Substation Structure Design Guide

Recommended Practice for Design and Use

Second Edition

Task Committee on Substation Structural Design

Sponsored by
the Task Committee on Substation Structural Design of the
Electrical Transmission Structures Committee of the
Special Design Issues Technical Administrative Committee of the
Technical Activities Division of the
Structural Engineering Institute of the
American Society of Civil Engineers

Edited by
George T. Watson, P.E.

Published by the American Society of Civil Engineers

Library of Congress Cataloging-in-Publication Data

Names: Watson, George T., editor. | American Society of Civil Engineers. Subcommittee on the Design of Substation Structures, sponsoring body.
Title: Substation structure design guide : recommended practice for design and use / Task Committee on Substation Structural Design, American Society of Civil Engineers ; edited by George T. Watson, P.E.
Description: Second edition. | Reston, VA : American Society of Civil Engineers, [2023] | Series: ASCE Manuals and Reports on Engineering Practice ; no. 113 | "Sponsored by the Task Committee on Substation Structural Design of the Electrical Transmission Structures Committee of the Special Design Issues Technical Administrative Committee of the Technical Activities Division of the Structural Engineering Institute of the American Society of Civil Engineers"--T.p. | Includes bibliographical references and index. | Summary: "MOP 113, second edition, documents electrical substation structural design practice and gives guidance and recommendations for the design of outdoor electrical substation structures"-- Provided by publisher.
Identifiers: LCCN 2023036758 | ISBN 9780784416174 (paperback) | ISBN 9780784485170 (pdf)
Subjects: LCSH: Structural design--Handbooks, manuals, etc.
Classification: LCC TA658.3 .S83 2023 | DDC 621.31/26--dc23/eng/20230812
LC record available at https://lccn.loc.gov/2023036758

Published by American Society of Civil Engineers
1801 Alexander Bell Drive
Reston, Virginia 20191-4382
www.asce.org/bookstore | ascelibrary.org

***Errata:* Errata, if any, can be found at https://doi.org/10.1061/9780784485170.**

ISBN 978-0-7844-1617-4 (print)
ISBN 978-0-7844-8517-0 (PDF)

Manufactured in the United States of America.

27 26 25 24 23 1 2 3 4 5

MANUALS AND REPORTS ON ENGINEERING PRACTICE

(As developed by the ASCE Technical Procedures Committee, July 1930, and revised March 1935, February 1962, and April 1982)

A manual or report in this series consists of an orderly presentation of facts on a particular subject, supplemented by an analysis of limitations and applications of these facts. It contains information useful to the average engineer in his or her everyday work, rather than findings that may be useful only occasionally or rarely. It is not in any sense a "standard," however, nor is it so elementary or so conclusive as to provides a "rule of thumb" for nonengineers.

Furthermore, material in this series, in distinction from a paper (which expresses only one person's observations or opinions), is the work of a committee or group selected to assemble and express information on a specific topic. As often as practicable the committee is under the direction of one or more of the Technical Divisions and Councils, and the product evolved has been subjected to review by the Executive Committee of the Division or Council. As a step in the process of this review, proposed manuscripts are often brought before the members of the Technical Divisions and Councils for comment, which may serve as the basis for improvement. When published, each manual shows the names of the committees by which it was compiled and indicates clearly the several processes through which it has passed in review, so that its merit may be definitely understood.

In February 1962 (and revised in April 1982), the Board of Direction voted to establish a series titled "Manuals and Reports on Engineering Practice" to include the manuals published and authorized to date, future Manuals of Professional Practice, and Reports on Engineering Practice. All such manual or report material of the Society would have been refereed in a manner approved by the Board Committee on Publications and would be bound, with applicable discussion, in books similar to past manuals. Numbering would be consecutive and would be a continuation of present manual numbers. In some cases of joint committee reports, bypassing of journal publications may be authorized.

A list of available Manuals of Practice can be found at https://ascelibrary.org/page/books/s-mop.

CONTENTS

PREFACE

The Subcommittee on the Design of Substation Structures under the Committee on Electrical Transmission Structures from the Structural Engineering Institute of ASCE developed this Manual of Practice (MOP). The subcommittee membership represented utilities, manufacturers, consulting firms, and general interest. The combined expertise of the subcommittee members contributed to make this a valuable substation structure design guide for the utility industry.

The primary purpose of this MOP is to document electrical substation structural design practice and to provide guidance and recommendations for the design of outdoor electrical substation structures. This MOP covers a review of structure types and typical electrical equipment. Guidelines for analysis methods, structure loads, deflection criteria, member and connection design, structure testing, quality control, quality assurance, connections used in foundations, detailing, fabrication, construction, and maintenance issues are presented. This second edition also includes a chapter on foundation types used in substations. A chapter on retrofitting existing structures was included along with a chapter on oil containment and barrier walls. Appendix D is included as a draft prestandard to show what this MOP might look like if it were advanced to become a standard. In addition to Appendix D, Appendixes A, B, and C were added to cover various examples of structure design (Appendix A), a detailed explanation of short-circuit forces application to substation structures (Appendix B), and supplemental information on application of seismic forces as it applies to substation structures. The recommendations presented herein are based on the professional experience of the subcommittee members. Although the subject matter of this manual has been thoroughly researched, its application should be based on sound engineering judgment.

The subcommittee wishes to thank the peer review committee for their assistance and contributions to this document.

BLUE RIBBON PANEL REVIEWERS

The subcommittee wishes to thank the peer review committee for their assistance and contributions to this document.

John Eidinger, G&E Engineering Systems
Tom Mara, CPP Wind Engineering Consultants
Robert Cochran, Seattle City Light
Benedict Chu, Pacific Gas & Electric
Kamran, Khan Trench Ltd.
William B Mills, FWT (retired)

The subcommittee thanks all the individuals who have contributed to the completion of this manual. Without their contributions, guidance, and dedication, this manual would not have been published. The following individuals have contributed to this manual, either as past subcommittee members or as corresponding members:

Florizel Bautista
Matt Flint
J. Casey Scoggins
Xiaosong Xiao
Vincent Chui
Kevin Hoang
Anthony Stone
Jin Yoon
John Dai
Gelberg Rodriguez
Reid Strain

ACKNOWLEDGMENTS

Task Committee on Substation Structural Design

George T. Watson, *Chair*
Majid R.J. Farahani, *Vice Chair*
Thomas Amundsen
Michael Clark
Kurt Edwards
Todd Gardner
Stefanie Gille
John Humphries
Chris Jayavendra
Mike Khavari
John Klotz
Paul Legrand
Michael Miller
Mark Nelson
Perumal Radhakrishnan
Michael Riley
Ross Twidwell
Josh Jordan, *Secretary*
Alan Blackwell
Marella Diokno
Eric Fujisaki
Jennifer Gemar
Mohammad Hariri
Mary Innamorato
Leon Kempner Jr.
Alex Kladiva
Brian Knight
Brian Low
Neil Moore
Anthony Pick
John Reed
Baker Tee
C. Jerry Wong

Committee on Electrical Transmission Structures

Otto J. Lynch, *Chair*
Frank W. Agnew
Ronald James Carrington
Timothy C. Cashman
Leon Kempner Jr.
Josh Sebolt, *Vice Chair*
Michael D. Miller
Vicki Schneider
George T. Watson

DEDICATION

Thomas A. Amundsen

This edition of the manual is dedicated to Thomas A. Amundsen. Tom provided 40 years of service to our industry and was instrumental in supporting, contributing, and mentoring this subcommittee's activities. His legacy lives on in this manual and with the engineers he developed, trained, and impacted over his life.

CHAPTER 1
INTRODUCTION

The purpose of this Manual of Practice (MOP) is to provide structural design guidance and a comprehensive resource for outdoor electrical substation structures and foundations. Engineers using this MOP may substitute or modify these recommendations on the basis of experience, research results, or test data. This MOP promotes the best utility practices for structural loads and design of electrical substation structures and acceptable electrical system performance. Electrical substations are an important component of the electrical system connecting the generation, overhead lines, and delivery points. The structure design loads provide an acceptably safe and reliable performance. This MOP uses the applicable ASCE 7-22 (ASCE 2022) wind, ice, and seismic loading in conjunction with overhead line tensions, and electrical clearances, as the basis of substation structure design. The application of these loads to substation structures is according to the procedures of this MOP.

The 113 committee has referenced many documents and included the specific year of publication where it was appropriate. As the various documents are revised, the reader should investigate what portion was changed and determine whether the changes impact the design of the substation structure. The term *latest edition* was avoided when referencing most documents.

Chapter 2, "Definitions, Electrical Equipment, and Structure Types," provides an overview of electrical equipment, identifies the various components and structure types, and describes structure configurations. With the exception of Chapter 11, "Retrofit of Existing Substation Infrastructure," the recommendations herein apply to new substation structures and foundations that support electrical equipment, rigid bus, and other conductors. Utilities should develop structure and foundation retrofit design criteria.

The electrical equipment supported by substation structures or foundations can be of significant weight, be subjected to substantial forces, and have attachments of porcelain or composite components. Guidelines for structural loads, deflection limits, analysis, design, foundations and anchorage, fabrication (quality assurance/quality control), maintenance, and construction of substation structures are presented. Oil containment and barrier walls are covered in Chapter 12, "Oil Containment and Barrier Walls." The structural loads and load combinations provided in this MOP can be regarded as minimum requirements. The selection of appropriate loads and load combinations for specific applications is the responsibility of the owner.

This MOP addresses steel, concrete, wood, aluminum, and porcelain or composite insulators used for the design of substation structures. Design equations are provided when references to existing structural design standards and codes (e.g., ACI, American Institute of Steel Construction, American Institute of Timber Construction, and ASCE) are not applicable.

Some figures (i.e., maps and graphs) are shown for information; the user of these figures can consult the original reference for more detail.

In the past, the utility industry has used both the allowable strength design (ASD) and ultimate strength design (USD) methods for substation structural design. ASD is a method of proportioning structural members such that elastically computed stresses produced in the members by nominal loads do not exceed specified allowable stresses. USD is a method of proportioning structural members such that the computed forces produced in the members by the factored loads do not exceed the member design strength [also called Load and Resistance Factor Design (LRFD)].

A significant issue discussed during the revision of this MOP was the direction it should take with respect to the design of substation structures using either ASD or USD concepts. It was decided that USD is the preferred approach for the structural analysis and design of structures. However, ASD may be used for foundation–soil interaction design if the limiting soil parameters are specified accordingly (e.g., allowable soil bearing pressure) and seismic qualification according to IEEE 693-2018 (Chapter 6, "Design," Section 6.5.5.2, "Allowable Strength Design according to IEEE 693"). In this MOP, the terms *ASD loads* and *service loads* are used interchangeably relative to deflection loading and foundation design.

Guidelines for the determination of substation structure loads for wind, ice, seismic forces, short circuit, line tensions, equipment reactions, operation, construction, maintenance, structure testing, and regulatory codes (e.g., IEEE 2023, PUC 2017) are provided in Chapter 3, "Loading Criteria for Substation Structures." The specific recommendations are based on structure type, such as dead-end structures, equipment, and bus supports. Recommended load factors and load combinations are presented.

The seismic load section complements IEEE 693 (IEEE 2018), which addresses the seismic design of electrical equipment and their supports (dedicated or intermediate supports). Dedicated supports are designed to exclusively support a single piece of equipment such as a pedestal for a current transformer (CT) or the concrete foundation under a power transformer. An intermediate support is a structural member or subassembly between the equipment and the primary substation structure (i.e., substation rack structure). This MOP references IEEE 693 (IEEE 2018) and provides seismic requirements for structures not covered by that reference.

Substation structures and the electrical equipment they support should be considered as a system. Excessive movement of structure supporting electrical equipment can cause the electrical equipment to experience mechanical damage, operational difficulties, and electrical faults. Recommended deflection limits and structure classes are defined in Chapter 4, "Deflection Criteria."

Analysis techniques and structural modeling concepts as they relate to substation structures are discussed in Chapter 5, "Method of Analysis." Both static and dynamic analyses are covered. Guidelines are given for selecting the appropriate analysis method for different structural behaviors, such as large versus small displacement theory, and nonlinear analysis.

Chapter 6, "Design," lists design standards for various materials of construction, and considerations for the design of rigid bus, insulators, base plates, structural connections, and other special considerations that include, but are not limited to, wind and seismic loading. This MOP, in general, notes only exceptions to these design standards and references other documents for the appropriate information.

The owner should designate each electrical installation (substation or circuit pathway) as either essential or nonessential based on its relative criticality to the power system. Installations of specific equipment designated as essential are those that are vital to power delivery and cannot be bypassed in the system or are undesirable to lose because of economic or operational considerations. Essential items range from equipment to support structures. Equipment that can be bypassed for short-term emergency operations may be considered nonessential.

Foundation design is discussed in Chapter 7, "Foundations," which reviews the types of foundations that are commonly used to support substation equipment and structures and outlines the geotechnical, seismic, durability, and other considerations for foundation design. The design of substation structure anchorage connection to foundations is presented in Chapter 8, "Connections to Foundations." Many types of anchorages are used to connect substation structures to their foundations. The most common anchorage is anchor rods cast in reinforced concrete. This MOP provides design recommendations for this type of anchorage. Special design considerations for anchorages subjected to seismic loads are covered.

Guidelines for quality control and quality assurance programs for substation structures are presented in Chapter 9, "Quality Control and Quality Assurance." Recommendations on when it is appropriate to test a unique substation structure design concept or perform individual component testing are provided in Chapter 10, "Construction, Maintenance, and Testing." Requirements for seismic testing are covered in IEEE 693 (IEEE 2018). Chapters have been added to this revision to address the retrofit of existing substations (Chapter 11, "Retrofit of Existing Substation Infrastructure,") and oil containment and barrier walls (Chapter 12, "Oil Containment and Barrier Walls").

Appendix D has a substation design draft pre-standard that has been added to this edition. The ASCE Task Committee on Substation Structure Design envisions the need for a design standard for substation facilities. The appendix presents the recommendations of this committee written in a prescriptive form for review and comment.

REFERENCES

ASCE. 2022. *Minimum design loads and associated criteria for buildings and other structures.* ASCE 7-22. Reston, VA: ASCE.

IEEE (Institute of Electrical and Electronics Engineers). 2018. *Recommended practice for seismic design of substations.* IEEE 693-2018. Piscataway, NJ: IEEE.

IEEE. 2023. *National electrical safety code.* ANSI C2-2023. Piscataway, NJ: IEEE.

PUC (Public Utility Commission). 2017. *State of California rule for overhead electric line construction.* General Order 95. Sacramento, CA: Public Utility Commission.

CHAPTER 2

DEFINITIONS, ELECTRICAL EQUIPMENT, AND STRUCTURE TYPES

2.1 PURPOSE

Substation and switchyard structures are used to support above-grade components and electrical equipment such as cable bus, rigid bus, and strain bus conductors; switches; surge arresters; insulators; and others. Substation and switchyard structures can be fabricated from angles that form chords, trusses, and braced frames. Wide flanges, tubes (round, square, and rectangular), pipes, and polygonal tubes (straight or tapered) are also used in substation structures. The common materials that are utilized are concrete, steel, aluminum, and wood.

This chapter provides an overview of electrical equipment, identifies the various components and structure types, and describes structure configurations. Photographs of selected substation structures and equipment are also included in this section. The photographs are shown for reference and pictorial purposes only, and inclusion of any equipment is not meant to be an endorsement of any manufacturer. The structures shown are not necessarily representative of all engineering applications and support types to be utilized.

2.2 DEFINITIONS, SUBSTATION TYPES, AND COMPONENTS

2.2.1 Substation

Substations are commonly defined as facilities where power voltage or other characteristics of power are transformed, redirected, or to adjust power quality in the electric transmission system. Larger substations may contain control enclosures, transformers, interrupting and switching devices, and surge protection (Figure 2-1). A substation may include one or more step-up or step-down power transformers operating at multiple voltages. Voltages may be transformed for transmission or distribution purposes.

2.2.2 Switchyard or Switching Station

The term *switchyard* is typically applied to the assemblage of switches, power circuit breakers, bus, and auxiliary equipment that are used to redirect power from the generators of a power plant and distribute it to the transmission lines at various points. Switchyards do not include the use of power transformers to transform voltages. Most electrical substations

Figure 2-1. Substation aerial view.
Source: Courtesy of Brian Low.

have a combination of substation and switchyard functionality. Switching stations are similar to switchyards, except that switching stations are not attached to a power plant. The power is only switched among connected transmission lines.

As far as structures are concerned and throughout the remainder of this Manual of Practice (MOP), the terms *substation* and *switchyard* will be used interchangeably. Switchyards are designed according to the substation design criteria.

2.2.3 Unit Substation

For lower voltages (typically 69 kV and lower), metal-enclosed *unit substations* (also referred to as *metalclad switchgear*) are sometimes used to house switches, fuses, circuit breakers, instrument transformers, and controls. They are usually mounted on a reinforced concrete pad or drilled shaft foundations.

2.2.4 Transmission Line

Transmission lines are power lines, typically with voltages at 69 kV and above. Lines carrying voltages lower than 69 kV are usually referred to as *sub-transmission* or *distribution lines*. Transmission line circuits can be protected from the effects of lightning strikes by shield wires or surge arresters. Shield wires on transmission lines entering substations are used for protecting the substation equipment. Shield wires may also be used for communication. Optical ground wires are shield wires using optical fibers, surrounded by layers of steel and aluminum, for communication and lightning protection.

The design of transmission lines and transmission line structures is not covered in this MOP. However, the design of transmission lines and shield wires entering the substation should be coordinated with the substation design. Transmission lines entering the substation may terminate on dead-end structures and these structures must be capable of supporting the wire configurations and wire tensions of the transmission line. Substation dead-end structures may be installed along with the transmission line construction, prior to the remaining substation structures. The owner may require the substation dead-end to be designed with a higher capacity than the transmission structures outside the substation. The intent would be to contain failure outside the substation where structure replacement is easier than inside the substation.

2.2.5 Air-Insulated Substation

An *air-insulated* substation has the insulating medium of air and is the most common type of installation. The high-voltage bus is a bare metallic tubing or cable, supported by insulators, and insulated from adjacent conductors, grounded structures, and substation grade by air. These types of substations use disconnect switches, which when open, depend on the air for insulation across the switch's open gap. Bushings, which are conductors insulated by porcelain or composite materials, are used to route electrical energy into the body of equipment (like circuit breakers and transformers) and prevent arcing between the conductors and the outside surface of the equipment, which is usually made of metal as well.

A variety of structure and foundation types are utilized to support the equipment in an air-insulated substation. Air-insulated substations typically require larger overall footprints than gas-insulated substations (GIS) of the same electrical configuration because of the larger electrical clearances necessary with air as the insulating medium. Where the footprint size is a constraint, a gas-insulated substation, among other alternatives, may be a consideration.

2.2.6 Gas-Insulated Substation, GIS

An alternative to the air-insulated substation is a *sulfur hexafluoride* (SF_6) GIS (Figure 2-2). High-voltage conductors are housed inside a metallic sheath filled with SF_6 gas under pressure. The metallic sheath is at ground potential and can be placed at or near the ground level. Disconnect switches are located inside the metallic sheath and use SF_6 gas for phase-to-ground, phase-to-phase, and open-gap insulation. Transmission lines and their associated equipment

Figure 2-2. Gas-insulated substation.
Source: Courtesy of Brian Low.

(surge arresters, wave traps, coupling capacitor voltage transformers, and line disconnect switches) are air-insulated. Air-to-gas bushings are used to connect gas-insulated equipment with air-insulated equipment.

Similar to an air-insulated substation, gas-insulated substations utilize a variety of structure and foundation types to support equipment. Unique structural designs are often required for the design of GIS structures. When SF_6 is used as the insulating medium, the electrical clearances that are required are minimized and a smaller overall substation footprint may be used, as opposed to an air-insulated substation of the same electrical configuration.

2.2.7 Electrical Clearance

Electrical clearances provide the physical separation needed for phase-to-phase, phase-to-structure, and phase-to-ground air gaps to provide safe working areas and to prevent flashovers. Minimum electrical clearances are specified in the National Electrical Safety Code (NESC) (IEEE 2023) or California's General Order 95 (PUC 2017).

2.2.8 Buswork System

The *buswork system* in a substation is the network of conductors that interconnect transmission lines, transformers, circuit breakers, disconnect switches, and other equipment. The term "buswork system" includes the conductors and the material and equipment that support these conductors.

The buswork system is selected on the basis of a desired switching arrangement and is configured to provide an orderly, efficient, reliable, and economic layout of equipment and structures. Three types of buswork systems that are commonly used are rigid, strain, and cable bus.

2.2.8.1 Rigid Bus System. A *rigid bus conductor* is an extruded metallic conductor. The conductor material is usually an aluminum alloy, but it can also be copper. Support locations in a rigid bus system may depend on extreme fiber stress and deflection of the bus under load. The cantilever strength of the supporting insulators must also be considered. Refer to IEEE 605-2008 (IEEE 2008) and Chapter 6 of this MOP, "Design," for recommendations on the design of rigid bus systems. Chapter 5, "Method of Analysis," Section 5.7.2, "Rigid Bus Analysis Methods Discussion," briefly discusses the dynamic analysis of short-circuit events in rigid bus systems, and Chapter 6, "Design," Section 6.9, "Rigid Bus Design," provides additional considerations.

2.2.8.2 Strain Bus System. A *strain bus conductor* is a stranded wire conductor installed under tension. When utilizing strain bus within a substation, the design of the wire system and supporting structures must satisfy required serviceability limits under all anticipated weather and operating conditions and tensions. Adequate electrical clearances must be maintained from other phases, structures, and ground under operating conditions such as blowout from wind or maximum operating temperatures. See Chapter 3, "Loading Criteria for Substation Structures," Section 3.5, "Serviceability Considerations," for additional discussion on serviceability recommendations.

2.2.8.3 Cable Bus System. *Cable bus conductors* are low-tension, stranded conductors supported on station post insulators. Similar to a strain bus, cable bus systems must maintain similar serviceability conditions under all anticipated weather and operating conditions.

2.2.9 Short-Circuit Force

Short-circuit forces are structure loads that are caused by short-circuit currents. Short-circuit currents are the result of electrical faults caused by equipment or material failure, lightning, or other weather-related causes and accidents.

2.2.10 Dead-End Structures

Dead-end structures (Figure 2-3) (also called *takeoff structures, pull-off structures, termination structures, anchor structures, gantry frames,* or *strain structures*) are designed to resist tension from phase conductors and shielding wires. In addition, they may support switches or other electrical equipment. They can support a single bay [three-phase alternating current (AC), two-pole direct current (DC)] or multiple bays. The first dead-end structure inside the substation is typically designed to support the transmission line conductors either at full tension or at reduced tension (slack span).

Dead-end structures are typically the tallest structures within the substation. They must be designed to allow flexibility in the transmission line design and support the required wire tension loads and substation equipment. Two types of dead-end structures that are commonly used include A-frame and H-frame dead-end structures. Line tensions, structure size, and foundation volume are some of the factors to consider when deciding between an A-frame, an H-frame, or a lattice dead-end structure. See Chapter 3, "Loading Criteria for Substation Structures," Section 3.1.4, "Wire Tension Loads," for additional design considerations.

Figure 2-3. Dead-end structure.
Source: Courtesy of US Department of Energy.

2.2.10.1 A-Frame Dead-End. *A-frame dead-end structures* are commonly used when the design requires the full tension of the transmission line to be supported by the substation dead-end structure or for other high-tension applications. An A-frame structure may be ideal in this situation because the geometry of the structure will convert large overturning moments, created by the wire tensions, into axial force couples, which may reduce the steel sections required and result in smaller foundations. Because A-frames convert large overturning moments into an axial force couple, uplift must always be considered when designing the foundations. Practical A-frame designs set the longitudinal leg spacing at approximately one-third of the conductor attachment height. The transverse leg spacing is a function of the conductor phase spacing.

2.2.10.2 H-Frame Dead-End. *H-frame dead-end structures* are more frequently used when reduced tensions, or slack-spans, are anticipated for the transmission line entering or exiting the substation. Designing for a reduced tension into the substation may require a transmission dead-end structure to be installed in the transmission line near the substation to resist the full tension of the transmission line. H-frame structures may provide a more practical design when used to resist slack-span line tensions because the weight of the structure is less than that of an A-frame, and fewer foundations will be required to support the structure.

2.2.11 Box-Type Structure

Box-type structures (Figure 2-4) (single-bay or multiple-bay space frames, also referred to as *rack* or *lattice steel structures*) can be used to support rigid bus conductors, switches, and other equipment.

Figure 2-4. Box-type structure.
Source: Courtesy of US Department of Energy.

2.2.12 Shielding Mast

This structure (also referred to as a *shield wire mast* or *static wire mast*) shields equipment in the substation from direct lightning strikes. These structures (Figure 2-5) have overhead wires attached to enhance protection and dampen mast vibration.

The height and location of shielding masts and shield wires should be coordinated with the engineer designing the structure (including determining wire tensions) and the electrical engineer performing the lightning protection study. The height and location of structures, as well as the sag of the shield wires, should be considered in the lightning protection study. Structures and wires should be located to provide adequate protection of the substation equipment from lightning strikes.

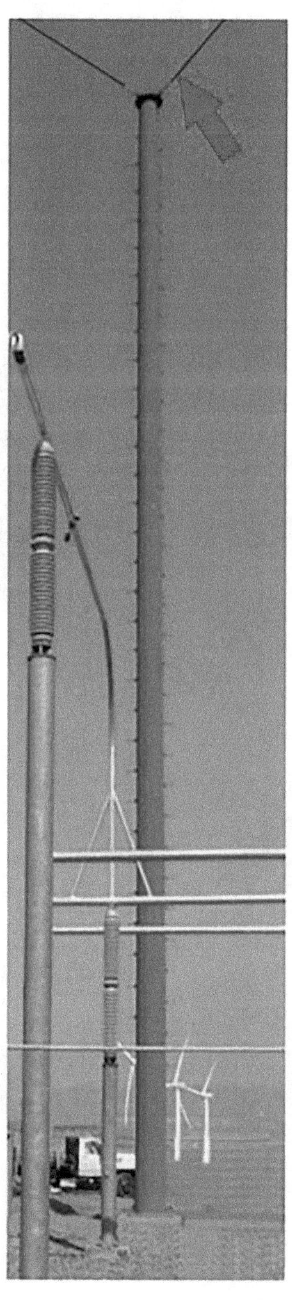

Figure 2-5. Shielding mast with wires.
Source: Courtesy of US Department of Energy.

Shielding masts may experience wind-induced vibrations because of the height and flexibility of the structure. Refer to Chapter 6, "Design," Section 6.10.2, "Vortex-Induced Oscillation and Vibration," for further discussions on wind-induced vibrations.

2.2.13 Lightning Mast

This free-standing structure (also referred to as a *ground mast*) shields equipment in the substation from lightning strikes. These structures (Figure 2-6) do not have overhead ground wires attached.

Figure 2-6. Lightning mast.
Source: Courtesy of Michael Miller.

The height and location of the structure should be coordinated with the engineer designing the structure and the electrical engineer performing the lightning protection study. Lightning masts may experience wind-induced vibrations because of the height and flexibility of the structure. Refer to Chapter 6, "Design," Section 6.10.2, "Vortex-Induced Oscillation and Vibration," for further discussions on wind-induced vibrations.

2.3 ELECTRICAL EQUIPMENT AND SUPPORTS

This section provides an overview of the typical types of electrical equipment support structures and a brief description of the electrical equipment that they support. An understanding of the functions, operations, and relationship of electrical equipment to the support structures is a prerequisite for a good structural design.

Foundation types may vary based on equipment type, loading, and site-specific soil conditions. Engineering judgment and knowledge of each of the previously mentioned factors should be considered when determining the appropriate foundation type. For additional foundation recommendations, refer to Chapter 7, "Foundations." All sites are unique, and the engineer should be aware of the site-specific soil conditions before selecting a suitable foundation type.

2.3.1 Power Transformer and Autotransformer

The *power transformer* and *autotransformer* are devices used to provide a connection between power systems of different voltage levels to permit power transformation from one voltage to another. Power transformers and autotransformers may be three-phase or single-phase units containing a core and windings that are submerged in insulating oil. Transformers are often expensive, long lead time items and are critical to the function of a substation.

Support: At transmission-level voltages, the power transformer and autotransformer are supported directly on a foundation.

2.3.2 Shunt Reactor

The *shunt reactor* (Figure 2-7) is, like a power transformer, a device containing a core and windings that are submerged in insulating oil used to compensate for the shunt capacitance and the resulting charging current drawn by a transmission line. If no compensation for charging current is provided, the voltage at the receiving end of a long transmission line can exceed the sending end voltage by as much as 50% under light loading or load rejection.

Support: The shunt reactor is supported directly on a foundation unless it is an air core reactor.

2.3.3 Current-Limiting Inductor or Air Core Reactor

A *current-limiting inductor* (Figure 2-8) is a device that adds inductive reactance to the system and limits the amount of current that can flow in a circuit. Current-limiting inductors are typically dry inductors and are usually not provided with magnetic shielding. These inductors have a magnetic field surrounding them under normal conditions, which increases in intensity when carrying short-circuit current. This magnetic field interacts with the magnetic field of

Figure 2-7. Oil-filled shunt reactor.
Source: Courtesy of US Department of Energy.

other inductors if they are spaced too closely. The magnetic field induces eddy currents in metallic objects placed in the inductor's magnetic field. However, currents are induced in any closed conductive loops located within the inductor's magnetic field. An *air core reactor* is a construction technique that may be used on several types of applications. Filter reactors, current limiting reactors/inductors, shunt reactors, neutral grounding reactors, and line traps are among the applications of air core reactors.

Air core reactors are typically not supplied with magnetic shielding and thus have a magnetic field surrounding them when carrying current. The necessary precautions regarding the magnetic field are given in Chapter 6, "Design," Section 6.10.1, "Precautions regarding the magnetic fields of air core reactors."

Air core reactors are supported by insulators and may have additional members to provide the necessary ventilation, or magnetic or personal clearance.

Support: Dry current-limiting inductors are usually supported on insulators (unless it is an air core reactor) and aluminum or fiberglass pedestals to maintain the manufacturer's recommended magnetic clearance, as well as the required personnel clearance. The supporting pedestals are anchored directly to the foundation. If independent foundations are adopted under individual porcelain support posts, the porcelain posts should be designed for additional stresses as a result of any differential movements between the foundations.

Figure 2-8. Current-limiting inductor or air core reactor.
Source: Courtesy of Kamran Khan.

2.3.4 Line Trap

A *line trap* (Figure 2-9) (also referred to as a *wave trap*) presents high impedance to carrier frequencies and negligible impedance to a normal 60 Hz (50 Hz outside the United States) line current. It is a blocking filter that is used to restrict the carrier signal to the transmission line on which it is installed and to prevent the carrier signal from entering substation equipment. Line traps are preferably mounted vertically to promote convection cooling. In some historical applications, where the line trap has a single layer of windings, a horizontal mounting is possible.

A line trap may be supported by either a single insulator element or multiple such elements. Normally, insulating elements are station post insulators, but it is possible to use a coupling capacitor or a coupling capacitor voltage transformer (CCVT) as an insulating element.

Support: Line traps are mounted vertically or horizontally on either a single- or multiple-pedestal support structure. They are also mounted with one end of the line trap supported by a coupling capacitor (CC) or CCVT and the other end by an insulator. The line trap can also be suspension-mounted from a structure. A suspension-mounted line trap (including suspension-mounted CC or CCVT) is not recommended in seismic areas as defined in Chapter 3, "Loading Criteria for Substation Structures," unless the attached bus work is suitably designed to accommodate seismic motions without imparting excessive impact loads on adjacent equipment. An excessive impact load is a load that exceeds the 50% of the rated strength of a porcelain component. However, if a suspension mount is used in seismic areas, the provisions of IEEE 693 (IEEE 2018) should be followed to add restraint to the system.

Figure 2-9. Line trap (Wave Trap) underhung on lattice truss.
Source: Courtesy of CenterPoint Energy.

2.3.5 Coupling Capacitor Voltage Transformer

Coupling capacitor voltage transformers (Figure 2-10) (formerly called *coupling capacitor potential devices*) are capacitance voltage dividers used to obtain voltages in the 66 V to 120 V range for relaying and metering. When supplied with carrier accessories, they are used to couple a carrier signal to a transmission line.

A *coupling capacitor* is a capacitance device with carrier accessories used to couple the carrier signal to the line conductor. It is similar to the CCVT, except that it does not have the voltage transformer.

Support: The CCVT is usually supported on a single pedestal or three-phase structure. At higher voltages, such as 500 kV, this equipment can be mounted at ground elevation within a fenced area.

2.3.6 Disconnect Switch

Disconnect switches (Figure 2-11) (when open) are used to electrically isolate a transmission line, a circuit breaker, or other electrical equipment (e.g., a transformer) and to provide a physical air gap to ensure the isolation. Disconnect switches are opened after the circuit has been de-energized to provide a visual air gap for safety and maintenance purposes. They can be manually or motor operated. Motor operators are applied to disconnect switches when physical operating requirements dictate and when automatic or remote-controlled operation of the switches is required.

Disconnect switches may use a vertical break, center break (including V switches), single side break, or a double side break. The side break requires larger phase spacing than the vertical break. The switches are equipped with buggy whip, gas blast, and vacuum-interrupting devices

Figure 2-10. Coupling capacitor voltage transformer.
Source: Courtesy of US Department of Energy.

to give the switch some load current (magnetizing current) or line-charging interrupting ability. Disconnect switches can be provided with opening and closing resistors. The disconnect switch can be provided with a maintenance grounding switch or high-speed grounding switch on either the hinge end or the jaw end of the switch. Grounding switches require a separate operating mechanism and can be interlocked with the main switchblade. The plane of motion of the grounding switchblade can be either parallel to the plane of the main switchblade or perpendicular to it. If the grounding switch operates in a plane perpendicular to the main switchblade, physical clearance from the grounding switchblade in the open position must be checked.

Figure 2-11. Disconnect switch on a truss-type support structure.
Source: Courtesy of US Department of Energy.

Support: The three phases (AC) of a disconnect switch are usually supported on a common structure for voltages less than 500 kV. At 500 kV and higher voltages, individual structures are used for each phase. The structure legs support the operating mechanism and control junction box. The structure should have adequate rigidity to permit proper switch operation. Each switch support may have a switch operating platform or two operating platforms if the switch has a grounding blade.

2.3.7 Circuit Switcher (Load Interrupter Switch)

A *load circuit switcher* (Figure 2-12), also referred to as a *load interrupter switch* or *line circuit breaker,* is a device that combines a disconnect switch with an SF_6 interrupter. It operates under electrical load and has limited fault current interrupting capability. It provides isolation and limited fault interrupting capability in a single device.

Support: The three phases (AC) of circuit switchers are usually supported on a common structure for voltages lower than 500 kV. At 500 kV and higher voltages, an individual structure is used for each phase. The structure legs support the operating mechanism and control junction box.

The circuit switcher imparts a dynamic load on opening or closing, and the structure should have adequate rigidity to permit proper switching operation.

Figure 2-12. Circuit switcher (load interrupter switch) on a frame-type support structure. Source: Courtesy of CenterPoint Energy.

2.3.8 Circuit Breaker

Circuit breakers (Figures 2-13 and 2-14) are used for electrical load switching and fault current interruption. They must be able to interrupt the fault current of the circuit in which they are applied. Oil, compressed air, SF_6 gas, and vacuum can be used as the insulating and interrupting media in the circuit breaker.

The circuit breaker tank is the chamber that houses the interrupter mechanism. The two types of tanks are live and dead. A *live tank breaker* (Figure 2-13) is one whose "tank" or interruption chamber is at line potential and is supported by an insulating column or columns. A *dead tank circuit breaker* (Figure 2-14) is one whose "tank" or interruption chamber is at ground potential. Dead tank circuit breakers typically have a current transformer at the base of one or more bushings.

Operational loads should be considered in the design of the breaker support and foundation. Operational loads may be considered negligible; however, the manufacturer should provide this information and it must be considered during design.

Support: Circuit breakers, including their supporting frames, are anchored directly on the foundation. In seismic areas, various types of damper units may be incorporated between the support frame and the foundation.

Figure 2-13. Live tank circuit breaker.
Source: Courtesy of US Department of Energy.

Figure 2-14. Dead tank circuit breaker.
Source: Courtesy of US Department of Energy.

Figure 2-15. Potential transformer.
Source: Courtesy of US Department of Energy.

2.3.9 Potential and Current Transformers

Potential transformers (PTs) (Figure 2-15) and *current transformers* (CTs) (Figure 2-16) are instrument transformers that change the magnitude of the primary circuit voltage to a secondary value that is suitable for use with relays, meters, or other measuring devices. The PT measures voltage, and the CT measures current.

Support: PTs and CTs are usually supported on a single pedestal or lattice stand structure. A single structure may also be used to support multiple phases.

2.3.10 Capacitor Bank

A grouping of capacitors is used to maintain or increase voltages in power lines and to improve system efficiency by reducing inductive losses. After the capacitor bank is installed, the outer periphery of the bank should be enclosed inside a fence for protection of personnel if electrical clearance is not provided.

Support: Capacitor banks are generally supported on an aluminum frame or insulator stacks and bolted directly to the foundation. The design of the support frames is generally provided by the manufacturer of the unit.

2.3.10.1 Shunt Capacitor. A *shunt capacitor* (Figure 2-17) is an installation of fused or fuseless capacitors and associated equipment (usually located in substations) and used to provide reactive power to increase system voltage and to improve the power factor at the point of delivery.

Figure 2-16. Current transformer.
Source: Courtesy of US Department of Energy.

The shunt capacitor improves the power transmission efficiency by adding capacitance, volt-amperes reactive (VAR), to the system. They are used to stabilize voltage drops on the system resulting from generation jump start-up or to stabilize long radial line feeds.

Support: The shunt capacitor is supported on a single frame-type structure. Shunt capacitor banks for low voltages are in an enclosed cabinet. Porcelain support insulators should be designed to sustain differential motions between independent foundations.

2.3.10.2 Series Capacitor. A *series capacitor* (Figure 2-18) is an installation of capacitors with fuses and associated equipment in series with a transmission line. Series capacitors are used (typically at 230 kV and above) to improve power transfer capability by compensating for the voltage drop along a transmission line. Usually located near the center of a line (but they can be located at any point), they are used to increase the capability of interconnections and in some cases to achieve the most advantageous and economical division of loading between transmission lines operating in parallel. Series capacitors can also force more power to flow over the transmission line with larger conductors when parallel lines have different conductor sizes.

Support: The support is provided by a metal platform. Because the series capacitors are at line potential, the platform must be mounted on insulators. Porcelain support insulators should be designed to sustain differential motions between independent foundations.

Figure 2-17. Shunt capacitor.
Source: Courtesy of US Department of Energy.

2.3.11 Surge Arrester

Surge arresters (Figure 2-19), sometimes called *arresters*, protect power equipment from overvoltage caused by switching surges or lightning. They are typically used near transmission line connection terminations near dead-end structures.

Support: Surge arrester support structures are usually single-phase supports, but could be three-phase supports, depending on the voltage. Surge arresters can be supported on a single pedestal or a lattice stand structure or directly mounted on a transformer.

2.3.12 Neutral Grounding Resistor

The *neutral grounding resistor* provides resistance grounding of the neutral transformers to limit ground-fault current to a value that does not damage generating, distribution, or other associated equipment in the power system, yet allows sufficient flow of fault current to operate protective relays to clear the fault.

Figure 2-18. Series capacitor.
Source: Courtesy of US Department of Energy.

Support: Resistors are sometimes mounted on separate structures but are usually mounted on the transformer tank.

2.3.13 Cable Terminator

The *cable terminator* (Figure 2-20) (also called a *pothead*) is used to change from a bare overhead conductor to a dielectric insulated cable. Other types of insulated cables such as gas- or oil-filled may also be used. Exposed cables may require protection from ultraviolet rays.

Support: Support structures of cable terminators of individual phases can be columns resting on a foundation. A structure supporting three phases can also be used. The foundation design must consider the underground cables and their allowable bend radii to avoid conflicts.

2.3.14 Insulator

Insulators electrically isolate energized conductors or components from supporting structures. They can be either suspension or station post insulators (Figure 2-21). Suspension insulators transfer tension forces from the conductor to the structure. Station post insulators can transfer tension forces, compression forces, and bending moments to the structure. They have various strength ratings and lengths for different basic impulse levels. Porcelain, glass, and composite materials are used for suspension and post insulators. The rated strength is generally established by test, and the allowable load under seismic, wind, or ice loading is typically 50% of the rated strength.

Support: Insulators can be supported on a single-phase or three-phase structure. Single-phase structures are usually supported by a support column (Figure 2-21), whereas

Figure 2-19. Surge arrester.
Source: Courtesy of US Department of Energy.

three-phase structures can be supported by a single structure, but this becomes impractical and uneconomical at 500 kV and above because the phase spacing is greater than 25 ft (7.62 m).

2.3.15 Bus Duct

There are three types of *bus ducts*. One is an isolated (iso) phase bus duct that connects the generator to the main power transformer and unit auxiliary transformers. The isolated phase bus duct consists of three separate conductors, each in its own grounded metal enclosure.

The second type is a nonsegregated phase bus duct that connects switchgear to switchgear and station auxiliary transformer to station auxiliary transformer. This type of bus duct consists of three bus bars in a common grounded metal enclosure.

The third type is a segregated bus duct and is the same as the nonsegregated phase bus duct except that there are grounded metal segregation barriers between bus bars.

Support: When installed inside an enclosure, an iso phase bus duct may be supported by a series of steel structures or anchored directly onto a foundation. When installed outside of an enclosure, the three phases of bus ducts are usually supported on separate structures.

Figure 2-20. Cable terminator or pothead.
Source: Courtesy of Brian Low.

Figure 2-21. Insulator and rigid bus conductors.
Source: Courtesy of US Department of Energy.

2.3.16 Fire Barriers

Power transformers and circuit breakers may contain large amounts of oil to provide insulation and a cooling medium. When installed in electrical equipment, this oil possesses properties to become a fire hazard. Minimum separation is required from transformers to equipment enclosures housing controls and between adjacent transformers. Where minimum separation requirements are not able to be met, a fire barrier is required (also called a *firewall*).

Fire barriers should be constructed of noncombustible materials such as concrete block, brick, sheet steel, reinforced concrete, or composites. Chapter 12, "Oil Containment and Barrier Walls," discusses the design of fire barriers.

Support: A variety of structures are available for fire barriers and include precast concrete, steel, or a combination of materials to form walls supported on foundations.

2.3.17 Control Enclosures

Control enclosures contain switchboard panels, relaying and controls, batteries, battery chargers, and other equipment for metering and communications. Control enclosures may be designed by the supplier as part of the equipment contained within or by using rules similar to non-building structures. Control enclosures are not covered in this manual.

2.3.18 Transformers

Transformers (Figure 2-22) change voltage from one level to another.

Support: Substation power transformers are supported on a foundation that is typically incorporated with oil spill containment.

Figure 2-22. Transformer.
Source: Courtesy of US Department of Energy.

2.4 DEFINITION OF RESPONSIBILITIES

2.4.1 Owner

The *owner* is defined as the responsible party that operates an existing or proposed substation. The owner may also include the owner's designated representative, who may be a consulting engineer, general contractor, or another entity such as the owner's engineer. Large utilities (owners) may have engineers on staff, whereas smaller utilities may use a contractor to act on behalf of the owner for engineering decisions.

2.4.2 Structure Designer

The *structure designer* is the party responsible for the design of the structures in the substation, who may be an agent of the owner or fabricator. The term "engineer" will be used in this MOP interchangeably with structure designer ("designer").

2.4.3 Supplier or Fabricator

The *supplier* or *fabricator* is the producer of the substation structures as directed by the owner or the structure designer.

REFERENCES

IEEE (Institute of Electrical and Electronics Engineers). 2008. *Guide for design of substation rigid-bus structures*. IEEE 605-2008. Piscataway, NJ: IEEE.

IEEE. 2018. *Recommended practice for seismic design of substations*. IEEE 693-2018. Piscataway, NJ: IEEE.

IEEE. 2023. *National electrical safety code*. ANSI C2-2023. Piscataway, NJ: IEEE.

PUC (Public Utilities Commission). 2017. *State of California rule for overhead electric line construction*. General Order 95. Sacramento, CA: PUC.

CHAPTER 3

LOADING CRITERIA FOR SUBSTATION STRUCTURES

All substation structures should be designed to withstand applicable loads that consider dead load, wind, ice, line tensions, earthquakes, short circuit, construction, maintenance, electrical equipment operating loads, and other specified or unusual service conditions.

This chapter discusses guidelines for developing substation structure loading criteria. Loads and load combinations recommended in this Manual of Practice (MOP) are considered appropriate for providing reliable substation structures described in Chapter 2. The selection of appropriate loads and load combinations using this MOP is the responsibility of the utility owner. Engineers may substitute or modify the recommendations on the basis of experience, research, or tests.

It should be noted that thermal loads in restrained structural elements can be significant. For example, a restrained rigid bus with a large temperature differential caused by electrical current will expand its length and induce loads on the bus and its connections. Structures and connections should be detailed to eliminate or minimize these thermally induced loads. When such detailing is not feasible, the engineer should consider the forces resulting from thermal loads. This MOP does not explicitly address methods for computing thermal loads.

3.1 BASIC LOADING CONDITIONS

3.1.1 Dead Loads

Structure, support equipment, and accessory weights should be included in the dead loads applied in conjunction with applicable design loads. Dead loads should include all masses associated with the equipment in operational condition, such as insulating oils, coolants, and appendages. Substation structures, in general, have well-defined dead loads such that dead load factors greater than 1.1 are typically not used.

3.1.2 Equipment Operating Loads

Operation of equipment, such as switches and circuit-interrupting devices with mechanized operations, can create loads on support structures. Loads resulting from operation should be combined with other loads if the equipment must be operable when weather conditions are most severe. The equipment manufacturer should be consulted regarding the application and magnitude of such loads. Equipment manufacturers should also be consulted for operational deflection requirements specific to the particular equipment. Recommended deflection limits are listed in Chapter 4.

The engineer should be involved in the equipment procurement review process to request and assess design information, such as equipment center of gravity, weights, equipment attachment points, seismic qualification, component sizes for wind surface area, and operating forces (both nominal and fault conditions).

3.1.3 Terminal Connection Loads for Electrical Equipment

The primary purpose of terminal connections on substation equipment is to serve as electrical conduits to the equipment. They are not intended as a primary load support system. Terminal connection loads are forces and moments that are created by flexible or rigid bus connectors to the electrical equipment. Connectors to equipment should be designed to accommodate sufficient movement or slack between equipment. The amount of movement required depends on the equipment connected and the loading that produces the relative movement.

Bus connections to the terminal pads of electrical equipment use fittings that typically vary from 6 to 18 in. (150 to 460 mm) long. These terminal connectors, which are either flexible or rigid bus, can act as a lever arm to create additional moments on equipment terminal pads. Substantial forces and moments can be transmitted to the terminal pads of the equipment by the terminal connector either by the loads transferred from the connected equipment or the load effects of the connector itself. Test terminals, for use by operations and maintenance personnel, may also be installed between the terminal connector and the equipment terminal pad.

Terminal pad connection capacities for disconnect switches are specified in IEEE C37.30.1 (IEEE 2011) for circuit breakers in IEEE C37.04 (IEEE 2018d) and for transformer bushings in IEEE C57.19.01 (IEEE 2000) and IEEE C57.19.100 (IEEE 2012). Moment capacities are typically not provided by these documents. When it is determined that the terminal pads may be subjected to a significant moment, the engineer should consult with the equipment manufacturer for the limiting terminal pad moment capacity. Refer to IEEE 1527-2018 (IEEE 2018b) or IEEE 693-2018 (IEEE 2018c) in the event of any seismic interconnection movement.

The following information is recommended for inclusion in specifications for purchasing electrical equipment:

- Type of terminal connectors (flexible or rigid bus) to the electrical equipment;
- Drawings of terminal connection hardware, if available;
- Preferred orientation of terminal connections; and
- Unfactored forces (axial, shear, moment) that act on the equipment terminal connections.

3.1.4 Wire Tension Loads

Structures supporting wire (conductor and shield wire) into and out of a substation are called *substation dead-end structures*. Dead-end structure wire tensions can be based on transmission line wire tensions or a specified reduced tension, called *slack-span tension*. The use of a slack-span tension (40% to 60% of the transmission line tension) separates the high tension loads of the transmission line with relatively long spans from the substation dead-end utilizing relatively short spans to lower the load demand. When a dead-end slack-span structure is used, the adjacent transmission line tower is typically designed as a full-tension dead-end structure. Substation dead-end structures that support wires that extend outside the substation yard should meet the requirements of the IEEE (2023), in addition to the requirements of this MOP.

The strain bus conductor (also called *cable bus* in this chapter) is composed of flexible conductors suspended above the ground between supporting structures within the substation using strain insulators. The span length is typically 300 ft (91 m) or less and takes the shape of a catenary or parabolic curve. Most stringing programs assume that the conductor spans the entire length between support structures, has the shape of a catenary or parabolic curve, and ignores the effect of insulators on the conductor sag.

Ice, wind, or lower temperatures increase the tension load in the wire beyond its original installation tension and should be considered when computing the wire tension load. These effects are discussed in detail in Sections 3.1.5 and 3.1.6.

With typical strain bus spans, porcelain or glass insulator weight may contribute to a substantial portion of the sag, and therefore, its impacts should be assessed. Also, some older stringing programs assume that the conductor loading is uniform and do not have provisions for concentrated loads. Concentrated loads along the strain bus span are typically created by taps (vertical jumpers) from the strain bus conductor to a lower bus conductor or electrical equipment. Depending on the tap location and magnitude, they may significantly affect the tension and sag of the conductor. A sag tension program that can accurately model these conditions should be used for strain bus design.

Taps should have adequate slack to allow unrestrained upward movement of the overhead conductor during cold temperatures or other loading conditions. Taps with insufficient slack may cause the tension in the conductor to become larger than the design tension on the structure, or the tap may transmit loads to the lower bus conductor or equipment that is not designed to resist this additional load. For long strain bus spans with taps, the use of mechanical springs inline with the strain bus conductor can reduce the effects of variations in sag caused by temperature and loading conditions. Finite-element programs can model the effects of insulators and concentrated tap loads.

Taps create vertical concentrated loads on strain bus conductors, which can significantly increase wire tensions. Some computer programs that calculate wire tensions and sags use uniform wire loads and do not have provisions for concentrated loads. Dividing the concentrated load by the span length and adding it to the uniform wire load may lead to an underestimation of the wire tension, particularly if the taps are located at midspan. A structural analysis program with cable elements can properly calculate the wire tension for the concentrated loads. If this type of program is not available, the effects of concentrated loads can be approximated using the general cable theorem (Figure 3-1).

The general cable theorem (Norris and Wilbur 1960) states,

> At any point on a cable acted upon by vertical loads, the product of the horizontal component of cable tension and vertical distance from that point to the cable chord equals the bending moment which would occur at that section, if the loads carried by the cable were acting on a pin-supported beam of the same span length as that of the cable.
>
> This theorem is applicable to any set of vertical loads, whether the cable chord is horizontal or inclined.

For a concentrated load at midspan, the bending moment is $PL/4$ (where P is the concentrated load and L is the horizontal span length) and is equal to the product of the horizontal tension and the sag at midspan (Figure 3-1). For a uniform load, the bending moment is $W_E L^2/8$, and it is also equal to the product of the horizontal tension and the sag at midspan. Therefore, $PL/4$ may be set equal to $W_E L^2/8$ and solved for W_E, which is the equivalent uniform load increase to account for the concentrated load. The equivalent uniform load, W_E, is then added

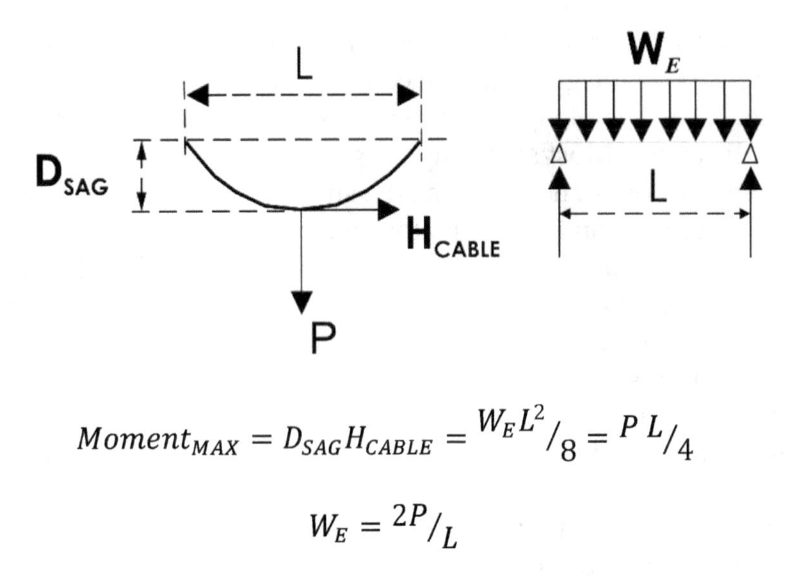

Figure 3-1. General cable theorem.
Source: Norris and Wilbur (1960).

to the conductor uniform load to calculate the tension and sag of the conductor. However, this example is valid only for concentrated loads that create the maximum bending moment at midspan.

The application of wire tension loads for dead-end structures should be chosen to allow for flexibility in routing transmission lines into and out of substations. This flexibility can be accounted for by selecting a maximum pull-off line angle (such as $\pm$ 15 degrees), reasonable wind span for transverse loads, wire sag for vertical loads, and wire tensions for longitudinal loads. The minimum phase-to-phase distance may limit the pull-off horizontal line angle.

Uplift loads can be applied to the substation dead-end structure when the wire attachment points of the first transmission line structure are higher in elevation than the attachment points of the substation dead-end structure. Design uplift loads can be specified for dead-end structures. Normally, the uplift capacity of the members and connections is sufficient, assuming that vertical wire loads can act either up or down and additional uplift load cases will usually not govern the design. However, the uplift reactions need to be accounted for in foundation design and may lead to substantial foundation sizes.

3.1.5 Extreme Wind Loads

Wind loads on substation structures, equipment, and conductors (bus and wire) should be applied in the direction that generates the maximum loading or maximum load effects on structure components. This section provides two equations for determining the wind loads on substation structures supporting equipment and/or overhead wires. Equations (3-1) and (3-2) can be applied to structures and supported electric equipment. The wind equation parameters are represented in Figure 3-2. Equation (3-1) can be used on structures other than lattice. Equation (3-2) can be used on lattice-type structures (such as line terminal and rack structures). For substation structures supporting wire loads, the longitudinal winds (in the direction of the wires) may also produce structure loading and should therefore be considered in the load calculation.

The wind force in the direction of the wind (WD) can be determined using the following equations:

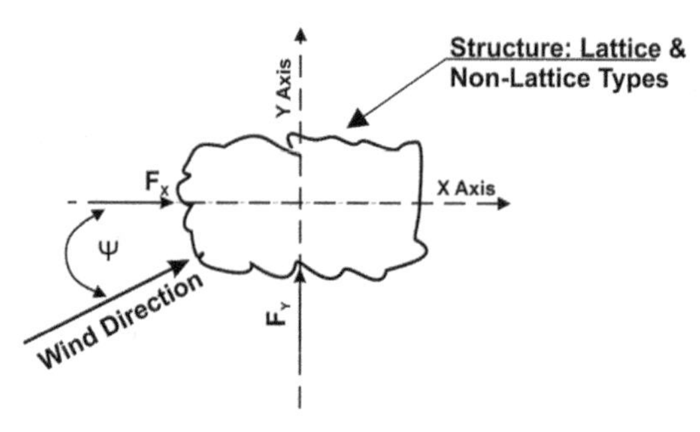

Figure 3-2. Wind equation parameters.

Non-lattice-type structures:

$$F_{\mathrm{WD}} = QK_zK_dK_{zt}(V_{\mathrm{MRI}})^2G_{\mathrm{RF}}C_{fWD}A_{fWD} \tag{3-1}$$

Lattice-type structures:

$$F_{\mathrm{WD}} = QK_zK_dK_{zt}(V_{\mathrm{MRI}})^2G_{\mathrm{RF}}(1+0.2\sin^2 2\Psi)(C_{fx}A_{fx}\cos^2\Psi + C_{fy}A_{fy}\sin^2\Psi) \tag{3-2}$$

where

F_{WD} = Wind force in the direction of wind (WD) [pounds (newtons)];
Q = Air density factor, default value = 0.00256 (0.613 SI), defined in Section 3.1.5.1;
K_z = Terrain exposure coefficient, defined in Section 3.1.5.2;
K_d = Wind directionality factor, defined subsequently;
K_{zt} = Topographical factor, defined subsequently;
V_{MRI} = Basic wind speed [mph (m/s)] defined in Sections 3.1.5.3 and 3.1.5.4;
G_{RF} = Gust response factor (structure G_{SRF} and wire G_{WRF}), defined in Section 3.1.5.5;
C_{fWD} = Force coefficient for face in the wind direction (WD), defined in Section 3.1.5.6;
C_{fx} = Force coefficient in x-axis, defined in Section 3.1.5.6;
C_{fy} = Force coefficient in y-axis, defined in Section 3.1.5.6;
A_{fWD} = Projected wind surface area normal to the wind direction WD [ft^2 (m^2)] for an entire structure or parts of the structure if warranted;
A_{fx} = Projected wind surface area normal to the x-direction [ft^2 (m^2)] for an entire structure or parts of the structure if warranted;
A_{fy} = Projected wind surface area normal to the y-direction [ft^2 (m^2)] for an entire structure or parts of the structure if warranted; and
Ψ = Wind direction angle measured from the x-axis.

The wind force in the direction of the wind can then be decomposed into forces in the x- and y-axes, respectively, to assist with the design of each direction

$$F_x = F_{\mathrm{WD}}\cos\Psi \tag{3-3a}$$

$$F_y = F_{\mathrm{WD}}\sin\Psi \tag{3-3b}$$

For simple equipment support structures such as instrument transformers and disconnect switches, and overhead shield wire supports, the application of Equation (3-1) is typically in the principal axes of the support structure ($\Psi = 0$ degree or 90 degrees). If a skewed wind is applied to simple equipment supports, the appropriate force coefficient ($C_{f\Psi}$) and effective wind area (the area projection normal to the wind) in the direction of wind should be used. For $C_{f\Psi}$ values, see ASCE 74 (ASCE 2020) and ASCE 7-22 (ASCE 2022).

The wind force calculated from Equations (3-1) and (3-2) is based on the selection of the wind speed, terrain exposure coefficient, wind directionality factor, topographical factor, gust response factor, wind area, and force coefficient. These parameters are discussed in subsequent sections. For structures supporting wire(s), the wire tension corresponding to the wind loading should be calculated using the temperature that is most likely to occur at the time of the extreme wind-loading events.

The wind loads recommended in this section are primarily based on the provisions of ASCE 74 (ASCE 2020) and ASCE 7-22. The topographical factor, K_{zt}, can be used to account for speed-up effects for unique terrain upwind from a substation. If the engineer determines that this effect is significant, the K_{zt} factor can be calculated using ASCE 74 (ASCE 2020) or ASCE 7-22; otherwise, K_{zt} can be assumed to be 1.0.

The wind directionality factor K_d accounts for the likelihood of the design wind speed occurring from the wind direction causing the maximum wind load on the structure; refer to ASCE 7-22. K_d can be assumed to be 1.0 for substation structures. The use of K_d with a value less than 1.0 for substation structures should be justified by conducting a site-specific wind study.

3.1.5.1 Air Density Factor, Q. The air density factor, Q, converts the kinetic energy of moving air into the potential energy of pressure. For a standard atmosphere, the air density factor is 0.00256 (or 0.613 SI). The standard atmosphere is defined as a sea-level pressure of 29.92 in. (101.32 kPa) of mercury with a temperature of 59 °F (15 °C). The wind speed, V, in Equations (3-1) and (3-2), should be the 3-second gust wind speed expressed in terms of miles per hour when using the constant 0.00256 and meters per second when the constant 0.613 is used.

An air density factor of 0.00256 should be used, except where sufficient elevation data are available to justify a different value. The air density factor will decrease as the altitude or temperature increases, and vice versa. Variations of air density for other air temperatures and elevations that are different from the standard atmosphere are given in ASCE 74 (ASCE 2020) or ASCE 7-22.

3.1.5.2 Terrain Exposure Coefficient, K_z. The terrain exposure coefficient, K_z, accounts for the wind speed variation with height as a function of terrain type (Table 3-1). It is recognized that wind speed varies with height because of ground friction and that the amount of friction varies with ground roughness. The ground roughness is characterized by the various exposure categories described in Section 3.1.5.2.1.

Table 3-1. Terrain Exposure Coefficient, K_z.

Height above ground z (ft)	K_z Exposure B	K_z Exposure C	K_z Exposure D
0–15	0.57	0.85	1.04
30	0.69	0.98	1.17
40	0.74	1.04	1.23
50	0.79	1.09	1.28
60	0.83	1.13	1.32
70	0.86	1.17	1.35
80	0.90	1.20	1.38
90	0.92	1.23	1.41
100	0.95	1.25	1.44

Note: 1 ft = 0.3048 m.

3.1.5.2.1 Exposure Categories. Three exposure categories, B, C, and D, are recommended for use with this MOP. The substation maximum structure height (*H*) can be conservatively used in the determination of the distances for exposure categories and transitions between categories. The maximum substation structure height (*H*) includes all structures within the substation that contribute to the delivery of electric power and excludes shielding masts or free-standing and guyed microwave communication structures. One exposure category is intended for all structures within the substation.

Exposure B: This exposure is classified as urban and suburban areas, well-wooded areas, or terrain with numerous closely spaced obstructions having the size of single-family dwellings or larger. Use of this exposure category should be limited to those areas for which terrain representative of Exposure B roughness prevails in the upwind direction for a distance of at least 2,600 ft (792 m) or 20 times *H*, whichever is greater, for *H* over 30 ft (9.1 m). For an *H* less than 30 ft (9.1 m), Exposure B roughness prevails upwind for a distance of 1,500 ft (457 m) (Figure 3-3).

Exposure C: Open terrain with scattered obstructions having height generally less than 33 ft (10 m). Category C includes flat open country and grasslands but excludes shorelines in hurricane-prone regions. Exposure Category C should be used where the conditions of Exposure Category B or D are not satisfied. Exposure C is the default for this MOP.

Exposure D: Flat, unobstructed areas and water surfaces. The category includes smooth mud flats, salt flats, and unbroken ice. Exposure D applies to winds flowing across a distance greater than 5,000 ft (1,524 m) or 20 times *H*.

Shorelines in Exposure D include inland waterways, the Great Lakes, and the coastal areas of California, Oregon, Washington, and Alaska, and shorelines in hurricane-prone regions. This exposure should apply to those structures exposed to the wind blowing from over the water or other smooth surfaces. Exposure D extends a distance of 600 ft (183 m) or 20 times *H*, whichever is greater (Figure 3-4). Note: Hurricane-prone regions were moved from Exposure C in prior editions of ASCE 7 to Exposure D in ASCE 7 (ASCE 2016) and ASCE 7-22.

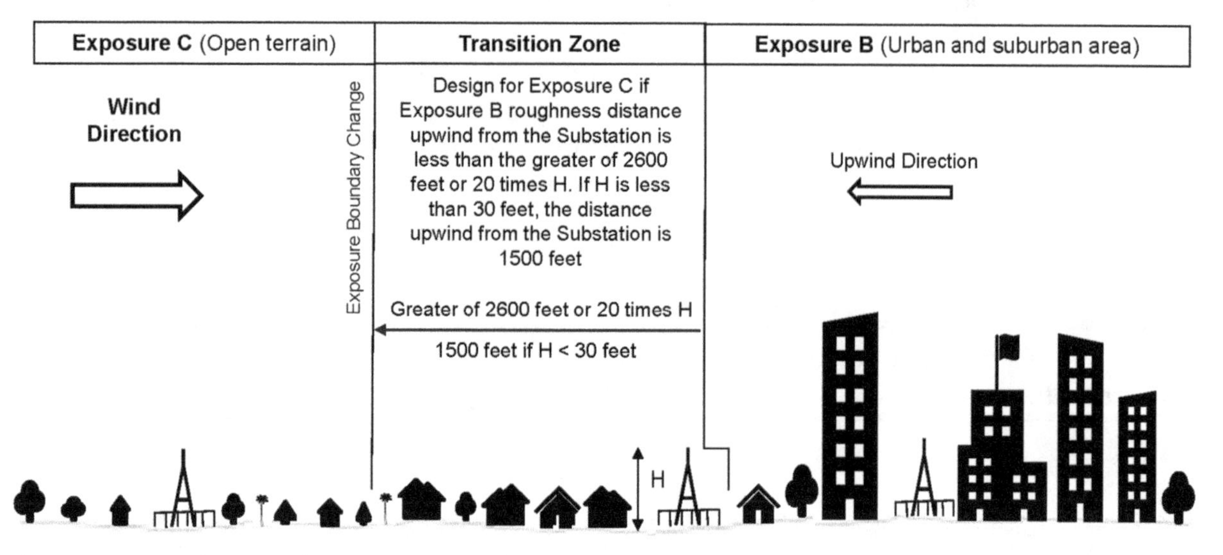

Figure 3-3. Surface conditions required for the use of Exposure Category B and the transition from Exposure C to Exposure B.

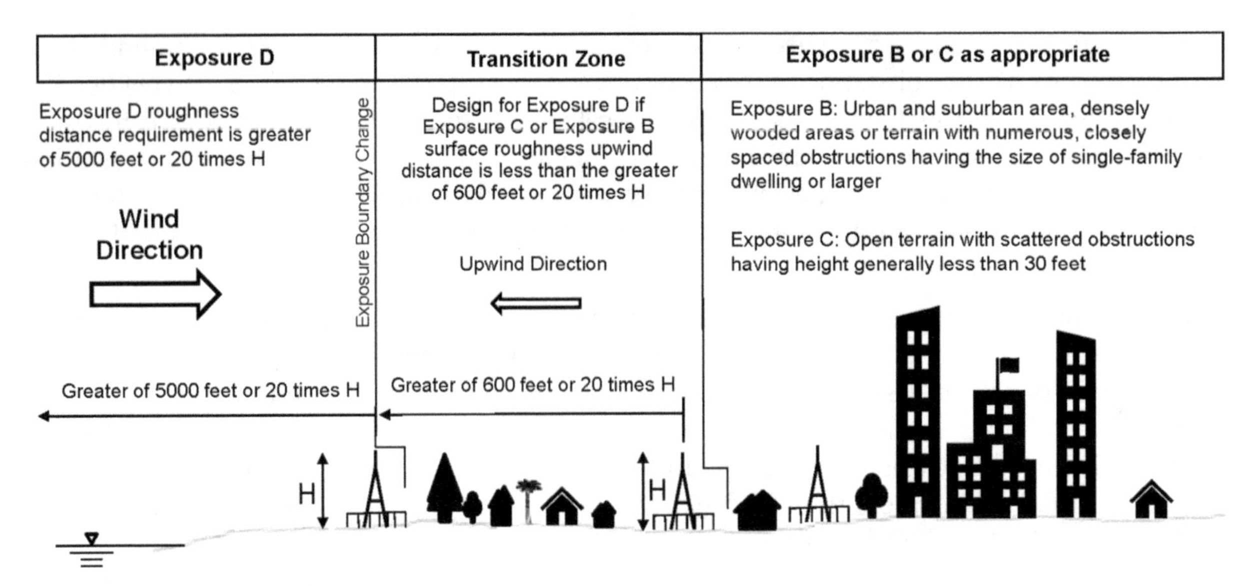

Figure 3-4. Surface conditions required for the use of Exposure Category D and the transition from Exposure D to Exposure C or Exposure B.

Values of the terrain exposure coefficient K_z are listed in Table 3-1 for heights up to 100 ft (30.5 m) above ground. Values for K_z are not given in Table 3-1, and for heights greater than 100 ft (30.5 m), they can be determined using Equation (3-4)

$$K_z = 2.41\left(\frac{z}{z_g}\right)^{2/\alpha} \quad \text{for } 15\,\text{ft}\ (4.6\,\text{m}) \leq z < z_g \tag{3-4}$$

where

z = Height at which the wind is being evaluated,
z_g = Gradient height (Table 3-2),
α = Power law coefficient for a 3-second gust wind (Table 3-2).

Note: The constant, 2.41, for K_Z was changed from the previous version of this MOP (where it was 2.01). The constant 2.41 was first implemented in ASCE 7-22.

If the structure height is less than 15 ft (4.6 m), K_z is taken as the value at 15 ft (4.6 m). The effects of terrain roughness on the wind force are significant. It is essential that the appropriate exposure category be selected after a careful review of the surrounding terrain. Exposure C should be used, unless the engineer has determined with good engineering judgment that other exposures are more appropriate. The power law constants α and z_g for the different terrain categories are given in Table 3-2.

Table 3-2. Power Law Constants.

Exposure category	α Power law coefficient for 3-second gust wind	z_g Gradient height [feet (m)]
B	7.5	3,280 (1,000)
C	9.8	2,460 (750)
D	11.5	1,935 (590)

Note: These table values changed from MOP 113, 1st edition, to these values that are taken from ASCE 7-22 (ASCE 2022).

3.1.5.2.2 Effective Height. An effective height, z, is used for the selection of the terrain exposure coefficient, K_z, the structure gust response factor, G_{SRF}, and the wire gust response factor, G_{WRF}. Sections 3.1.5.5.1 and 3.1.5.5.2 define the location for the effective height for the structure and wire. The effective height for the wire-supporting structures is two-thirds the vertical distance from grade to the top of the structure being analyzed. The effective height includes any foundation projection, any elevated support, and the height of the equipment itself. The effective height for an overhead wire is the vertical distance above grade to the wire attachment point at the structure.

3.1.5.3 Basic Wind Speed, V_{MRI}. The basic wind speed, to be used in Equations (3-1) and (3-2), is the 3-second gust wind speed at 33 ft (10 m) above ground in a flat and open country terrain (Exposure Category C) and associated with a recommended 300-year mean recurrence interval (MRI), shown in Figure 3-5 for USD. The basic wind speed recommended for ASD loads is a 100-year MRI, Exposure Category C, at 33 ft (10 m) above ground. The wind speed values for a 100-year MRI can be obtained from the online ASCE Hazzard Tool application. See Section 3.1.5.4 for making adjustments to the basic wind speed MRI.

There are certain regions in the country, such as mountainous terrain, where topographical characteristics may cause significant variations in wind speed over short distances (ASCE 7-22) (ASCE 2022). In these regions, local meteorological data should be collected or site-specific wind studies performed to establish the design wind speed. General guidelines of developing local wind data are recommended in ASCE MOP 74 (ASCE 2020).

3.1.5.4 Mean Recurrence Interval Wind Speed. The MRI is approximately the reciprocal of the annual probability of occurrence. For example, a 300-year MRI signifies approximately a 0.33% annual probability of wind loading that will exceed or equal the design value. For critical substation structures that require a higher level of reliability, a longer MRI may be considered (Table 3-3).

The selection of a longer MRI provides a method of adjusting the level of structural reliability. The use of a 300-year MRI does not imply that the structures are not important. Rather, it represents a good understanding of the probabilities of failure and required structural reliability. It is the responsibility of the owner's engineer of record to select an appropriate MRI for their substation structures. Longer MRI (and corresponding higher wind speeds) are provided in ASCE 7-22, Section 26.5.1. It is not recommended that an MRI less than 100 years be used for extreme wind design for hurricane zones.

3.1.5.5 Gust Response Factor. The gust response factors, G_{SRF} and G_{WRF}, account for the dynamic effects of gusts on the wind response of structures and wires, respectively. The gust response factors originally presented by Davenport for transmission line towers have been modified for compatibility with current ASCE 7-22 wind-loading procedures [see ASCE 74 (2020), Appendix F, for a complete discussion].

3.1.5.5.1 Structure Gust Response Factor G_{SRF}. The structure gust response factor, G_{SRF}, accounts for the response of the substation structure (dead-ends, equipment and bus support structures, lightning, and shield wire masts) to the wind gust. It is used for computing the wind loads acting on substation structures.

G_{SRF}, for structures supporting equipment and bus (rigid and flexible), but not overhead wires, can be assumed as a constant value of 1.0.

G_{SRF}, for overhead wire–supporting structures (dead-end and line termination structures) and other non-overhead wire–supporting structures, including freestanding shielding masts

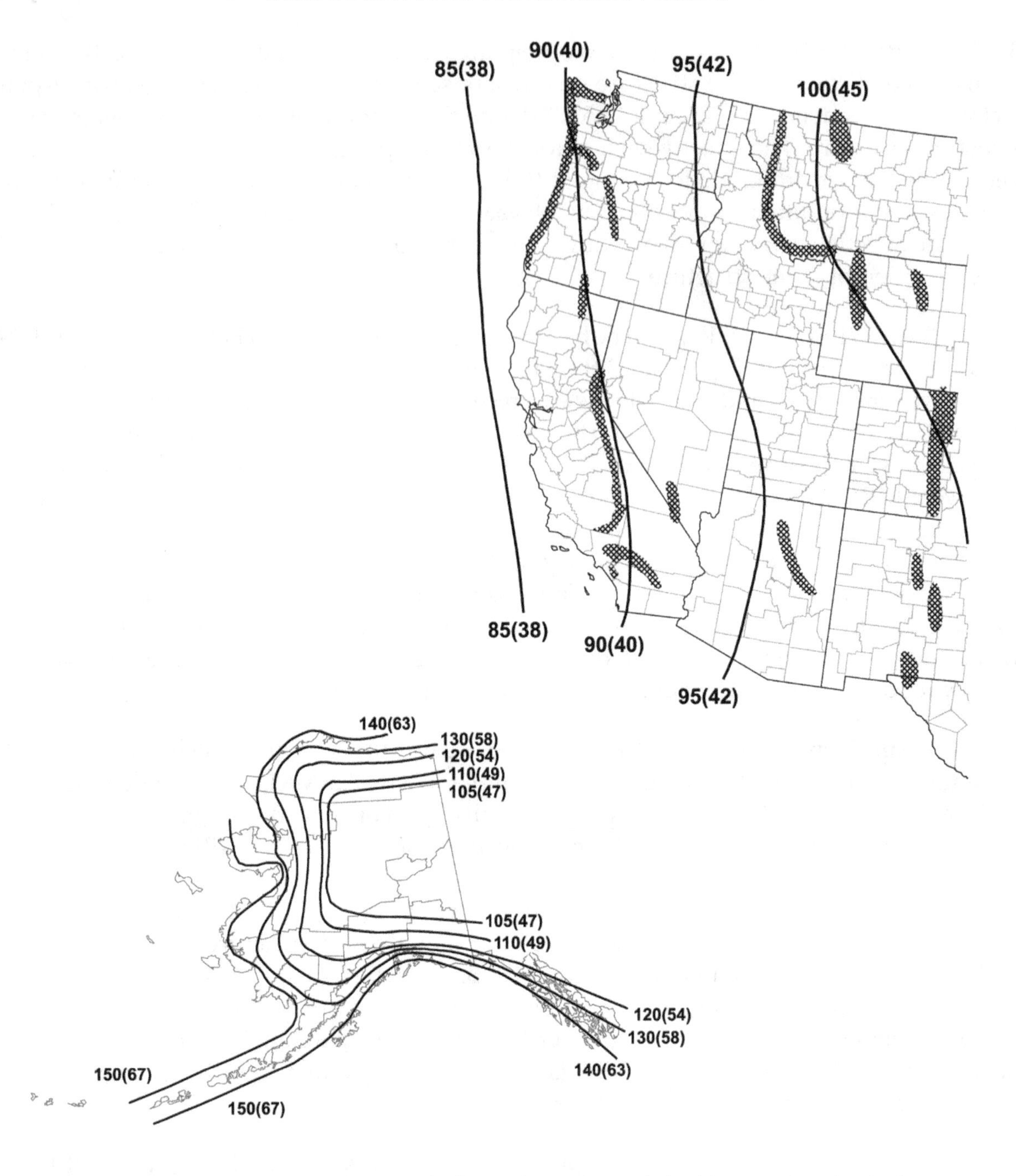

Figure 3-5. 300-year MRI 3-second gust wind speed map in mph (m/s) at 33 ft (10 m) above ground in Exposure C, ASCE 7-22, for USD load combinations in Table 3-18.
Notes: (1) Values are nominal design 3-second gust wind speeds in mph (m/s) at 33 ft (10 m) above ground for Exposure Category C. (2) Linear interpolation is permitted between contours. Point values are provided to aid with interpolation. (3) Islands, coastal areas, and land boundaries outside the last contour shall use wind speed contour. (4) Mountainous terrain, gorges, ocean promontories, and special wind regions shall be examined for unusual wind conditions. (5) Wind speeds correspond to an approximately 15% probability of exceedance in 50 years (annual exceedance probability = 0.00333, MRI = 300 years). (6) Location-specific basic wind speeds can be determined using the ASCE 7 Hazard Tool website.

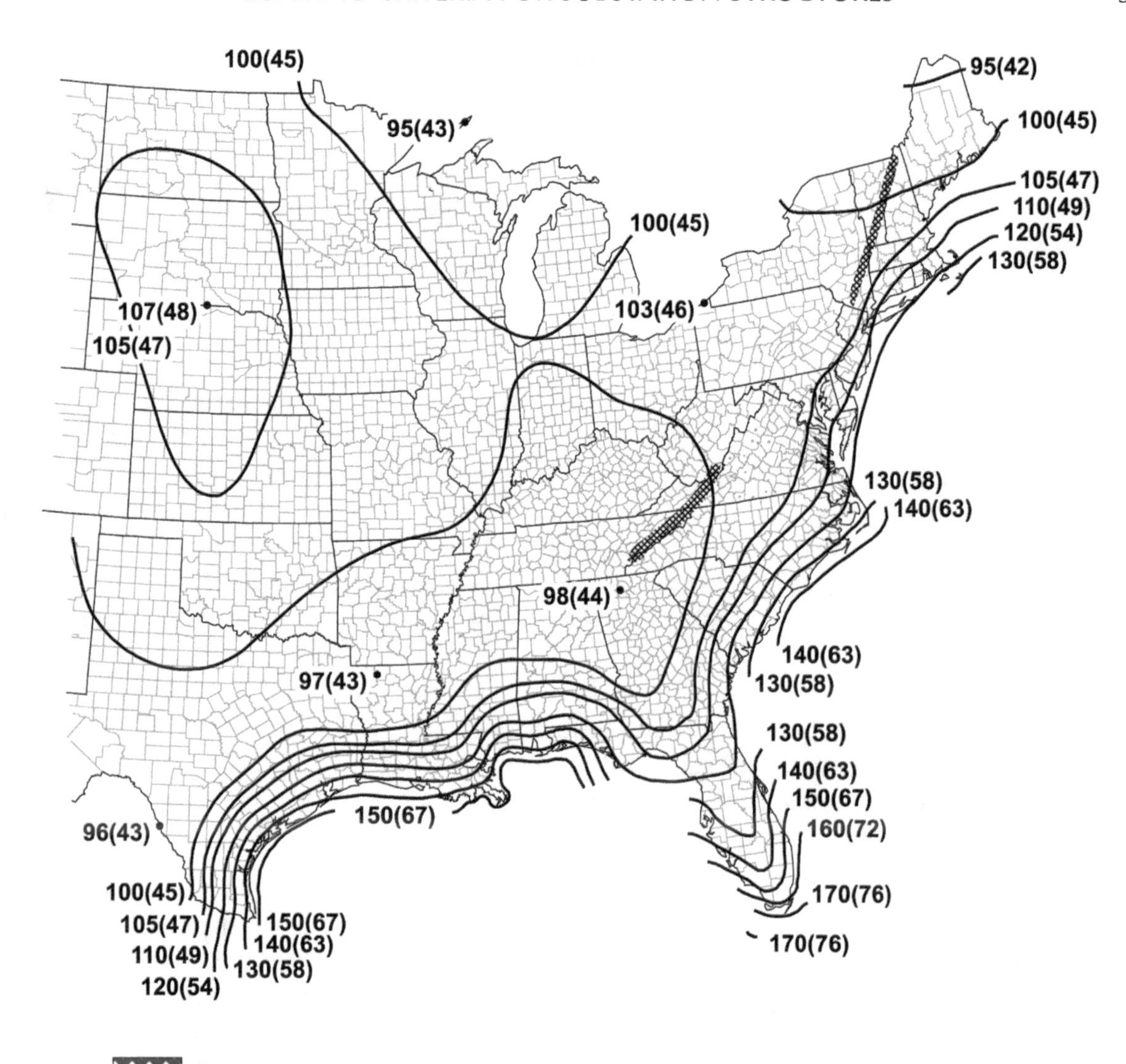

Figure 3-5. (Continued)

(with fundamental frequency less than 1 Hz), are based on Davenport's (1979) wind load model. The gust response factors used in this MOP for these structure types do not include the dynamic resonant response of the structure or wires. This concept is consistent with ASCE 74 for transmission line structures.

Table 3-3. Exceedance Probability for Various MRIs.

Typical conditions	MRI (Years)	Probability that the MRI load is exceeded in any one year (%)	Probability that the MRI load is exceeded at least once in 50 years (%)
Reduced reliability	100	1.00	40
Recommended Reliability	300	0.33	15
Enhanced reliability	700	0.14	7

The wire-supporting structure and shielding mast gust response factor, G_{SRF}, can be calculated using Equation (3-5). The above ground height (h_s) of the substation structure should be used in this equation. A factor of 0.67 is included in the determination of I_{ZS} used in Equation (3-5). This factor assumes that the average wind force is equal to the force calculated at two-thirds of the structure height, defined as the structure effective height.

$$G_{SRF} = \left[\frac{1+6.1\varepsilon(I_{ZS})B_S}{1+6.1(I_{ZS})}\right] \tag{3-5}$$

where

$I_{ZS} = c_{exp}(33/0.67h_s)^{1/6}$, Turbulence intensity at effective height of the structure (ft);
$I_{ZS} = c_{exp}(10/0.67h_s)^{1/6}$, Turbulence intensity at effective height of the structure (m);
c_{exp} = Turbulence intensity constant, based on exposure (Table 3-5);
$\varepsilon = 0.75$, Wire-supporting structures (dead-end, line termination structures, shield wire masts);
$\varepsilon = 1.00$, Non-overhead wire–supporting structures such as lightning masts
(ε represents the separation factor when the structure supports the wire);
$B_S = [1/(1+0.375(h_s/L_s))]^{0.5}$ quasi-static background response, does not include a dynamic component
h_s = Total above ground height of the structure (minimum h_s listed in Table 3-5); and
L_s = Integral length scale of turbulence (Table 3-5).
Note: h_s and L_s must be in consistent units.
$G_{SRF} = 1.0$ for equipment support structures.
Note: 1 ft = 0.3048 m.

Structure and wire gust response factors, G_{SRF} and G_{WRF}, respectively, are determined using Equations (3-5) and (3-6). This simplified approach is applicable for most practical overhead wire–supporting substation structure types (dead-end and line termination structures) and shielding mast structures. The complete gust response factor equations in Appendix F of ASCE 74 (ASCE 2020) can be considered for the calculation of G_{SRF} and G_{WRF}. The use of the complete equations (wind background and resonant component responses) of Appendix F can be considered for tall [≥250 ft (76 m)] or unique wire-supporting structures, or shielding mast structures with fundamental frequencies of 1 Hz or less. Table 3-4 provides structure gust response factors for different substation structure types.

Table 3-5 contains parameters for Terrain Exposure Categories B, C, and D that are used to calculate the simplified gust response factors. The parameters from Tables 3-2 and 3-5 can be used to replace Table F-1 parameters in Appendix F of ASCE 74 (ASCE 2020) to obtain the complete gust factors G_{SRF} and G_{WRF} for substation dead-end structures and shielding masts.

Table 3-4. Structure Gust Response Factor for Different Substation Structure Types.

Type of structure	Structure gust response factor G_{SRF}
Overhead wire support structures: lattice or tubular dead-end, shield wire mast, box structure, and so on	Equation (3-5) or Table 3-6 with $\varepsilon = 0.75$
Rigid and flexible equipment support	1.0
Non-overhead wire support structures: lightning mast/pole	Equation (3-5) or Table 3-6 with $\varepsilon = 1.0$

Table 3-5. Parameters for Calculation of the Gust Response Factor by Exposure.

Exposure	c_{exp}	L_s (ft)	L_s (m)	h_s min (ft)	h_s min (m)
B	0.3	320	97.54	30	9.14
C	0.2	500	152.4	15	4.57
D	0.15	650	198.12	7	2.13

To help the user of this MOP, Table 3-6 provides structure gust response factors that can be used for structures up to 100 ft (30 m) in height, measured from ground elevation. The G_{SRF} values shown in the table vary by less than 2% over this height range. Equation (3-5) can be used to calculate G_{SRF} values in the table and for heights above 100 ft (30 m). When the substation structure height h_s is lower than the minimum for the exposure category given in Table 3-5, the minimum h_s is used in Equation (3-5).

3.1.5.5.2 Wire and Strain Bus Gust Response Factor, G_{WRF}. Substation structures supporting overhead wire and strain bus (both will be referred to as wire) should include the wire gust response factor (G_{WRF}) for determining the wind loads acting on the wire. This factor accounts for the response of the wire system to the wind gust.

G_{WRF} is a function of the exposure category, design wind span between structures, and the effective height. The wire gust response factor can be calculated using Equation (3-6). In Equation (3-6), the effective height for an overhead wire is the height above the ground to the wire attachment point at the structure.

$$G_{WRF} = \left[\frac{1+4.6(I_{ZW})B_W}{1+6.1(I_{ZW})}\right] \tag{3-6}$$

where

$I_{ZW} = c_{exp}(33/h_w)^{1/6}$, Turbulence intensity at effective height of the wire (ft);
$I_{ZW} = c_{exp}(10/h_w)^{1/6}$, Turbulence intensity at effective height of the wire (m);
c_{exp} = Turbulence intensity constant, based on exposure (Table 3-5);
$B_W = \sqrt{1/(1+0.8(L/L_s))}$, quasi-static background response (does not include a dynamic component);
L = Wire horizontal span length;

Table 3-6. Structure Response Factor (G_{SRF}) for Structure Heights $\leq$ 100 ft (30 m) for Equipment Support Structures and Wire Support Structures

Height	Exposure Category B		Exposure Category C		Exposure Category D	
(ft)	G_{SRF} equip. supports	G_{SRF} wire supports	G_{SRF} equip. supports	G_{SRF} wire supports	G_{SRF} equip. supports	G_{SRF} wire supports
10 < 45	1.00	0.83	1.00	0.85	1.00	0.87
46 < 100	1.00	0.82	1.00	0.86	1.00	0.88
> 100	Use Equation (3-5)					

Notes:

1. Equation (3-5) can be used to calculate G_{SRF} values [in the table and above 100 ft (30 m)].
2. G_{SRF} = 1.00 for equipment support structures.
3. G_{SRF} for wire support structures where $\varepsilon = 0.75$.

h_w = Wire and strain bus attachment height on the structure; and
L_s = Integral length scale of turbulence (Table 3-5).

Note: h_w and L_s must be in consistent units.
Note: 1 ft = 0.3048 m.

Equation (3-6) is taken from ASCE 7 (ASCE 2022) and excludes the resonant response terms. This simplified approach is applicable for most practical overhead wire installations supported on substation structures. Table 3-7 gives the wire gust response factor for Terrain Exposure Categories B, C, and D with a span length range of up to 750 ft (229 m) and a wire effective height of up to 100 ft (30 m). The G_{WRF} values shown in the table vary by less than 5% over the selected height and span length ranges. Equation (3-6) with the appropriate exposure category constants given in Table 3-5 can be used to determine G_{WRF} values in the table and outside the table parameter ranges.

3.1.5.6 Force Coefficient. The force coefficient, C_f, in the wind force formula, Equations (3-1) and (3-2), is the ratio of the resulting wind force per unit area in the direction of the wind to the applied wind pressure. This coefficient takes into account the effects of the structural member's wind characteristics (such as shape, size, and orientation with respect to the wind, solidity, shielding, and surface roughness). It is also referred to as a drag coefficient, pressure coefficient, or shape factor.

The ratio of a member's length to its diameter (or width) is known as the aspect ratio. Shorter members have lower force coefficients than longer members of the same shape. The force coefficients given in the tables of this section are applicable to members with aspect ratios greater than 40. Correction factors for members with an aspect ratio of less than 40 may be used to determine a correction C_f' that is substituted with C_f in Equations (3-1) and (3-2), as follows:

$$C_f' = (c)C_f \tag{3-7}$$

where c is the correction factor from Table 3-9, and C_f is the force coefficient from Tables 3-8 and 3-10.

3.1.5.6.1 Lattice Structure Force Coefficients. The term "yawed wind" is used to describe winds whose angle (Ψ) of incidence with a structure is other than perpendicular. The maximum effective wind on square base lattice structures occurs at a yaw angle of slightly less than 45 degrees (Bayar 1986, BEAIRA 1935). The effect of these wind loads is typically 12% to 15% greater than the calculated perpendicular wind loads.

Table 3-7. Wire Gust Response Factors (G_{WRF}), Effective Heights up to 100 ft.

Wind exposure category	Span length $L < 100$ ft	Span length $100 \leq L < 250$ ft	Span length $250 \leq L < 500$ ft	Span length $500 \leq L \leq 750$ ft
B	0.85	0.80	0.75	0.70
C	0.88	0.85	0.82	0.78
D	0.89	0.88	0.85	0.82

Notes:

1. Equation (3-6) can be used to calculate G_{WRF} values in the table and outside the table parameter ranges.
2. Values in the table were generated for a 100 ft wire height and the minimum span listed in the range. (3) 1 ft = 0.3048 m.

An important factor that influences the force coefficient of lattice truss structures is the solidity ratio of the frame. The force coefficient for the total structure is dependent on the airflow resistance of individual members and on the airflow pattern around the members. For lattice tower structures that are less than 200 ft (61 m) in height, the solidity ratios for the various tower panels over the height of the transverse and longitudinal faces may be averaged to simplify the wind load calculation. Solidity ratio, Φ, is defined as the ratio of the area of all members in the windward face to the outline area of the windward face of a latticed structure.

When two members are placed in line with the wind, such as in a lattice structure, the leeward frame is partially shielded by the windward frame. The shielding factor is influenced by the solidity ratio, Φ, spacing between frames, and yaw angle. For all wind directions considered, the area, A, consistent with the specified force coefficients in Equation (3-2) should be the solid area of a tower face projected on the plane of that face for the tower segment under consideration. The specified force coefficients are for towers with structural angles or similar flat-sided members. Wind forces should be applied in the directions resulting in maximum member forces and reactions.

Tables 3-8 and 3-10 provide the force coefficients recommended by ASCE 74 (2020). Equation (3-8a) gives a correction factor, C_c, for converting the C_f for flat-sided member values in Table 3-10 to C_f values for round-section members. These values are assembled from the latest boundary-layer wind-tunnel and full-scale tests and from previously available literature. Other

Table 3-8. Force Coefficients, C_f, for Structural Shapes, Bus, and Members Commonly Used in Substation Structures.

Member shape	Force coefficient, C_f
Structural shapes (average value)	1.6
Structural circular pipes and round tubes	0.9
Hexadecagonal (16-sided polygonal)	0.9
Dodecagonal (12-sided polygonal)	1.0
Octagonal (8-sided polygonal)	1.4
Hexagonal (6-sided polygonal)	1.4
Square and rectangle structural tubes, square shapes	2.0
Round rigid bus, strain or cable bus, and stranded conductor	1.0
Insulators (station post, suspension, and strain)	1.0

Note: The 1.6 value has been commonly used for structural shapes such as angles, channels, and wide flange–type shapes. ASCE MOP 74 (ASCE 2020); see Appendix G for these and other structural shapes.

Table 3-9. Aspect Ratio Correction Factors.

Aspect ratio	Correction factor (c)
0–4	0.6
4–8	0.7
8–40	0.8
> 40	1

Table 3-10. Force Coefficients, C_f, for Normal Wind on Latticed Structures Having Flat-Sided Members.

Tower cross section	C_f
Square	$4.0\Phi^2 - 5.9\Phi + 4.0$
Triangle	$3.4\Phi^2 - 4.7\Phi + 3.4$

Note: Φ is the solidity ratio.

force coefficients can be used where justified by experimental data. Additional background information on force coefficients can be found in ASCE 7-22 and ASCE 74 (2020)

$$C_c = 0.51\Phi^2 + 0.57, \text{ but not greater than } 1.0 \quad (3\text{-}8a)$$

Note: Φ is the solidity ratio.

$$C_f\ (\text{round}) = C_f(\text{flat}) \times C_c\ (\text{correction}) \quad (3\text{-}8b)$$

3.1.5.7 Application of Wind Forces to Structures. The wind forces for lattice-type towers determined by Equation (3-2) using the recommended force coefficients of this MOP have accounted for both windward and leeward faces, including shielding. The wind forces calculated on a complete lattice truss system can be distributed to the panel points of the structure without further consideration.

Equipment, or portions of a structure, can be partially shielded from the wind by an adjacent item. This is known as "wind shading" or "wind shielding." A slight change in the wind direction will usually eliminate or reduce the amount of wind shielding. For these reasons, the effects of wind shielding are normally not considered for substation equipment support structures.

Where the lattice truss systems in a structure or individual tubular shaft members of a frame-type structure are separated, the windward faces and leeward faces should be considered as each being individually exposed to the calculated wind force with appropriate force coefficients.

It is up to the designer to choose the effective diameter for wind on substation post insulators. The outside diameter of the sheds has been used conservatively.

3.1.6 Ice Loads with Concurrent Wind

Glaze ice is the most common ice load. This MOP recommends a 100-year MRI ice thickness for USD loads and a 50-year MRI ice thickness for ASD loads. The map shown in Figure 3-6a is modified from ASCE 7-22 (500-year MRI) to obtain a 100-year MRI event by using the 0.6 conversion factor (Figure 3-6d) on the ice thickness values. ASCE 7-22 provides conversion factors to be used with the 500-year MRI map to obtain a 100-year MRI (0.60) or a 50-year MRI (0.50) glaze ice thickness value. The conversion factors shown in Figure 3-6d are taken from ASCE 7-22, Table C10.4-2. The maps shown in Figure 3-6a are for the continental United States. Ice thickness values for Alaska, Great Lakes, and Colombia River Gorge should be obtained from the ASCE 7 Hazard Tools website (500-year MRI) and converted to the 100-year MRI or 50-year MRI values using the conversion factors in Figure 3-6d.

Concurrent wind speed to be used with all return periods is shown in Figure 3-6b. Ambient temperatures to be used concurrent with glaze ice thicknesses caused by freezing rain are shown in Figure 3-6c. Specific local ice thickness values should be obtained from the ASCE 7 Hazard Tool website (500-year MRI) and converted to the 100-year MRI or 50-year MRI values using the conversion factors in Figure 3-6d.

In certain geographical areas, other types of ice loads, such as rime or in-cloud ice, wet snow, and hoarfrost, may become important in the design of substation structures. For information on non-glaze ice loads, meteorological and engineering studies can be conducted to properly account for non-glaze ice loads in design practice.

(a) West

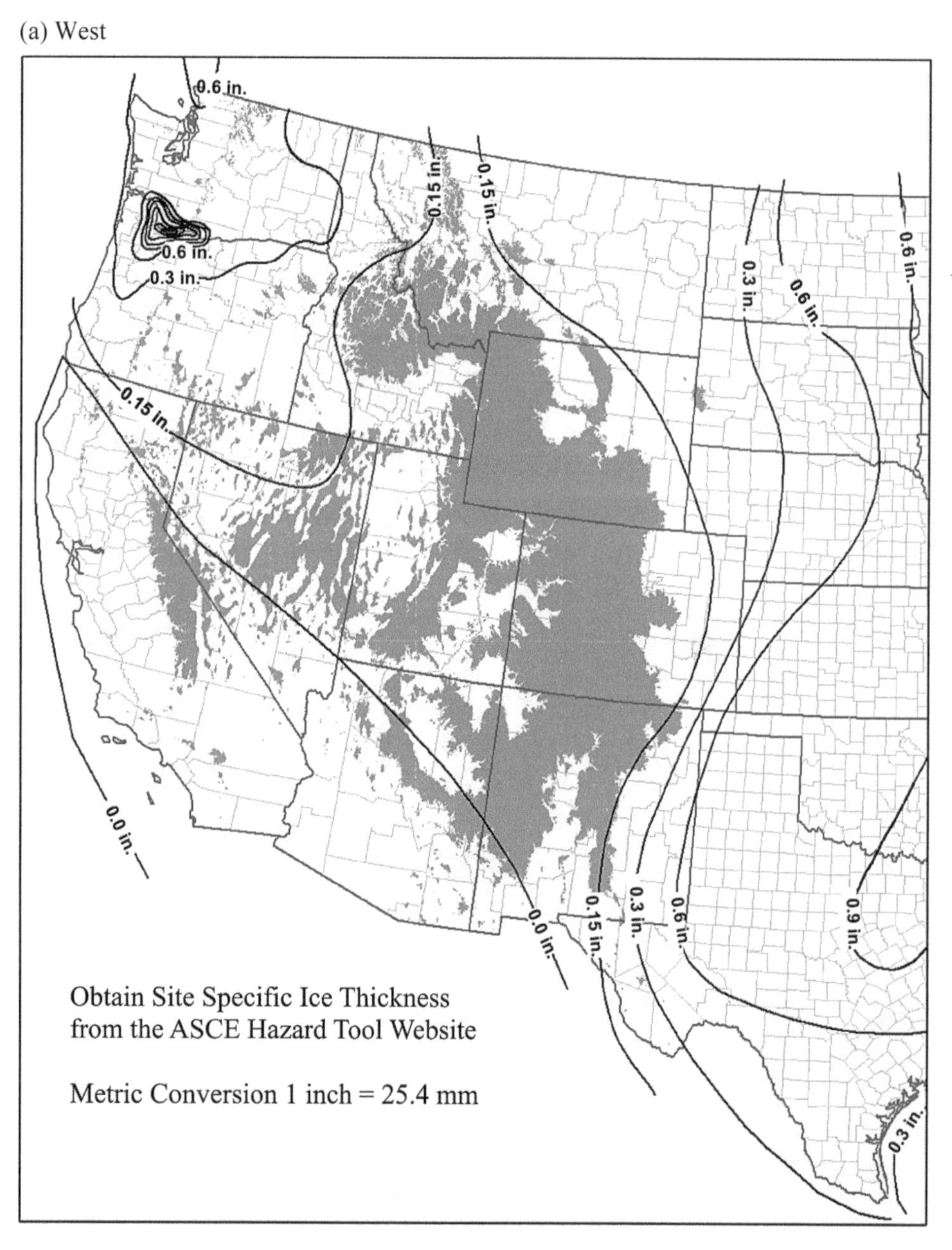

Figure 3-6. (a, West, East) 100-year MRI radial glaze ice thickness (in.) caused by freezing rain, to be used with (b) Concurrent wind speed, western United States, for USD load combinations in Table 3-18. (b) Concurrent wind speed 33 ft (10 m) above ground in Exposure C, for USD load combinations in Table 3-18 and ASD load combinations in Table 3-19, ASCE 7-22. (c) Temperatures concurrent with ice thicknesses caused by freezing rain, ASCE 7-22. (d) Ice thickness conversion factors for return periods other than a 500-year MRI.

(a) East

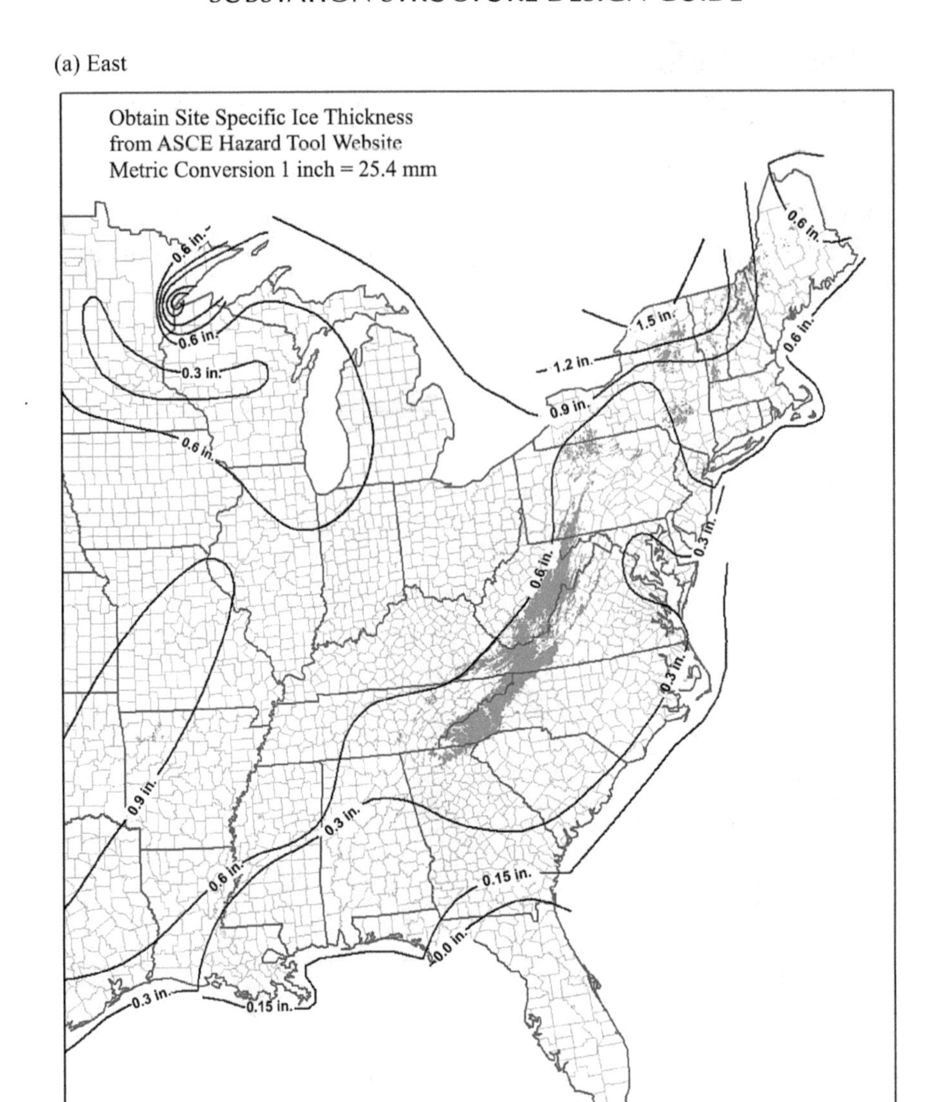

(b)

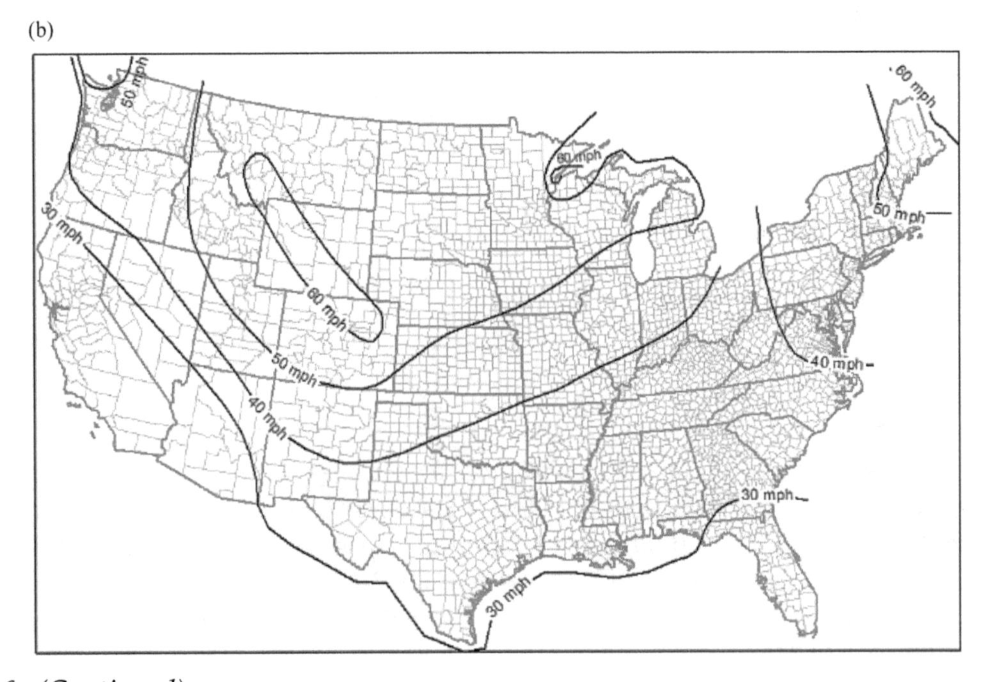

Figure 3-6. (Continued)

(c)

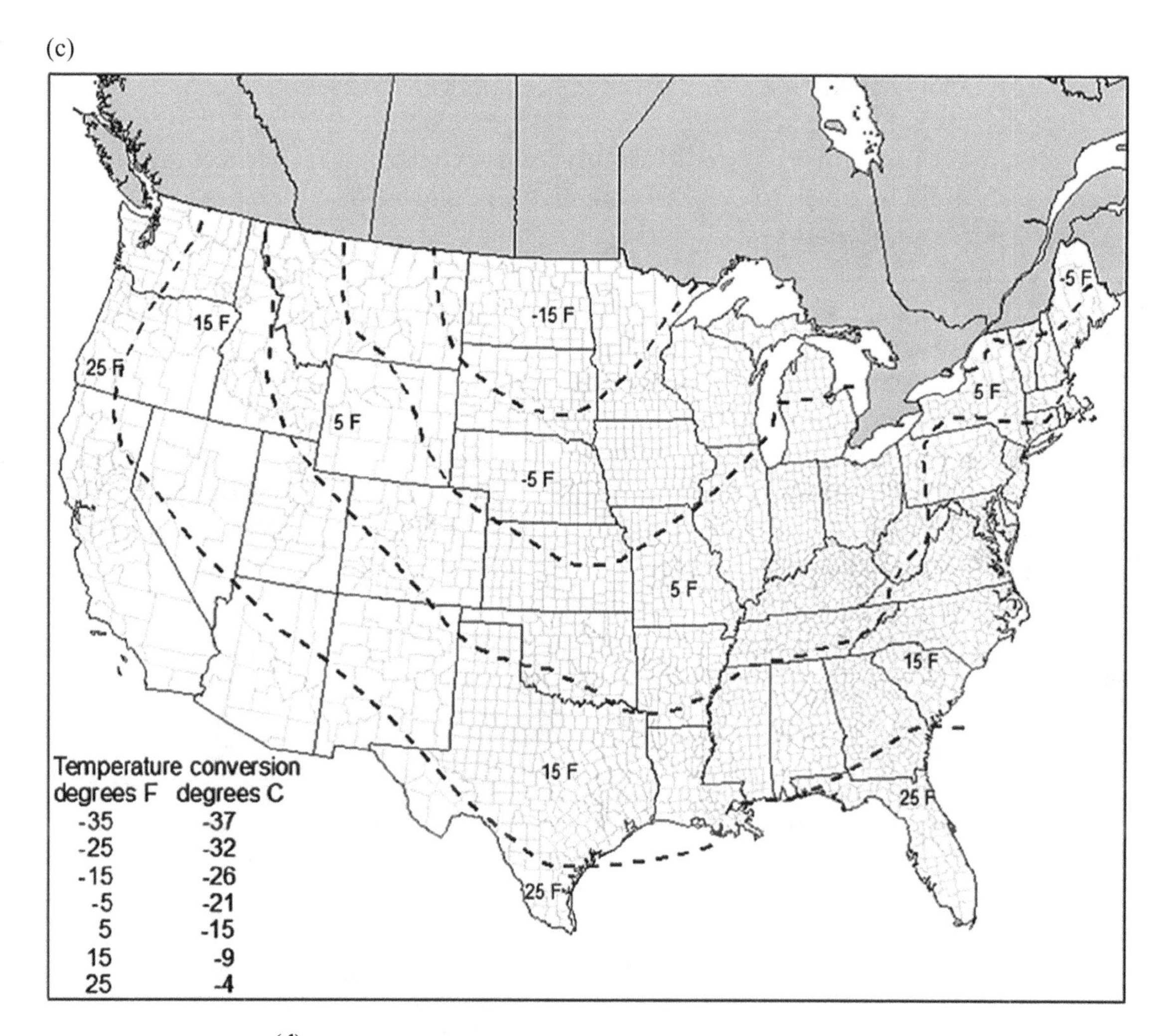

(d)

Mean Recurrence Interval (MRI)	Ice Thickness Conversion Factor
50 year	0.5
100 year	0.6
500 year	1.0

Figure 3-6. (Continued)

This MOP recommends a minimum of a 100-year MRI for glaze ice design. When necessary, a longer MRI can be selected on the basis of site conditions. The engineer is responsible for selecting, from available data, the most appropriate ice thickness to use for the location of the facility being designed. Utilities can conduct icing studies, with the assistance of a consulting meteorologist with ice expertise, to develop more accurate ice loading for substation sites in their service areas.

The ice thickness variation with height above ground is described in Section 3.1.6.2.

3.1.6.1 Effects of an Icing Event on Structures. The effect of ice includes the dead load of the ice in combination with the increased wind load caused by the added surface area. To calculate ice loads for the design of substation structures, glaze ice density is assumed to be a minimum of 56 lb/

ft^3 (pcf) (900 kg/m^3). Rime ice can accumulate to a much larger thickness than glaze ice adding to the surface area, but its density is lower than glaze ice and may be assumed to be 25 pcf (400 kg/m^3). Rime ice-density values range from 19 to 56 pcf (300 to 900 kg/m^3) for soft rime and hard rime.

Engineers need not consider ice loads on every structure or structural component. Considerations should be given to only ice-sensitive structures for which the load effects from atmospheric icing in combination with wind control the design of a part or an entire structural system. For example, in a dead-end structure design, the ice load on the conductor is included in the design, but the ice load on the structure is often ignored because of the relatively small effect on the structure.

The effect of ice includes the dead load of the ice and the wind loading on the ice-covered members. In a substation, ice-sensitive structures include equipment and rigid bus conductors and supports. Considerations should be given to only ice-sensitive structures. In addition, ice loads may be applied to only selected components in ice-sensitive structures. The engineer should verify that the effect of ice on the structural members can indeed be ignored, especially for lattice-type structures where the increase in the projected area of structural members with ice for wind loading can be significant.

Glaze ice should be considered to occur on all surfaces exposed to precipitation. For openings or perforations smaller than the specified glaze ice thickness, the area can be considered as being continuous.

3.1.6.2 Ice Thickness Variation with Height. The amount of ice that accretes on a wire or structure depends on the temperature and wind speed at the wire or structure height. Design thicknesses of glaze ice t_z for heights z above ground can be obtained from Equation (3-8c, d).

$$t_z = t_{\mathrm{MRI}}\left(\frac{z}{33}\right)^{0.10} \quad \text{for } 0 < z < 900\,\text{ft} \tag{3-8c}$$

$$t_z = t_{\mathrm{MRI}}\left(\frac{z}{10}\right)^{0.10} \quad \text{for } 0 < z < 275\,\text{m} \tag{3-8d}$$

where

t_z = Design ice thickness at height z above ground,
t_{MRI} = Nominal ice thickness (e.g., t_{100} for a 100-year MRI), and
z = Height above ground (feet or meters).

The design ice thickness t_z may be used for Exposures B, C, or D. Structures and wires less than 33 ft (10 m) height above ground may conservatively use the ice thickness t_{MRI}.

3.1.7 Seismic Loads

Earthquakes can interrupt the delivery of power in several ways. Low-level ground motion may trip equipment relays but without causing any long-term damage, which may require remote or a manual re-energizing of relays. Moderate-level ground motion may cause minor repairable equipment damage. Major-level ground motion may cause equipment damage, destruction, or both. Earthquake hazards are primarily determined from data obtained from the United States Geological Survey (USGS). Earthquakes may also be induced by hydraulic fracturing, filling of large reservoirs, and other human activities. Such effects are currently under study by USGS, and the owner should consider such effects on a case-by-case basis.

Seismic loads, unlike wind loads, are not loads that are applied directly on the structure. Seismic loads include five phenomena: ground motion, liquefaction, landslide, fault offset, and tsunami. This MOP provides guidance on only how to address ground motion. Available USGS national seismic hazard maps are typically adequate to determine horizontal ground motion. Seismic hazards are quantified by parameters such as peak ground acceleration (PGA), peak ground velocity (PGV), and response spectra (RS). Using the principles of dynamics, PGA, PGV, and RS can be converted into equivalent loads that are applied on the structure for design purposes to approximate the structural effects (deformations, strains, stresses) caused by earthquake ground motion. Structures respond to earthquake ground motion by acceleration. The acceleration of the structure's mass creates forces that impart deformations, strains, and stresses. The amplitude and acceleration of ground motion at the base of the structure is a function of its location (latitude, longitude, proximity to faults), soil types (above bedrock), and assumed return period for the earthquake. The response of the structure to this ground motion is dependent on structural characteristics such as stiffness, distribution of mass, fundamental frequencies, and damping. Soil–structure interaction can impact structure demands by reducing the effective fundamental frequency or increasing the total structure displacement. The effects of the soil–structure interaction on the structure could be considered in the design where applicable.

3.1.7.1 Purpose and Scope. This section presents recommendations for seismic loads for substation structure types not covered in IEEE 693 (IEEE 2018c) but that are commonly encountered in substation design practice. IEEE 693 (IEEE 2018c) covers dedicated substation equipment support structures (1.4.3). IEEE 693 (IEEE 2018c) also covers primary substation structures (1.4.4) that have an intermediate support for equipment not qualified on a dedicated support (5.10.7) and qualified on a dedicated support but installed on a primary substation structure (5.10.5 and 5.10.7). The recommendations of this section are intended for new designs. This section primarily provides guidance for the seismic design of substation structures, as described in Section 6.7.2, such as dead-ends, racks, bus supports, and shielding masts. The seismic performance of existing structures and alterations to these structures is discussed in Chapter 11. Substation structures and associated foundations with the sole purpose of supporting a particular substation equipment type, such as instrument transformer, capacitors, and disconnect switch, are covered by IEEE 693 (IEEE 2018c). In addition, the design of anchor rods for IEEE 693 (IEEE 2018c) substation structure types is covered by this MOP. The structural design of the bus supports (whether rigid bus, strain bus, or cable bus) that interconnect equipment in the switchyard is covered by this MOP. The seismic load from the bus must be resisted by any attached equipment.

The seismic design loads provided in this MOP are obtained by factoring the maximum considered earthquake (MCE) (Section 3.1.7.2) loads by two-thirds [Equations (3-9) and (3-10)] and by further dividing by the response modification factor $R \geq 1$, [Table 3-13, Equation (3-12)]. This will result in forces in structural elements that are less than that corresponding to a structure that remains in a linear-elastic state of stress when subjected to the MCE. Consequently, a structure designed to this MOP when subjected to the MCE may result in structural elements that yield, buckle, or exhibit inelastic behavior. The yielding of structural elements and the inelastic deformations will reduce the structural stiffness (increase deflections) and the corresponding effective response frequency, which for most structures will reduce the structural demand. Further reduction in structural demand is enabled by energy dissipation from the inherent damping prior to structural yielding and subsequent hysteretic damping with yielding and inelastic deformations. This approach of designing for loads less than those corresponding to the MCE will provide an acceptable structural performance when the structure has adequate ductility and a continuous load path. It is important to ensure that the connections have a higher capacity than the members to maintain structural integrity when they perform in the inelastic range. This is achieved by designing these connections to the prescribed seismic forces amplified by the overstrength factor Ω_0 ($\Omega_0 > 1.0$) (Table 3-13). Substation structures designed in accordance with Table 3-13 are not required to meet the special seismic provisions in AISC 341-16 (AISC 2016).

Substation structures designed in accordance with this MOP typically are not controlled by earthquake-induced inertia forces, as other traditional loads such as extreme wind, with and without ice, and wire tensions provide adequate capacity to account for the demands of an earthquake. Structures located in low seismic hazard areas with $S_{DS} < 0.167$ and $S_{D1} < 0.067$ need not be designed to the seismic provisions in Section 3.1.7, where S_{DS} and S_{D1} are determined in accordance with Section 3.1.7.2.

Substation equipment and their supports should be designed in accordance with IEEE 693 (IEEE 2018c). Seismic loads for the design of foundations for equipment and their supports should be as specified in IEEE 693 (IEEE 2018c). Seismic loads for the design of anchorages of equipment and their supports should be as specified by IEEE 693 (IEEE 2018c) and as modified by the overstrength factors provided by this MOP.

Figure 3-7 shows the USGS Relative Seismic Hazard map. This map is shown only for informational purposes. Spectral response acceleration maps obtained from the ASCE 7 Hazard Tool website can be used to determine the design acceleration level.

3.1.7.2 Seismic Ground Motion Acceleration Parameters. The site seismic ground motion acceleration parameters can be determined from the ASCE 7 Hazard Tool website. The ground motion spectral response acceleration for 5% critical damping corresponding to a short period is referred to as S_{MS}, and the 1.0 s period value is referred to as S_{M1} (long period). The USGS maps are based on risk targeted MCE for 1% probability of collapse within a 50-year period, which is acceptable for substation structures that are typically located within a restricted fenced yard and not accessible to the public.

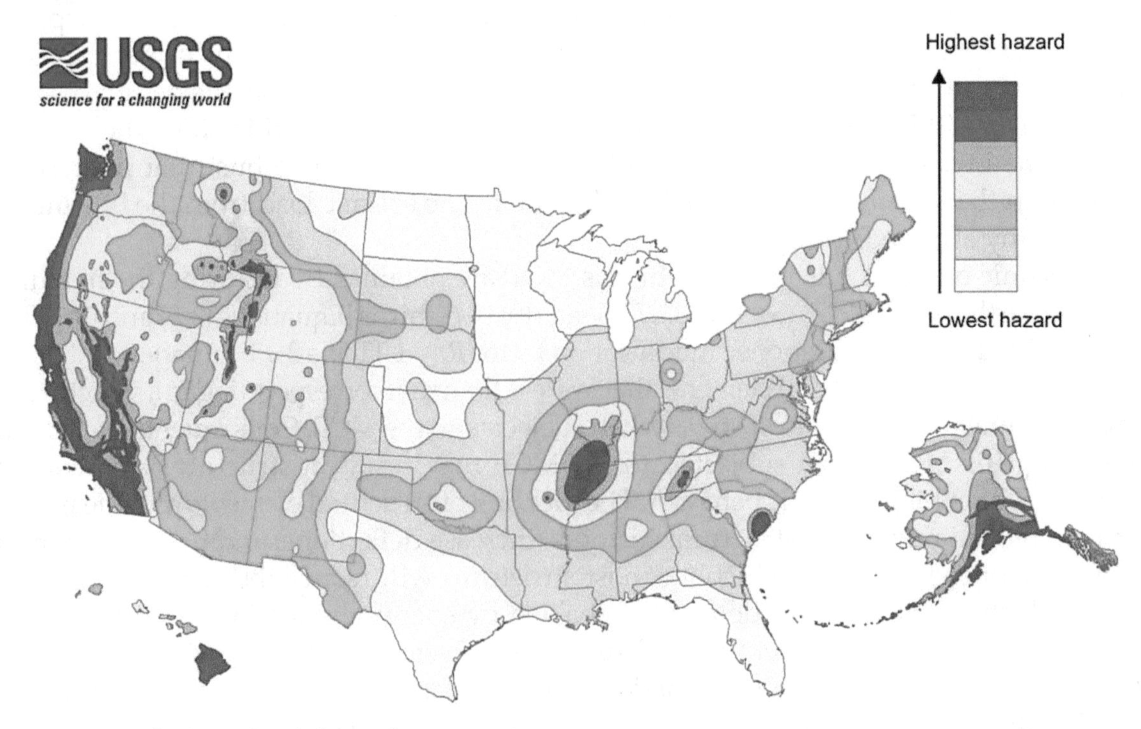

Figure 3-7. Relative seismic hazard map.
Source: USGS (2018).

Table 3-11. Soil Site Class and the Associated Soil Type.

Soil site class	Soil description
A	Hard rock
B	Medium hard rock
BC	Soft rock
C	Very dense sand or hard clay
CD	Dense sand or very stiff clay
D	Medium dense sand or very stiff clay
DE	Loose sand or medium stiff clay
E	Very loose sand or soft clay
F	Very poor soil (determined by a geotechnical engineer in accordance with ASCE 7-22, Chapters 20.2.1 and 21.1)

The ground motion spectral response acceleration values (S_{MS} and S_{M1}) are adjusted on the basis of soil site classes. The soil site classes are based on different soil types and associated shear wave velocity (Vs_{30}) measured in the upper 100 ft (30 m) of the soil profile. Table 3-11 shows different soil site classes based on soil description. Table 3-12 shows the shear wave velocity for associated soil site classes.

Where the soil properties are not known in sufficient detail to determine the soil site class, spectral response accelerations shall be based on the most critical spectral response acceleration at each period of Site Class C, Site Class CD, Site Class D, and Site Class DE, unless the geotechnical data determine that Site Class E or F soils are present at the site, subject to the limitations described in Section 3.1.7.2.1. A geotechnical engineer should be consulted for determining site-specific soil conditions. Additional information can be found in the commentary to ASCE 7-22 and FEMA P-2082/1 (FEMA 2020).

3.1.7.2.1 Site-Specific Ground Motion Procedures. The definitions of near-fault sites given in ASCE 7-22, Chapter 11.4.1, should be used when applicable: 9.5 mi (15 km) or less from the surface projection of a known active fault capable of producing M_w 7 or larger events, or 6.25 mi (10 km) or less from the surface projections of a known active fault capable of producing M_w 6 or larger events, but smaller than M_w 7.

Table 3-12. Soil Site Class–Associated Shear Wave Velocity.

Soil site class	Shear wave velocity Vs_{30} (ft/s)	Shear wave velocity Vs_{30} (m/s)
A	$>$ 5,000	$>$ 1,524
B	$>$ 3,000–5,000	$>$ 914–1,524
BC	$>$ 2,000–3,000	$>$ 610–914
C	$>$ 1,450–2,000	$>$ 442–610
CD	$>$ 1,000–1,450	$>$ 305–442
D	$>$ 700–1,000	$>$ 213–305
DE	$>$ 500–700	$>$ 152–213
E	$\leq$ 500	$\leq$ 152
F	Determined by a geotechnical engineer in accordance with ASCE 7-22, Chapters 20.2.1 and 21.1	Determined by a geotechnical engineer in accordance with ASCE 7-22, Chapters 20.2.1 and 21.1

Two exceptions are applied to identify near faults: (1) Faults with estimated slip rates less than 0.04 in. (1 mm) per year shall not be used to determine whether a site is a near-fault site, and (2) the surface projection used to determination a near-fault site classification shall not include portions of the fault at depths of 6.25 mi (10 km) or greater.

Site-specific ground motion hazard should be performed for structures on Site Class F in accordance with ASCE 7-22, Chapter 21.1, unless exempted in accordance with Chapter 20.3.1. A ground motion hazard analysis in accordance with ASCE 7-22, Section 21.2, can be used to determine ground motions for any structure. When the procedures of either ASCE 7-22, Section 21.1 or 21.2, are used, the design response spectrum shall be determined in accordance with ASCE 7-22, Section 21.3, the design acceleration parameters shall be determined in accordance with ASCE 7-22, Section 21.4, and, if required, the MCE_G peak ground acceleration parameter PGA_M shall be determined in accordance with Section 21.5.

3.1.7.2.2 Design Spectral Accelerations. The design spectral accelerations S_{DS} and S_{D1} are determined using Equations 3-9 and 3-10 or obtained directly from the ASCE 7 Hazard Tool website. S_{DS} and S_{D1} are adjusted for different soil site classes on the basis of shear wave velocity (Table 3-12). ASCE 7-22 and the ASCE 7 Hazard Tool website now include the site coefficients (soil class) used in the seismic map S_{DS} and S_{D1} values.

Short Period Design Spectral Acceleration:

$$S_{DS} = \frac{2}{3} S_{MS} \tag{3-9}$$

where S_{MS} is 5% damped, spectral response acceleration parameter at short period. The S_{DS} value can be obtained directly from the ASCE 7 Hazard Tool website.

Long Period (1.0 s) Design Spectral Acceleration:

$$S_{D1} = \frac{2}{3} S_{M1} \tag{3-10}$$

where S_{M1} is 5% damped, spectral response acceleration parameter at a period of 1 s.

The S_{D1} value can be obtained directly from the ASCE 7 Hazard Tool website (for 5% damping).

When damping values other than 5% are desired, IEEE 693-2018 may be used to develop the damped design spectral accelerations.

3.1.7.3 Design Response Spectra. Where a design response spectrum is required for substation structures in which the primary function is not to support substation equipment, the design response spectrum should be developed as described in ASCE 7-22, Chapter 11. Substation structures that primarily support substation equipment should use IEEE 693 (IEEE 2018c). The high seismic performance level response spectrum is shown in Figure 3-8. Time histories for analysis of these structures representing the IEEE 693 (IEEE 2018c) spectra are available on the IEEE website. The 1 g time history can be used to adjust the specific ground motion PGA selected for performing the analysis of the substation support structure.

3.1.7.4 Seismic Design Coefficients/Factors. Seismic coefficients depend on the type of lateral force resisting system. Recommended seismic coefficients, R, Ω_0, and C_d, for substation structures that are not designed in accordance with IEEE 693 (IEEE 2018c) are given in Table 3-13 and defined as follows:

R = Response modification coefficient,

Ω_0 = Overstrength factor, and

C_d = Deflection amplification factor.

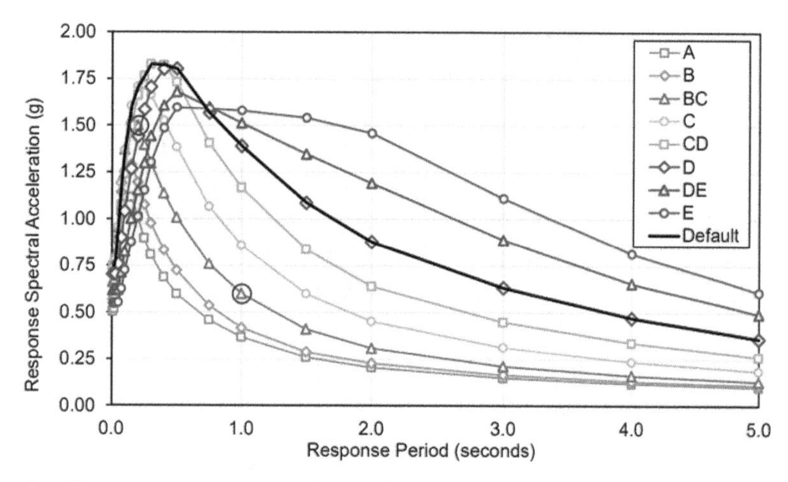

Figure 3-8. Example of a multiperiod response spectrum.

For combinations of different types of structural systems within the same structure along the same loading axis (horizontal and vertical), the *R* factor value (Table 3-13) used for design in that direction should not be greater than the least value of any of the systems used in that same direction. *R* factor values for structural systems and materials not given in Table 3-13 can be found in ASCE 7-22. Substation steel structures designed in accordance with the recommendations of this MOP need not meet the provisions in AISC 341-16, except as noted in this section.

Table 3-13. Seismic Design Coefficients.

Structure or component type	R	Ω_0	C_d	Figures	Remarks/notes
Steel moment frame, wire supporting	2	2	2	2-3, 2-12	Note 4
Steel lattice portal frame/truss, wire-supporting	3	1.5	3	2-4, 2-9	Notes 1 and 4
Steel-braced frame/truss, wire-supporting	1.5	1	1.5	2-11	Notes 1 and 4
Frame-type rigid bus or equipment support	2	2	2	2-12	Notes 2 and 4
Cantilever rigid bus support	1	1	1	2-21	Note 4
Poles wire supporting: steel, wood, and concrete	1.5	1.5	1.5	2-5	Note 4
Poles without wires: steel, wood, and concrete	1.5	1.5	1.5	2-6	Note 4
Fire or sound barrier walls	1.25	2	2.5	12-2	Note 4
Rigid bus conductor, insulators	—	—	—	2-21	Notes 2 and 4 (Section 3.1.7.11)
Structures not otherwise covered herein	1.25	2	2.5	__	Notes 3 and 4

Notes:

1. Latticed and braced structures composed of axial members that resist lateral loads primarily in compression and tension.
2. When more than 50% of the mass (including support structure and high-voltage components such as equipment, insulators, and bus) is concentrated at the top of a single column, such as a rigid-bus support, the moment demand of the column connection at the top should be designed for at least one-half of the moment demand at the base of the column.
3. The user of this MOP can select an *R* Factor using their engineering judgment for structures or components not listed in Table 3-13. The use of *R* factors in Table 3-13 should follow the recommendation of this section.
4. Users of Table 3-13 may use $R = 1$, $\Omega_0 = 1$, and $C_d = 1$ for all structures.

Many substation structures are essential facilities and play a critical role in maintaining reliability and operation of the electric grid. Because of this, substation structures designed in accordance with this MOP should consider using $R = 1$ to ensure that steel and aluminum structures are designed to respond within elastic limits. The first edition of this MOP provided recommended values for $R > 1$ for inelastic design of ASCE 113 structures to be used at the discretion of the designer, where appropriate, for the limited substation structure types not covered by IEEE 693 (IEEE 2018c). To maintain consistency with the previous version, this MOP also includes expanded provisions for $R > 1$ for inelastic design of ASCE 113 structures to be used at the discretion of the designer or owner. If values of $R > 1$ are used for design, special attention must be provided to account for inelastic displacements (member deflections and structure drift) where the structure interacts with attached systems including, but not limited to, bus, cables, conduits, and equipment. Substation structure separation and interaction effects must also be investigated to prevent damage to adjacent structures. In addition, special attention should be provided to the structure connections and anchorage to ensure reliable behavior of the structure. Appendix C provides additional information on the concept of the response modification factor (R). The utility should select the use of the R value on the basis of the performance objectives for the substation during and after the earthquake event.

The advantages of a design with $R = 1$ are an elastic design that is more predictable and that reduces structure damage during a seismic event. Also, member forces and displacements do not need to be magnified by Ω_O or C_d. The advantage of a design with $R > 1$ is an inelastic design that reduces seismic loads and therefore the weight of the structure. The disadvantages of using $R > 1$ are that the inelastic behavior of the structure is less predictable than elastic behavior and higher deformations could result in damage to the structure and attached rigid bus and flexible cable connections.

This MOP does not recommend R factors greater than 3. If R factors greater than 3 are used, the structure should be designed and detailed in a manner that allows significant development of inelastic energy dissipation mechanisms before any instabilities (e.g., buckling) or weaker non-ductile failure modes occur. If R factors are selected from ASCE 7-22, then all provisions associated with the selection should be considered to apply, including the detailing requirements of AISC 341-16 and the limitations related to the structure height or seismic hazard at the substation site.

If the designer determines that the structure does not have the capability to develop the necessary inelastic behavior and/or energy dissipation mechanisms for using a response modification factor (R), the designer should consider using an R factor equal to 1, with $\Omega_0 = 1$ and $C_d = 1$; see *Note* 4 in Table 3-13. Appendix C provides additional information on the concept of the response modification factor (R).

For the structures and components listed in Table 3-13 and in accordance with Section 3.1.7.8.2, all structural connections in the seismic load path, including column-to-base plate connections and anchorages to foundations, should be designed for seismic forces factored by Ω_0. There are exceptions to this criterion:

1. Structures with $0.167 < S_{DS} < 0.33$ and $0.067 < S_{D1} < 0.133$, unless the structure is a cantilever column system such as a cantilever support, or pole type; or
2. Ductile anchor elements designed to the ductile anchorage provisions in ACI CODE-318-19(22) (ACI 2019); or
3. IEEE 693 (2018), *Qualified Equipment and Support Anchorages;* see Section 3.1.7.10.1 for applicable provisions.

Table 3-14. Importance Factor for Seismic Loads (I_e).

Structures and equipment essential to operation	1.25
All other structures and equipment	1.0

3.1.7.5 Importance Factor. The importance factors for seismic loads, I_e, recommended by this MOP are given in Table 3-14. The owner should designate each electrical installation (substation or circuit pathway) as either essential or nonessential on the basis of its relative criticality to the power system. Installations or specific equipment designated as essential are those that are vital to power delivery and cannot be bypassed in the system or are undesirable to lose because of economic or operational considerations.

The selection of an appropriate importance factor is the responsibility of the owner, and various importance factors may be used within the substation depending on the structure and equipment. The importance factors (I_e) specified in this section are the same as the recommended values for (I_p), used in IEEE 693 (IEEE 2018c) for foundation design ($I_e = I_p$).

3.1.7.6 Seismic Analysis. The equivalent lateral force (ELF) method of analysis is, in general, appropriate for seismic analysis and design. However, specific systems may require a dynamic analysis to provide a better understanding of seismic structural response, including a more accurate distribution of seismic forces.

The structure designer should use judgment in deciding whether a dynamic analysis is needed. Consideration should be given to the fact that the formulations in the ELF method are typically empirical and are based primarily on building structures. Analysis by either ELF or dynamic analyses should be performed using linear elastic methods.

3.1.7.6.1 Selection of Analysis Procedures. When performing a seismic analysis of substation structures, two options are commonly used: (1) equivalent lateral force procedure (3.1.7.6.2), and (2) dynamic analysis (3.1.7.6.3).

The selection of the analysis method is based on the structural system, dynamic properties, regularity, and economy. For most cases, the ELF method is appropriate to determine lateral forces and their distribution. However, dynamic analyses may be performed for unusual structures that have structure irregularities in geometry, mass, or stiffness. Table 3-15 provides

Table 3-15. Substation Structure Configurations for Possible Dynamic Analysis.

Example photo	Sketch of substation structure	Description
Figure 3-9a	Figure 3-9b	Type 1: Rack structure with geometric or torsional irregularity caused by a nonsymmetrical lateral force–resisting system
Figure 3-9c	Figure 3-9d	Type 2: Steel braced frame/truss with stiffness irregularity caused by an unbraced panel
Figure 3-9e	Figure 3-9f	Type 3: Distribution rack structure with mass irregularity caused by a difference in the mass of equipment. Mass irregularity exists if the total equipment weight at any elevation W_i is $>$ 1.5 times that of any other elevation (W_j), and/or the distance to the center of mass from the centerline is $> B/6$

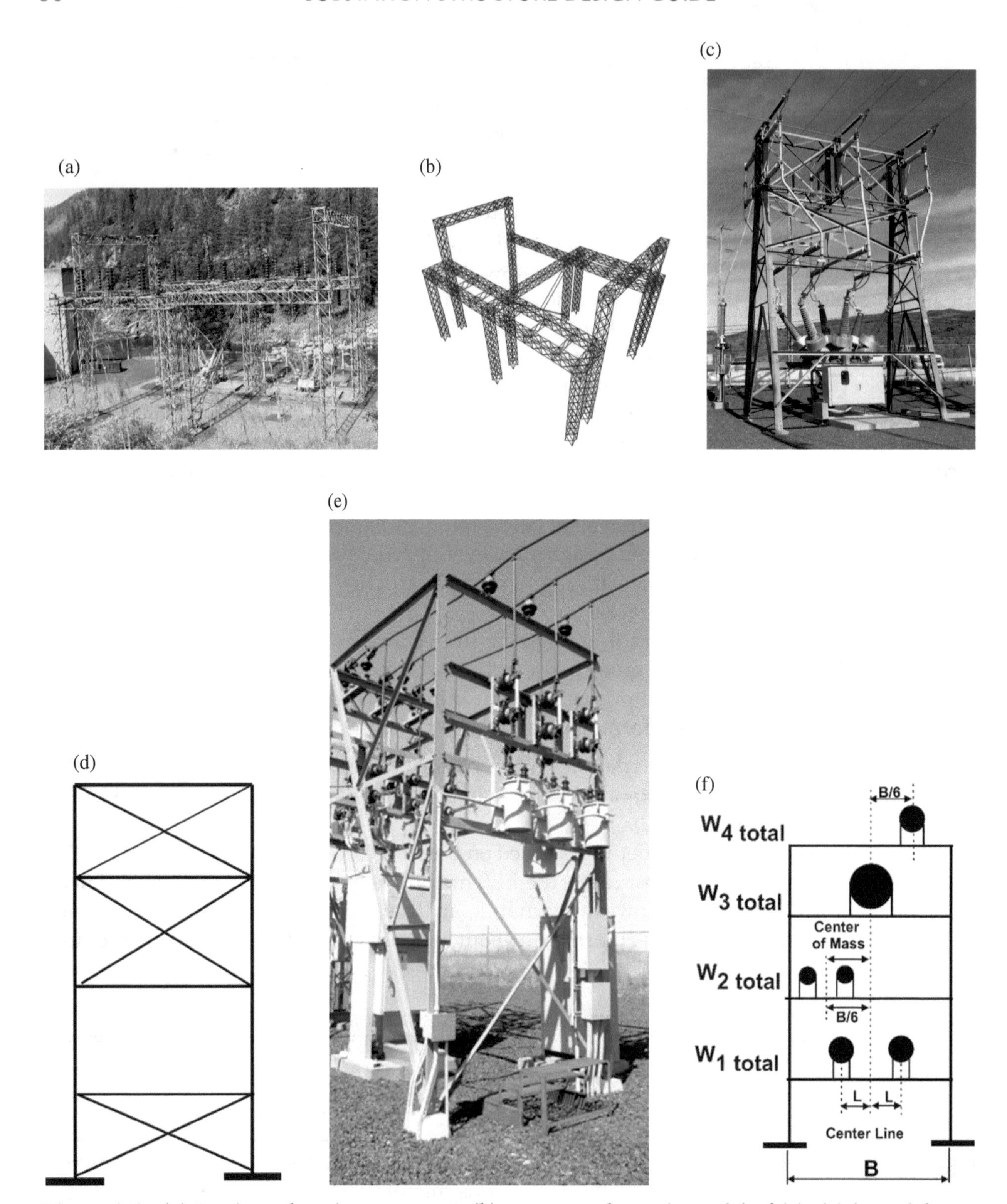

Figure 3-9. (a) Lattice substation structure, (b) computer dynamic model of (a), (c) braced frame, (d) sketch for (c), (e) Rack structure with differential mass, (f) computer model of (e) (the terms equipment mass and weight are interchangeable).
Source: (a) Courtesy of Brian Low, (e) courtesy of the US Department of Energy.

examples of configurations common to substations where a dynamic analysis is suggested at the discretion of the structure designer. As seen in Table 3-15, mass irregularity exists if any mass W_i on a given elevation (typically locations of significant concentrated mass, points of connection of horizontal members or substructures) varies by more than a factor of 1.5 of the mass W_j of any other elevation; or If the horizontal position of the summation of masses (center of mass) on any one elevation varies from the structure support centerline by more than $B/6$, where B is the support structure width. Further descriptions on the types of irregularities are given in Section 3.1.7.6.2.

3.1.7.6.2 Equivalent Lateral Force Procedure. The ELF procedure described in this section is suitable for determining seismic loads for most substation structures that respond as single-degree-of-freedom systems. This procedure is based on ASCE 7-22, Chapter 15, "Seismic Design Requirements for Nonbuilding Structures." In cases where the contribution of higher modes of vibration may be significant, the engineer should apply I_{mv} as described in Equation (3-11). The dynamic analysis methods described in Sections 3.1.7.6.3 and 5.5 are appropriate for determining seismic loads for such structures and may be used as an alternative to the ELF procedure. Refer to Chapter 5 in this MOP for static and dynamic analysis procedures. The seismic base shear, V, in a given direction should be determined by

$$V = C_s W(I_{mv}) \tag{3-11}$$

where

W = Total effective seismic weight, which includes the dead load of permanent equipment above the base of the structure;
I_{mv} = 1.5 if multimode contributions are significant, otherwise I_{mv} = 1.0; an example structure with multimode contributions is a box-type structure (Figure 2-4) with nonsymmetrical stiffness and mass. Whereas a structure without significant multimode effects would be a cantilever-type structure such as a shielding mast (Figure 2-5) or a single-phase bus support (Figure 2-21); and
C_s = Seismic response coefficient determined by using the flowchart shown in Figure 3-10 and either equation (3-12, 3-13a, 3-13b, 3-14, or 3-15):

$$C_s = \frac{S_{DS}}{R / I_e} \tag{3-12}$$

where

S_{DS} = Design spectral response acceleration from the ASCE 7 Hazard Tool website,
R = Response modification factor listed in Table 3-13, and
I_e = Importance factor listed in Table 3-14.

C_s need not exceed the following:

$$C_s \leq \frac{S_{D1}}{T\left(\dfrac{R}{I_e}\right)} \tag{3-13a}$$

$$C_s \leq \frac{S_{D1}\,T_L}{T^2\left(\dfrac{R}{I_e}\right)} \tag{3-13b}$$

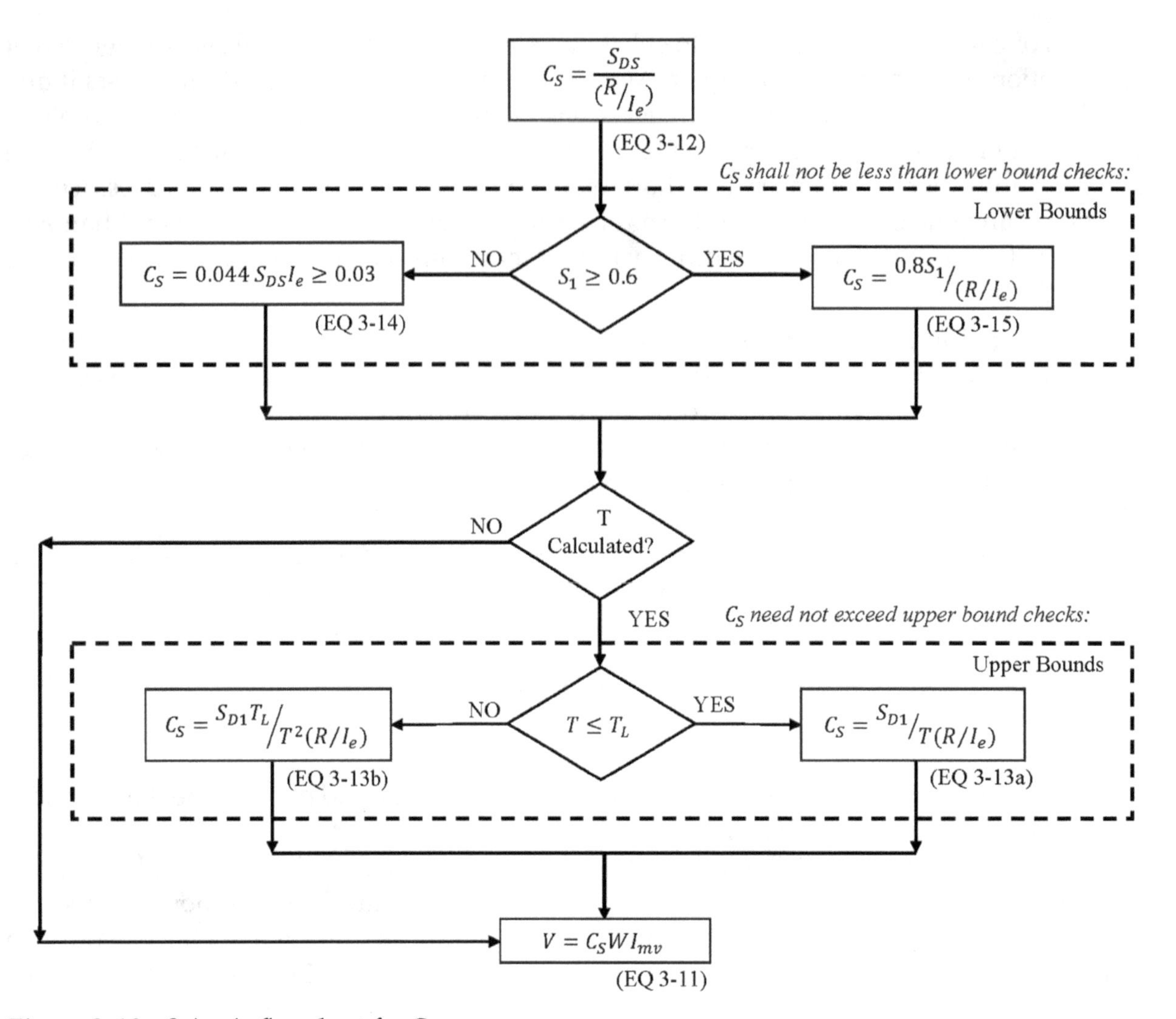

Figure 3-10. Seismic flowchart for C_s.

where

S_{D1} = Design spectral response acceleration from the ASCE 7 Hazard Tool website;
T = Fundamental period of vibration of the structure, calculated as recommended in Chapter 5 or by modal analysis; and
T_L = Long-period transition period.

C_s should not be less than

$$C_s = 0.044 S_{DS} I_e \geq 0.03 \tag{3-14}$$

In addition, for $S_1 \geq 0.6$ g, C_s should not be less than

$$C_s \leq \frac{0.8S_1}{R/I_e} \tag{3-15}$$

where S_1 is the MCE_R spectral response acceleration, at 5% damping, from the ASCE 7 Hazard Tool website.

If T is not computed, the coefficient C_s may conservatively be assumed as calculated in Equation (3-12). When damping values other than 5% that are available in the ASCE 7 Hazard Tool website are desired, such as 2% for substation structures, IEEE 693 (IEEE 2018c) may be used to develop the damped design spectral accelerations.

3.1.7.6.2.1 Vertical Distribution of Seismic Forces. The vertical distribution of the total lateral seismic force should be proportional to the height and effective seismic weight of the structure and computed with the following equations:

$$F_x = C_{vx}V \tag{3-16}$$

and

$$C_{vx} = \frac{w_x h_x^k}{\sum_{i=1}^{n} w_i h_i^k} \tag{3-17}$$

where

C_{vx} = Vertical distribution factor;
V = Total design lateral force at the base of the structure;
w_i and w_x = Effective seismic weight at the elevation i or x being considered (typically locations of significant concentrated mass, points of connection of horizontal members or substructures, or as appropriate for discretization of the structure model);
h_i and h_x = Height above the base to elevation i or x;
k = exponent dependent on the substation structure fundamental frequency (f) as follows:
$k = 1, f \geq 2$ Hz;
$k = 2, f \leq 0.4$ Hz;
$k = 2.25 - 0.625f$, when f is between 0.4 and 2.0 Hz;
n = number of mass vertical locations (elevations) selected.

The method described in this section is applicable to most substation structures commonly encountered in practice.

3.1.7.6.2.2 Horizontal Distribution of Seismic Forces. For structures with horizontal bracing (horizontal truss) connecting vertical trusses or frames of the lateral load–resisting system, the horizontal distribution of lateral seismic force at any elevation should be based on the relative lateral stiffness of the vertical trusses or frames of the lateral load–resisting system. For structures without horizontal bracing, the horizontal distribution of seismic force to the vertical trusses or frames of the lateral load–resisting system should be based on the tributary weights to each vertical truss or frame. In Table 3-15, mass irregularity exists if any weight W_{eq1}, W_{eq2}, W_{eq3}, or W_{eq4} varies by more than a factor of 1.5 of the weight of any one elevation, or if the horizontal distribution of weight on any one elevation varies from the structure support centerline by more than $B/6$, where B is the support width.

When distributing seismic forces, the concurrent effects of a rigid bus within switch racks (4 kV and higher) should be considered. If there is inadequate slack in the flexible bus, those impact forces should also be considered.

3.1.7.6.2.3 Rigid Substation Structures. For rigid substation structures that have a fundamental period, T, less than 0.03 seconds (frequency $\geq$33 Hz), the lateral force may be computed as

$$V = 0.30 S_{DS} W (I_e) \tag{3-18}$$

where

S_{DS} = Design spectral response acceleration from the ASCE 7 Hazard Tool website,
W = Total effective seismic weight, and
I_e = Importance factor listed in Table 3-14.

3.1.7.6.3 Dynamic Analysis Procedure. The dynamic analysis procedure may be used for all substation structures with or without irregularities. An analysis should be conducted to determine the fundamental modes of vibration for the structure. The analysis should include a minimum number of modes to obtain a cumulative modal mass participation of at least 90% of the actual mass in each orthogonal horizontal direction of response considered in the model.

The value of each force-related design parameter of interest for each mode of response should be computed using the properties of each mode and the design response spectrum defined in Section 3.1.7.3 divided by the quantity R/I_e.

The value of each parameter of interest calculated for the various modes should be combined using the methods discussed in Section 3.1.7.8. The three methods include (1) a time-history analysis with the 100/40/40 percent rule, (2) the modal superposition square root of the sum of the squares (SRSS), and (3) the complete quadratic combination method, in accordance with A.1.4.4 IEEE 693 (IEEE 2018c). The first method uses three analyses, each combining the three orthogonal earthquake components. Each analysis uses 100% of one of the orthogonal earthquake components with the sum of 40% of the other two components, as provided in Equations (3-20) to (3-22) and (3-27) to (3-29).

The second and third methods use three separate analyses of each of the individual orthogonal earthquake components. The results are combined using the SRSS or CQC method. Equations (3-23) and (3-30) are used for the SRSS method. The SRSS method assumes that the vibration modes are independent, well separated, and therefore does not account for cross-coupling vibration mode effects. For structures with coupled modes of vibration, structures with unsymmetrical stiffness or mass, the SRSS method can result in an underestimation of the dynamic response. When using the SRSS method, the response of vibration modes whose frequencies differ by less than 10% should be first summed using absolute values.

The CQC method also assumes independent modes but has the advantage of accounting for the cross-coupling effects of vibration modes. If there are no cross-coupling effects (mode frequencies differ by great than 10%), the CQC and SRSS will give similar results. Equations (3-24) and (3-31) are used for the CQC method. Where the combined response for the modal base shear (V_t) is less than 100% of the calculated base (V) using the ELF procedure, the forces shall be multiplied by (V/V_t), where V is the equivalent lateral force procedure base shear calculated in accordance with Section 3.1.7.6.2 and V_t is the base shear from the required modal combination.

3.1.7.6.3.1 Dynamic Analysis Recommendations. This section along with Section 5.6.2, "Dynamic Analysis," provides guidance for choosing the type of seismic analysis procedure for substation structures assuming that the least complicated type of analysis is the first choice of analysis type. This guidance is not a recommendation that a more complicated type of analysis cannot or should not be used. It is important for the structure designer to carefully choose the type of seismic analysis for a structure based on all contributing factors including, but not limited to, structural irregularities, fundamental natural period, structure height, and economics.

Table 3-15 provides some structural configurations where a dynamic analysis is recommended. In cases where the participation of higher modes will significantly influence the seismic load effect, E, dynamic analysis should be performed.

3.1.7.7 Vertical Seismic Load Effect. The seismic design force for vertical seismic load effect (E_v) is determined as follows:

$$E_v = (0.80)S_{DS}D \tag{3-19a}$$

where S_{DS} is the design spectral acceleration at short period from the ASCE 7 Hazard Tool website; and D is the dead load.

Or when the design spectral response vertical acceleration S_{av} is known, it can be used to determine E_v:

$$E_v = S_{av} D \tag{3-19b}$$

Either value of E_v is acceptable.

EXCEPTION:

The vertical seismic load effect, (E_v), is permitted to be taken as zero when determining the demands on the soil–structure interface of foundations.

3.1.7.8 Seismic Load Effects and Combinations. Structures, foundations, and anchorage should be designed for the seismic load effects described in this section, including the combinations of horizontal and vertical (as applicable) earthquake components.

3.1.7.8.1 Basic Seismic Load Effect. Horizontal seismic loads effect, (E_h), should be determined as follows: $V = E_h$ from Equation (3-11). E_h should be applied in two orthogonal horizontal directions, and vertical seismic load effect, (E_v), should be considered to determine the controlling member and foundation design forces where

$$E = E_h \pm E_v \tag{3-19c}$$

Applicable structures should be designed for the earthquake component combinations described in one of the methods presented subsequently.

The seismic load vectors, E, in the load combinations described in Section 3.3 should be determined in the following load cases:

$$E = E_{h1} + 0.4E_{h2} \pm 0.4E_v \tag{3-20}$$

$$E = 0.4E_{h1} + E_{h2} \pm 0.4E_v \tag{3-21}$$

$$E = 0.4E_{h1} + 0.4E_{h2} \pm E_v \tag{3-22}$$

The aforementioned combination requires that the maximum component strength be used. Note that the aforementioned equations are a combination of orthogonal seismic load vectors and not algebraic equations.

If an SRSS earthquake component combination is used, the seismic load effect E should be determined from

$$E = \sqrt{E_{h1}^2 + E_{h2}^2 + E_v^2} \tag{3-23}$$

where E_{h1}, E_{h2} have a horizontal seismic load effect in mutually perpendicular directions from Section 3.1.7.6.2 (equivalent lateral force procedure) or 3.1.7.6.3 (dynamic analysis), and E_v is the vertical seismic load effect (acting upward or downward) from Section 3.1.7.7.

The anchor bolts and welded connections to the foundation should not be reduced by the effects of friction.

If a CQC earthquake component combination is used, the seismic load effect E should be determined from

$$E = \sqrt{\sum_{i=1}^{2}\sum_{j=1}^{2} E_{hi}^2 \alpha_{ij} E_{hj}^2 + E_v^2} \tag{3-24}$$

$$\alpha_{ij} = \frac{8\xi^2(1+\beta)\beta^{\frac{3}{2}}}{(1+\beta^2)^2 + 4\xi^2(1+\beta)^2} \tag{3-25}$$

$$\beta = \omega_i/\omega_j \text{ IF } (\omega_j > \omega_i) \tag{3-26}$$

where

E_{hi}, E_{hj} = Horizontal seismic load effect in mutually perpendicular directions from Section 3.1.7.6.3,
E_v = Vertical seismic load effect (acting upward or downward) from Section 3.1.7.7,
a_{ij} = Correlation coefficient with constant damping applied to all vibration modes,
ω_i, ω_j = Frequencies of the ith and jth modes of vibration,
ξ = Damping ratio, and
β = Vibration mode frequency ratios.

It is assumed in this MOP that seismic loads are applied during the condition of no wind, or ice, and at 60 °F (15.6 °C). The owner should determine whether it is appropriate to combine seismic loads with wind, ice, short circuit, and operating loads. When appropriate, wire loads should be applied to the structure (Section 3.1.7.12).

3.1.7.8.2 Seismic Load Effect with Overstrength Factor. The seismic load effect, E, described in Equations (3-20) to (3-23) should be modified by an overstrength factor Ω_0 for designing connections, as described in Table 3-13 and this section. The modified seismic load effect E_m is determined from

$$E_m = \Omega_0 E_{h1} + 0.4\Omega_0 E_{h2} \pm 0.4E_v \tag{3-27}$$

$$E_m = 0.4\Omega_0 E_{h1} + \Omega_0 E_{h2} \pm 0.4E_v \tag{3-28}$$

$$E_m = 0.4\Omega_0 E_{h1} + 0.4\Omega_0 E_{h2} \pm E_v \tag{3-29}$$

The structure may have different overstrength factors in the two orthogonal directions depending on the structure geometry. If the overstrength factors are different, the appropriate overstrength factor to each orthogonal direction should be applied accordingly.

If the SRSS earthquake component combination is used, the modified seismic load effect E_m should be determined from

$$E_m = \sqrt{(\Omega_0 E_{h1})^2 + (\Omega_0 E_{h2})^2 + E_v^2} \tag{3-30}$$

If the CQC earthquake component combination is used, the modified seismic load effect E_m is determined from

$$E_m = \sqrt{\sum_{i=1}^{2}\sum_{j=1}^{2}\Omega_0^2\left(E_{hi}^2 \alpha_{ij} E_{hj}^2\right) + E_v^2} \tag{3-31}$$

The overstrength factor Ω_0 is intended to provide a level of assurance that the connections in the structural seismic load path have sufficient strength to resist demand forces that exceed the basic seismic load effects that may have been reduced by an R factor (Table 3-13). Examples for applying the overstrength factor, Ω_0, to the structural seismic load path are

- Photograph of Structure Type 2 (Table 3-15), the complete structure with the exception of the lower diaphragm member supporting the standoff insulators;
- Photograph of Structure Type 1 (Table 3-15), the complete structure designed as the structural seismic load path system; and
- Photograph of Structure Type 1 (Table 3-15), selected portal trusses as the structural seismic load paths in the orthogonal directions.

3.1.7.9 Seismic Deflection Considerations. The effects of structure displacements resulting from seismic loads should be evaluated to ensure that structural stability is maintained and attached equipment, rigid bus connections, and conductor slack are not adversely affected. The displacements obtained from an equivalent lateral force analysis or dynamic analysis should be multiplied by the quantity (C_d/I_e). Refer to Table 3-13 for deflection amplification factor C_d for typical substation structures. Substation structures need not comply with drift limits as specified in ASCE 7-22, Chapter 12 (Table 12.12-1). Refer to Section 3.1.7.4 for deflection cautions when $R > 1$ is used.

3.1.7.10 Anchorage Design Forces. Seismic load effects should be considered in the design of substation structures and equipment anchorage as recommended in Sections 3.1.7.10.1 and 3.1.7.10.2 and Chapter 8 of this MOP.

3.1.7.10.1 IEEE 693 (IEEE 2018c)—Qualified Equipment and Supports Anchorage Design Forces. Anchors (anchor material, diameter, and welds) for IEEE 693 (2018) qualified equipment should be designed in accordance with the provisions in IEEE 693 (2018). The anchorage in the concrete foundation (anchor embedment in concrete) should be designed in accordance with the provisions in this MOP for anchor loads corresponding to the IEEE 693 (2018) seismic qualification, amplified by the overstrength factor Ω_{PL}, where appropriate, as specified in Equations (3-28) to (3-33).

Loads derived from IEEE 693 (2018) provide requirements that are based on post seismic operability of the equipment. Therefore, IEEE 693 (2018) is a performance-based document that can result in potentially higher demands on the supporting structure. The engineer needs to understand the following concepts of the IEEE 693 (2018) to provide adequate anchorage for structures supporting qualified equipment. Loads from an IEEE 693 (2018) qualification can be derived from two qualification approaches called *design-level* and *performance-level* qualification. The performance-level loads are derived only from shake table tests and represent loads that the equipment is expected to withstand under the ultimate seismic loading condition where the equipment may be approaching functional failure. The design level loads result from either qualification by analysis or shake table testing at 50% of the performance-level loading condition, using unfactored load combinations or the IEEE 693 (IEEE 2018c) required response spectrum. The LRFD seismic design–level load combinations given in Section A.2.1.1 of IEEE 693 specify a load factor of 1.4 on the seismic load demand. In all IEEE 693 qualifications, no credit is explicitly taken for inelastic energy absorption, and the response modification factor R is taken as unity.

To maintain the integrity of the intent of IEEE 693 on anchorage design, a discussion of the overstrength factors to be used is necessary with respect to ductile versus nonductile anchorage designs. A description of ductile vs nonductile anchors can be found in Section 8.3 "Anchor Arrangements and General Design Considerations" of this MOP. It should be noted that Ω_{PL} for the anchorage of IEEE 693 (2018)—qualified items is different in concept from the Ω_0 factor described in Section 3.1.7.4. The Ω_{PL} factor is intended to provide assurance that the maximum withstand capability of the equipment (e.g., performance-level loading) can be achieved without anchorage failure.

In contrast, the Ω_0 factor described in Section 3.1.7.4 is intended to amplify the design forces on critical structural elements (e.g. connections) that may experience higher demands because of inelastic behavior ($R > 1$) of the structure or overstrength of structural elements (e.g., actual material strength greater than the minimum specified) that deliver the demand forces to those critical elements.

1. Design-Level Qualifications

If a ductile anchorage design is provided, then the design-level (ASD) loads should be factored by 1.4 and the anchorage designed to resist this magnitude of load using LRFD methods.

If a nonductile anchorage design is provided, then the design-level (ASD) seismic loads should be factored by 1.4, and $\Omega_{PL} = 1.5$ should be applied only to the horizontal components of the earthquake, using any method of qualification approved by IEEE 693 (2018). This maintains the intent of the design level being projected to the performance level by achieving a factor of about 2 (1.4 load factor $\times$ 1.5 overstrength factor $= 2.10$).

The following strength design load combinations should be used for designing anchorages:

$$1.2D + \Omega_{PL}(1.4E_{DL}) \quad (3\text{-}32)$$

$$0.9D + \Omega_{PL}(1.4E_{DL}) \quad (3\text{-}33)$$

where

D = Dead load effects;
E_{DL} = Earthquake loads from IEEE 693 (2018) Design-Level (ASD) qualification;
Ω_{PL} = Overstrength factor for IEEE 693 (2018) qualified items, 1.0 for ductile anchorage designs, or 1.5 for nonductile anchorage designs, when forces from design-level qualifications are used.

2. Performance-Level Qualifications

For performance-level forces, $\Omega_{PL} = 1.0$ may be used. The equipment reactions from a performance-level test are intended to represent the highest levels of loading for which the equipment is qualified, resulting in a condition where the equipment may be close to failure. The likelihood of the equipment being exposed to such levels of motion depends on the seismic hazard at the installation site and other site conditions. Variability associated with the equipment reactions reported from a performance-level shake table test should be expected because of the methods used during testing and the interpretation of test data. These uncertainties and risks should be considered by the structure designer when selecting a ductile versus nonductile anchorage design and when considering whether additional margin is warranted for a nonductile design. In general, ductile anchorage designs are preferred to nonductile designs, and it is undesirable for the equipment anchorage to become the governing element in the overall design.

The following strength design load combinations should be used for designing anchorages:

$$1.2D + \Omega_{PL}(E_{PL}) \quad (3\text{-}34)$$

$$0.9D + \Omega_{PL}(E_{PL}) \quad (3\text{-}35)$$

where

D = Dead load effects;
E_{PL} = Earthquake loads from IEEE 693 (2018) Performance-Level qualification; and
Ω_{PL} = Overstrength factor for IEEE 693 (IEEE 2018c)-qualified items, 1.0 for ductile and nonductile anchorage designs when forces from the performance-level qualification are used.

3.1.7.10.2 Equipment to Structure Anchorage Design Forces. The seismic acceleration experienced by equipment installed on a structure can be several times the ground acceleration resulting from structural amplification. The corresponding amplified seismic forces need to be considered when designing the anchorage of the equipment to the structure. Seismic qualification of equipment to IEEE 693 (2018) by analysis or shake table testing may have been performed with its own dedicated support or on an intermediate support [Section 6.7.1, "Structures That Support Electrical Equipment are qualified for IEEE 693" (IEEE 2018c)], or it may have been qualified without a support to 2.5 times the acceleration levels specified in the standard. If the equipment has been seismically qualified in accordance with IEEE 693 (2018), and this qualification must be maintained for the as-installed position on the structure designed to the provisions of this MOP, then the structure designer should verify that the seismic forces experienced by the equipment on the structure will not exceed the IEEE 693 seismic qualification forces. A dynamic analysis of the equipment support structure may be required for this validation. See IEEE 693, Section 5.10.7. If the IEEE 693 qualification was performed for 2.5 times the acceleration levels (equipment only tested), the anchorage determined from the shake table test should be used for designing the equipment connection to the support structure. A dynamic analysis of the equipment/support structure system should be performed to obtain anchor loads used to design the foundation connections.

The equipment anchorage loads to the structure (or intermediate support) should be obtained from the IEEE 693 seismic qualification, as mentioned in the previous paragraph. The structure, as a complete system with the equipment, is designed according to the requirements of this MOP. When equipment qualified to the IEEE 693 provisions is installed on a structure designed to the provisions of this MOP, the connection of the equipment to the structure (or intermediate support) should be designed to resist the anchorage forces not less than those corresponding to the IEEE 693 seismic qualification and as modified by the overstrength factors as described in this MOP.

When the equipment is not qualified to the IEEE 693 provisions, the equipment to structure anchorage design forces should be determined in accordance with Section 3.1.7.11.

3.1.7.11 Seismic Demand on Other Components. The other components applicable to this section are rigid bus conductors, insulators, appurtenances, and equipment that is not subject to seismic qualifications in accordance with IEEE 693 (2018). The seismic demand as determined in this section should be used to design the component and anchorage to the support structure.

The horizontal seismic design force (F_p) determined in accordance with Equation (3-36) should be applied at the component's center of gravity and distributed relative to the component's mass distribution.

$$F_p = 1.6 S_{DS} I_e W_p \tag{3-36}$$

where

F_p = Seismic design force;
S_{DS} = Design spectral acceleration, short period, as determined from the ASCE 7 Hazard Tool website;
I_e = Importance factor according to Section 3.1.7.5; and
W_p = Total effective seismic weight, which includes the dead load weight.

The seismic design force for vertical earthquake motions should be determined from Section 3.1.7.7. The combination of seismic load effects from the horizontal and vertical components of an earthquake should be determined from Section 3.1.7.8.1.

For design of components such as cable trays, raceways, and cable supports inside control enclosures, refer to ASCE 7-22, Chapter 13. $I_p = 1.5$ should be assumed.

3.1.7.12 Seismic Forces on a Wire Bus (Strain Bus or Cable Bus). In most cases, the loads on wire support structures resulting from the seismic loading on flexible wires are not believed to be a dominant effect, and as a result, wire support structures have typically been designed without taking into consideration the flexible wire seismic loading. This is attributed to the relative light weight of the conductors, presence of sufficient slack, low frequency of vibration of the conductors, and the uncoupling of important modes of vibration of the conductor from the wire support structure. History has shown that even with the high levels of ground motion experienced in historical earthquakes, substation and transmission line wire support structures have performed well.

The rigorous treatment of the seismic loading on flexible wires involves the analysis of a complex mechanical system. Such an assessment would typically require the time-history analysis of a non-linear structural system comprising the support structures, flexible wires, and insulators with different damping values for the various elements in the load path. This section presents a simplified method that applies to intermediate and longer spans [greater than about 30 ft (9 m)] of flexible wires (strain bus or cable bus) connected to wire support structures with transmission line–type suspension and dead-end insulators.

The method presented is not intended for cases with short spans (jumpers) such as those used for connecting adjacent equipment to each other or to bus supports. The methods described in IEEE 1527 (IEEE 2018b) are more suitable for the seismic design of jumpers and other short-span flexible bus, and a flexible bus that is supported by post insulators.

Tension forces, during an earthquake, in the strain bus caused by wire dead load and seismic loading effects (T_{EA} and T_{EB}) may be mitigated by the flexibility of the wire support structures. Tensile forces in the strain bus may be counterbalanced by the resistance from conductors on the adjacent span for multi-span configurations. For single-span or end-span cases, the tensile forces in conductors will result in support structure deflection that reduces the tension in the conductor. Because of these effects, the procedure suggested in this section is expected to be conservative. A dynamic analysis of the wire span(s) and the wire support structures will provide a more representative seismic performance for the design of wire support structures.

The wire tension for the condition with seismic loading transverse to the strain bus support structures may be estimated as follows:

1. Determine the strain bus unit weight and profile (coordinates of end points, sag) for the ambient conditions of no wind, no ice, and a temperature of 60 °F. The utility may decide to use a different temperature, if appropriate, for consideration with the seismic loading.
2. Determine the strain bus lateral seismic loading per unit length as $S_{MS} \times$ strain bus unit weight, where S_{MS} is the maximum considered earthquake spectral response acceleration parameter for short period (S_{DS}) according to Equation (3-9) times 1.5 or obtained from the ASCE 7 Hazard Tool website.
3. Determine the strain bus tension at the support structures for the combined effect of the vertical and lateral loads described in Steps 1 and 2. This may be determined from sag/tension calculations in a manner similar to transverse wind loading on conductors.
4. A wire weight of 50% and 50% of the wire inertia load (Item 2 times span length) should be applied at the seismic tension load points on the structure.

The tension force determined from this procedure should be used in conjunction with the structure seismic loading for designing the structure. Strain bus tension loads should not be factored by Ω_0.

Forces applied to structures from strain bus seismic loading should consider the sag and tension in the strain bus. The displacement and accelerations of the structures at the strain bus support points and strain bus geometry (unit weight, sag, and stringing tension profile) will influence the tensions at the ends of the strain bus tension section. A strain bus tension section is assumed to have adequate seismic slack if the slack is greater than the inequality shown in Equation (3-37).

The dynamic interaction between the support structures and the strain bus in a seismic event can result in large tensile forces in the strain bus if all slack is used up by the displacement of the support structures, resulting in the strain bus reaching a taut condition. The additional tension in the strain bus caused by seismic effects is not believed to be significant if sufficient slack is provided. The structure designer should verify the sufficiency of slack provided by using the following equation:

$$\Delta_{\text{slack}} \geq \sqrt{(C_{d1}\Delta_{e1})^2 + (C_{d2}\Delta_{e2})^2} \tag{3-37}$$

where

Δ_{slack} = Conductor slack between two strain bus support structures provided in the installation under normal temperature conditions with no wind or ice,
Δ_{e1} = Calculated elastic seismic displacement of the first support structure parallel to the strain bus, with the structures unconnected,
Δ_{e2} = Calculated elastic seismic displacement of the second support structure parallel to the strain bus, with the structures unconnected, and
$C_{d1} = C_{d2}$ = Deflection amplification factor (Table 3-13).

3.1.8 Short-Circuit (Fault) Loads

The current flowing in a conductor causes a magnetic field. When currents in adjacent parallel conductors are large enough, their magnetic fields can interact, as shown in Figure 3-11a, b, which causes conductors to attract or repel. At normal operating currents, Figure 3-11a, the magnetic fields are sufficiently small, making these forces typically insignificant. However, during a short-circuit event, Figure 3-11b, the currents increase by orders of magnitude higher than normal operating conditions and can cause large forces in conductors.

Because the magnetic force is proportional to the product of the adjacent currents, the resulting force is two orders of magnitude greater than that at normal operating levels. The

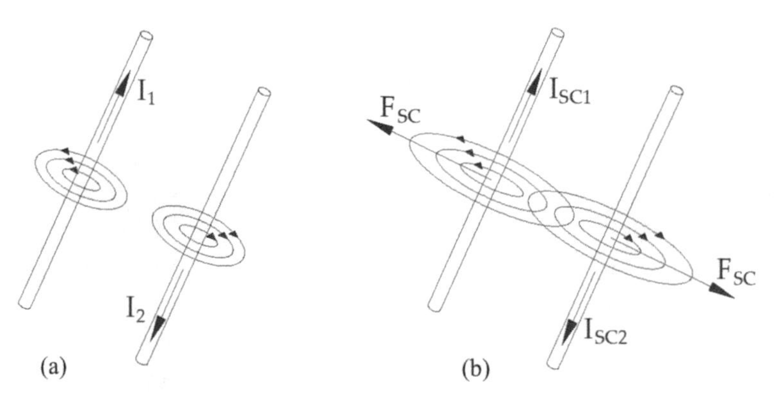

Figure 3-11. Magnetic fields generated by current flowing in adjacent conductors. (a, left) Typical operating currents resulting in inconsequential magnetic fields and lack of electromagnetic force, (b, right) currents during a fault resulting in large magnetic fields and electromagnetic force.

fault current may have a DC component that is more than twice the effective current when the fault occurs. This DC component decays with time based on the electrical parameters of the related power system. Because the fault current is comprised of both a decaying DC component and an AC oscillating component, the fault current oscillates with the frequency of the power system (60 Hz in the United States) and the resulting force on the conductor is made of components oscillating at twice the system frequency (from the AC contribution) and at the system frequency (from the DC component).

The primary electrical protection system can recognize a fault and open the circuit in 2 to 6 cycles (0.03 to 0.10 s), provided the circuit breaker and relays operate properly. The backup control system normally takes 10 to 30 cycles (0.17 to 0.5 s) to open the circuit. If a fault interrupts in the 2 to 6 cycle range, rigid bus conductors and supports may or may not have appreciable structural response to the fault, depending on their structural properties. If the primary control system fails, the backup system is required to operate, and rigid bus conductors and their support system will likely have sufficient time to structurally respond to the fault. The inertia of the rigid bus conductors and supports can be overcome, creating deflections and forces.

All of the factors described in this section point to the dynamic nature of the fault force.

Prior to the 1980s, bus design had been traditionally performed using simplified, static methods. After the 1980s, more accurate methods (such as the finite-element method) have been successfully applied, taking account of the dynamic nature of the short circuit force. Simplified methods remain largely in use today because of their simplicity of use. These simplified methods typically involve the use of beam tables to calculate the maximum conductor moments and support reactions under static loading for a single or continuous beam with equal spans. This is quite simplistic because of the complexity of the dynamic fault loading and structural response. The natural frequency of the bus span in high-voltage substations is often significantly lower than the frequency of the fault and the fault may be cleared before the conductor span reaches its peak deflection. Therefore, applying fault loading as a static force may yield results significantly different from the actual response.

There are several types of electrical faults. Each one has a different likelihood of occurrence. Listed from most likely to least likely, they are

1. Phase-to-ground fault,
2. Phase-to-phase fault, and
3. Three-phase fault.

In the following sections, the simplified equations for the short-circuit force are first presented. Appendix B presents an introductory treatment on fault current and the resulting forces so that the engineer can understand the issues involved before selecting a design procedure. This MOP references CIGRE Brochure 105: *The Mechanical Effects of Short-Circuit Currents in Open Air Substations* (CIGRE 1996), IEEE 605: *IEEE Guide for Bus Design in Air Insulated Substations* (IEEE 2008), and IEC 60865-1: *Short-Circuit Current – Calculation of Effects* (IEC 2011). This MOP is not intended to replace these documents. The engineer should familiarize themselves with these documents and others on the topic to fully understand the complex phenomenon of fault loading, particularly before pursuing more advanced analysis techniques.

The functions for current and force variation with time discussed in this MOP were largely gathered from CIGRE Brochure 105: *The Mechanical Effects of Short-Circuit Currents in Open Air Substation* (CIGRE 1996), with adjustments to phase-angle convention and nomenclature to match those symbols more commonly used in the United States. The discussions with regard to conductor short-circuit forces are limited to a parallel arrangement of the conductors. A nonparallel bus may also be subject to short-circuit forces, and the resolution of these forces is not covered in this MOP.

3.1.8.1 Simplified Static Short-Circuit (Fault) Force on Rigid Conductors. The equation for the peak fault force presented from IEEE 605-2008, Section 11.3, is reproduced as Equation (3-38). This force represents the peak force assuming full DC offset.

The equation for the basic maximum distributed force between two parallel, infinitely long conductors is

$$F_{sc} = \frac{\frac{\mu_0}{2\pi}\Gamma\left(2\sqrt{2}\,I_{sc}\right)^2}{D} \tag{3-38}$$

For SI units, this simplifies to

$$F_{sc} = \frac{16\Gamma\left(I_{sc}^2\right)}{10^7(D)} \tag{3-39}$$

For customary units, this simplifies to

$$F_{sc} = \frac{3.6\Gamma\left(I_{sc}^2\right)}{10^7(D)} \tag{3-40}$$

where

F_{sc} = Fault force in N/m or lb/ft,
I_{sc} = Symmetrical RMS fault current in amps,
D = Conductor center-to-center spacing in meters or feet,
Γ is 1.0 for phase-to-phase faults and 0.866 (middle conductor) or 0.808 (outer conductors) for three-phase faults, and
$\mu_0 = 4\pi \times 10^{-7}$ N/A^2 = 2.825×10^{-7} pounds/A^2 (magnetic constant – magnetic permeability in a classical vacuum).

This gives the maximum value of the force between the conductors, conservatively assuming that the current in the two conductors equals the RMS fault current simultaneously and also assuming that the AC component and decaying DC components of the fault are at their peaks simultaneously. However, the actual peak force occurs in approximately the first quarter to half cycle after fault initiation, when the AC component of the fault force is at its peak, but the DC component of the fault force has partially decayed. Therefore, a half-cycle decrement factor, D_f, should be used to determine the actual peak fault force.

The basic maximum fault force from Equation (3-41) is modified to obtain the peak force by multiplying by the square of the half-cycle decrement factor, D_f, as given in IEEE 605-2008.

$$F_{sc_{\text{corrected}}} = K_f F_{sc}\left(D_f^2\right) \tag{3-41}$$

where

$$D_f = \frac{1+e^{-(1/2fc)}}{2} \quad \text{(Decrement Factor)} \tag{3-42}$$

$$c = \frac{X}{R\times 2\pi f} \text{ Time constant of the circuit (also often given as } T_a \text{ in IEEE 605)} \tag{3-43}$$

f = Power system frequency (60 Hz in the United States),
K_f = Mounting flexibility factor (IEEE 605-2008, Figure 20),
X = System reactance (Ω), and
R = System resistance (Ω).

This corrected force given by Equation (3-41) is typically used in a simplified static analysis to design conductors, insulators, and supporting structures. The corrected peak force of Equation (3-41) is slightly less than the basic force of Equation (3-38) but is still potentially conservative when applied as a static force, because it does not account for the dynamic nature of the fault force and the dynamic structural response. The structural response may be greatly affected by the potentially large difference between the frequency of the fault force and the natural frequency of the bus system. The flexibility factor, K_f, is usually assumed to be 1.0 for three-phase mounting structures. Other values can be obtained from IEEE 605-2008, Figure 20.

3.1.8.2 Simplified Static Short-Circuit (Fault) Force on the Strain Bus. Each utility, as part of their Short-Circuit Fault (SCF) design criteria, may consider parameters for the application of short-circuit forces to the strain bus. If the SCF on the strain bus is to be considered, refer to IEEE 605-2008 and IEC 60865-1 (2011) for the force applications without or with jumpers, respectively.

Although the equations in IEEE 605-2008 and IEC 60865-1 (IEEE 2011) appear similar, there are some potentially substantial differences between the two, particularly in the determination of the stiffness norm, N, for the case of a "Strained conductor attached to the structure with two insulator chains (one on each side)." IEC 60865-1 includes the stiffness of the attachment points, S, whereas IEEE 605-2008 does not include this parameter. IEC 60865-1 also includes an effective conductor Young's modulus where IEEE 605-2008 does not. In addition, there is an apparent typographical error in Equation 56 in IEEE 605-2008. It appears that the number of subconductors, n, was inadvertently left out of this equation.

3.1.8.3 Short-Circuit Forces on Equipment. Where applicable, the equipment manufacturer should be contacted to provide both the magnitude and the direction of possible fault load limits on the equipment. Internal loads need not be given, but any additional short-circuit loads to the equipment's anchorage should be quantified.

3.1.8.4 Short-Circuit Force Additional Information. Additional information can be obtained from the references used in CIGRE Brochure 105: *The Mechanical Effects of Short-Circuit Currents in Open Air Substations* (CIGRE 1996), IEEE 605-2008: *IEEE Guide for Bus Design in Air Insulated Substations* (IEEE 2008), and IEC 60865-1: *Short-Circuit Current – Calculation of Effects* (IEC 2011). Appendix B in this MOP provides additional information on short-circuit force development.

3.1.9 Construction and Maintenance Loads

During construction and maintenance operations, it is sometimes necessary for workers to be supported from a component of the structure or to impose forces on the structure for fall protection anchors. Occupational Safety and Health Administration (OSHA 2020) requirements for worker safety should be reviewed. IEEE 1307 (IEEE 2018a) provides guidance for the determination of worker safety loads on substation structures. Each structure should be evaluated for the need to impose such loads for design. IEEE 1307 (IEEE 2018a) allows a worker to free-climb structures, whereas OSHA has eliminated the electric utility exemption and requires 100% fall protection for workers climbing utility structures. The structure designer should determine whether other unusual loads will be applied during construction and maintenance operations, and if so, their magnitude. These loads are usually not combined with other extreme climatic loads because it is unlikely that workers will perform construction or maintenance during severe weather conditions, and construction and maintenance loads are specified with higher load factors. Additional information on construction and maintenance considerations is provided in Chapter 10.

3.1.10 Wind-Induced Oscillations

Relatively low sustained wind speeds may occasionally produce an oscillating motion in an individual member or an overall structure. This phenomenon is referred to as vortex-induced oscillation (VIO) or vortex-induced vibration (VIV). Oscillations resulting from VIO may not be foreseen in the design stage but can be mitigated in the field by the utility if they develop. The prediction of VIO for a member or structure is difficult because of the number of parameters that can contribute to the phenomenon. These include

- Natural frequency (which is a function of mass and stiffness),
- Structural damping (the rate at which a structure will dissipate energy),
- Wind speed (affected by the local wind climate and upstream terrain),
- Turbulence characteristics (affected by upstream terrain or topography), and
- Orientation of the member or structure to the wind.

Oscillations caused by VIO can be excessive in situations where structures that have design parameters that make them vulnerable to the local wind conditions are built (i.e., common speeds, directions). Cases where amplitudes associated with VIO are not excessive may still contribute to low-stress high-cycle fatigue if not addressed. Mitigation approaches for high-amplitude and low-amplitude VIO are often similar but will vary in effectiveness.

Examples of structures that have been observed to be sensitive to VIO include shielding masts with no shield wires, rigid bus systems, and unloaded tubular dead-end structures (during the construction stage). Mitigation techniques such as external tie-downs to a structural member or internally suspended heavy steel chains encased in rubber can be effective in reducing such motions. These mitigation techniques, as well as others, improve the performance of structures sensitive to VIO by adding damping and/or mass to the system. Aerodynamic mitigation (i.e., helical strakes) is also possible but is often not the preferred solution because of feasibility reasons. Design considerations for wind-induced VIO are discussed in Section 6.10.2 of Chapter 6. Construction considerations are discussed in Section 10.1 of Chapter 10.

3.1.11 Loading Criteria for Deflection Limitations

Where the owner has not developed specific loading conditions for deflection analysis, the load conditions from Sections 3.1.11.1, 3.1.11.2, and 3.1.11.3 should be used with a load factor of 1.0. The owner should determine whether additional loads should be applied in combination with the recommended deflection load cases. The deflection limitations in Chapter 4, Section 4.1, "Structure Classifications and Deflection Limits," are to be used with the following load combinations.

3.1.11.1 Wind Load for Deflection Calculations. A 70 mph (31.3 m/s) wind speed is recommended for the velocity (V) in Equation (3-1) or Equation (3-2) to calculate wind load associated with deflection criteria for substations located outside hurricane-prone regions. Wind speeds used to determine deflections in hurricane-prone regions should be determined by the owner. Hurricane-prone regions are shown on the wind maps in Section 3.1.5.3 along the East coastline, the Gulf of Mexico, and are also defined in Chapter 26 of ASCE 7 (ASCE 2022). Without any specific input from the owner, an 80 mph (35.8 m/s) wind speed for V in Equation (3-1) or Equation (3-2) can be used to calculate wind load associated with deflection criteria for substations located in hurricane-prone regions. A load factor of 1.0 is recommended for all wind load deflection calculations. These wind speeds are recommended values to allow for equipment operability. The owner should evaluate the appropriate wind speed for their system on the basis of factors such as remote operability of switches and other equipment during a wind event.

3.1.11.2 Ice with Concurrent Wind Load for Deflection Calculations. A 5-year MRI peak ice thickness should be used to calculate the ice load with concurrent wind associated with deflection criteria. Table 3-16 provides conversion factors that should be used in conjunction with extreme ice thickness with concurrent wind obtained from Figure 3-6a.

3.1.11.3 Other Deflection Considerations. Typically, substation equipment is not operated during extreme ice, hurricane, and seismic events. The owner should determine whether the electrical equipment is expected to operate during an extreme event, and what the deflection limits are for the structure, equipment, and supporting foundation.

Loads resulting from short-circuit current or fault current need not be considered in deflection analysis. This loading condition is a short-duration extreme event that requires structural capability but not necessarily operational capability. Given the difficulty in predicting deflections under dynamic conditions and the limited need for equipment operation during the event, deflection analysis for this extreme event condition would be difficult, imprecise, and of questionable use.

3.1.12 National Electrical Safety Code Loads

The *National Electrical Safety Code* (IEEE 2023), Section 16, Paragraph 162.A, requires that substation structures supporting facilities (wires) that extend outside the substation fence comply with the loading and strength sections of the NESC. The *National Electrical Safety Code* (2023) is a minimum safety code and is not intended as a design specification.

3.1.13 State and Local Regulatory Loads

State and local regulatory loads should be reviewed for applying them to substation structures [e.g., California's General Order 95 (PUC 2017)].

Table 3-16. Conversion Factors for Ice Thickness and Concurrent Wind Load for Deflection Computations.

	100-Year to 5-year mean recurrence
Ice thickness conversion factor	0.2
Concurrent wind load conversion factor	1.0

3.2 APPLICATION OF LOADS

The following are recommended guidelines on the application of the structure-loading criteria presented in this MOP.

The following loading conditions should be considered for designing substation structures:

1. Dead loads (Section 3.1.1),
2. Equipment operating loads (Section 3.1.2),
3. Wire tension loads (Section 3.1.4),
4. Extreme wind loads (Section 3.1.5),
5. Combined ice and wind loads (Section 3.1.6),
6. Seismic loads (Section 3.1.7),
7. Short-circuit loads (combined with other load conditions when appropriate) (Section 3.1.8),
8. Construction and maintenance loads (Section 3.1.9), and
9. IEEE (2023) loads (and other state or local regulatory codes) (Sections 3.1.12 and 3.1.13).

Note: Structures that support wires and wire tension in appropriate weather conditions should be applied.

The following loading conditions should be considered for checking substation structure deflections:

10. Wind loads expected during substation equipment operation (Section 3.1.11.1),
11. Combined ice with concurrent wind loads expected during substation equipment operation (Section 3.1.11.2), and
12. Equipment operating loads (Sections 3.1.2 and 3.1.11.3).

Table 3-17 lists substation structure loading conditions that have the potential to control the design of the structure types listed. These load conditions should be considered, as well as other important load conditions identified by the owner, for designing substation equipment and support structures, and wire support structures.

3.3 LOAD FACTORS AND COMBINATIONS

The methods for estimating loads, especially those for weather-related events, are mostly based on statistical models. These models, although scientifically correct, have limitations on the precision of their prediction. To ensure structural reliability, load factors (or longer-return periods, MRI, for the event) are introduced to compensate for this uncertainty. Loads derived using this MOP should be factored in accordance with the load combinations listed in Table 3-18 (USD) and Table 3-19 (ASD). A design based on these loads is expected to withstand the effects of events having return periods (MRIs) that are consistent with current practice within the electric utility industry. The owner may choose a longer MRI than recommended for risk mitigation or more critical infrastructure.

The load factors and load combinations recommended in this section are selected on the basis of the unique characteristics of typical electrical substation structures. The recommended load factors and combinations are different from those in other documents such as the International Building Code (ICC 2018), ASCE 7-22 (ASCE 2022), or ACI CODE-318-19 (ACI 2019)

Table 3-17. Basic Loading Conditions.

Loading conditions	Wire-loaded substation structures	Switch and circuit switcher supports	Rigid bus supports	Other equipment supports
IEEE (2023)[a]	Y	N	N	N
Extreme wind	Y	Y	Y	Y
Combined ice with wind	Y	Y	Y	Y
Earthquake	Y[b]	Y[b]	Y	Y[b]
Short-circuit (fault)	N[c]	Y	Y	N[d]
Construction and maintenance	Y	Y	Y	Y
Equipment operation	N	Y	N	Y
Deflection	Y	Y	Y	Y

[a]IEEE (2023) or other state or local regulatory codes that may apply (e.g., PUC 2017). These codes may impose different load combinations.

[b]Substation equipment and their supports (Substation structures with the sole purpose of supporting a particular substation equipment type, such as instrument transformers, capacitors, or disconnect switches) should be designed in accordance with IEEE 693 (IEEE 2018c). Seismic loads for the design of foundations for equipment and their supports should be as specified in IEEE 693. Seismic loads for the design of anchorages of equipment and their supports should be as specified in IEEE 693 and as modified by the overstrength factors provided by this MOP. This MOP primarily provides guidance for the seismic design of substation structures, as described in Section 6.7.2 "Structures Not Covered by IEEE 693," such as dead-ends, racks, bus supports, and shielding masts. The seismic performance of existing structures and alterations to these structures is discussed in Chapter 11.

[c]Refer to Section 3.1.8.2.

[d]Short-circuit (fault) loads should be considered if the owner determines that this load effect is significant.

Note: For all structures and supports, include dead load and wire tension where appropriate.

that address loads mainly related to building-like structures. Electrical transmission line grids are distributed systems with multiple redundancies. Typically, multiple looped paths are provided from the generation sources to the point of service. The operational reliability of the substation and transmission line system is typically addressed by grid looping and circuit redundancies. Thus, the reliability of the electrical grid typically does not depend on one individual structure. The owner, however, should be cognizant of the importance of the structures staying operational during extreme events such as large ice and windstorms or large-magnitude earthquakes. The reliable delivery of electric power has been receiving increased attention from electric system owners and/or operators with the increasing importance of electric power to the functioning of modern society. The selection of appropriate levels of loading for structure design contributes to this end.

Furthermore, unlike failures of building-type structures, the failure of a substation structure during an extreme load event is likely to present a low hazard to utility personnel because almost all substations are unoccupied. For substation structures supporting wires that extend outside the substation fence, IEEE (2023) provides minimum requirements for public safety.

If loads derived using ASCE 7-22 are required to be used in design, they should be closely evaluated because the return periods can be significantly longer than those used as the basis of this MOP. If ASCE 7-22 loading is required by the owner, the structure designer is encouraged to discuss this issue with the owner. For example, this MOP uses wind maps with an MRI of 300 years. The loading combinations presented in ASCE 7-22 do not align well with the electric

Table 3-18. Ultimate Strength Design (USD) Load Combinations.

Case[a]	Combinations (all eight cases may not apply)
1[b]	$1.1\,D + 1.0\,W_{300} + 0.75\,SC + 1.1\,T_{W-300}$
2[b]	$1.1\,D + (1.0\,I_{100} + 1.0\,W_{WI-100}) + 0.75\,SC + 1.1\,T_{WI-100}$
3	$1.1\,D + 1.0\,\mathrm{SC} + 1.1\,T_{\mathrm{APP}}$
4[b]	$1.1\,D + 1.0\,E + 0.75\,SC + 1.1\,T_{EA}$
5[b]	$0.9\,D + 1.0\,W_{300} + 0.75\mathrm{SC} + 1.1\,T_{W-300}$
6[b]	$0.9\,D + (1.0\,I_{100} + 1.0\,W_{WI-100}) + 0.75\mathrm{SC} + 1.1\,T_{WI-100}$
7	$0.9\,D + 1.0\mathrm{SC} + 1.1\,T_{\mathrm{APP}}$
8[b]	$0.9\,D + 1.0E + 0.75SC + 1.1\,T_{EA}$

[a]Other load combinations may be required by the owner or be prudent to include

where

D = Structure and equipment dead load;

W_{300} = Extreme wind load (F_{WD}) obtained from Equation (3-1) or Equation (3-2) using a 300-year MRI wind map;

I_{100} = Extreme ice load from a 100-year MRI ice map (Section 3.1.6);

W_{WI-100} = Concurrent wind load in combination with ice from the 100-year MRI ice map;

T_{WI-100} = Wire tension resulting from the following loads acting simultaneously: weight of the wire; weight of ice corresponding to ice thickness in the 100-year MRI ice map; wind load on the iced wire corresponding to wind speed in the 100 MRI ice map shown in Figure 3-6a, and wire temperature shown in Figure 3-6c.

E = Seismic load as defined in Section 3.1.7.8;

T_{W-300} = Wire tension caused by wire weight acting simultaneously with the wind force corresponding to the wind speed from the 300-year MRI wind map at an ambient temperature determined by the owner;

SC = Short-circuit load; the load combination factors for short circuit were selected assuming that the engineer calculates the short-circuit forces as outlined in Section 3.1.8.1. That is, the engineer calculates loads on the rigid bus using the Simplified Short-Circuit Equations 3-38 to 3-41 with the simplifying assumptions of parallel, infinitely long conductors and the peak short-circuit force applied statically. If different procedures are used to determine short-circuit forces, such as considering the end and crossing effect or accounting for a dynamic structural response, the short-circuit load combination factors should be adjusted accordingly.

T_{APP} = Wire tension caused by the wire weight acting simultaneously with any appropriate ice weight and temperature as determined by the owner (under every-day or normal operational conditions);

T_{EA} = Wire tension corresponding to wire dead load acting simultaneously with the seismic loading in accordance with Section 3.1.7.12, at an ambient temperature determined by the owner.

[b]The combination of SC loads with extreme events listed in this table should be determined by the owner.

power industry, as ASCE 7-22 does not address issues such as fault currents, line tensions, or redundancy of the electric power system. However, if load factors and combinations in ASCE 7-22 are used to determine the structural loading demands, then the structural capacity should be calculated using the applicable material standards referenced in ASCE 7-22.

Loads derived from RUS 1724E-300 (USDA 2001) and IEEE (2023) should be factored and combined on the basis of the load factors within the respective document and structures designed using the material strength provisions in the corresponding standard.

Substation structural configurations are simple enough that dead weight can easily be taken into account, unlike that for building-type structures. The nature of substation structures

Table 3-19. Allowable Strength Design (ASD) Load Combinations (Service Level Loads).

Case[a]	Combinations (all eight cases may not apply)
1[b]	$1.0\,D + 1.0W_{100} + 0.5SC + 1.0T_{W-100}$
2[b]	$1.0\,D + (1.0I_{50} + 1.0W_{WI\text{-}50}) + 0.5SC + 1.0\,T_{WI-50}$
3	$1.0\,D + 0.7SC + 1.0T_{APP}$
4[b]	$1.0\,D + 0.7E + 0.5SC + 1.0T_{EB}$
5[b,c]	$0.6\,D + 1.0W_{100} + 0.5SC + 1.0T_{W-100}$
6[b,c]	$0.6\,D + (1.0I_{50} + 1.0W_{WI\text{-}50}) + 0.5SC + 1.0\,T_{WI-50}$
7[c]	$0.6\,D + 0.7SC + 1.0T_{APP}$
8[b,c]	$0.6\,D + 0.7E + 0.5SC + 1.0T_{EB}$

[a]Other load combinations may be required by the owner or be prudent to include
where

D = Structure and equipment dead load;

W_{100} = Extreme wind load (F_{WD}) obtained from Equation (3-1) or (3-2), using a 100-year MRI wind map;

I_{50} = Extreme ice load from a 50-year MRI ice map;

W_{WI-50} = Concurrent wind load in combination with ice from the 50-year MRI ice map;

T_{WI-50} = Wire tension resulting from the following loads acting simultaneously: weight of the wire; weight of ice corresponding to ice thickness in the 50-year MRI ice value obtained from the ASCE 7 Hazard Tool; wind load on the iced wire corresponding to wind speed from Figure 3-6b and: wire temperature from Figure 3-6c;

E = Seismic load as defined in Section 3.1.7.8;

T_{W-100} = Wire tension resulting from wire weight acting simultaneously with the wind force corresponding to the wind speed from the 100-year MRI wind map at an ambient temperature determined by the owner;

SC = Short-circuit load; the load combination factors for short circuit were selected assuming that the engineer calculates the short-circuit forces as outlined in Section 3.1.8.1. That is, the engineer calculates loads on the rigid bus using the simplified short-circuit Equations 3-38 to 3-41 with the simplifying assumptions of parallel, infinitely long conductors and the peak short-circuit force applied statically. If different procedures are used to determine the short-circuit forces, such as considering the end and crossing effect or accounting for a dynamic structural response, the short-circuit load combination factors should be adjusted accordingly.

T_{APP} = Wire tension resulting from the wire weight acting simultaneously with any appropriate ice weight and temperature as determined by the owner (under every-day or normal operational conditions);

T_{EA} = Wire tension caused by the wire weight acting simultaneously with any appropriate ice weight and temperature as determined by the owner (under every-day or normal operational conditions);

T_{EB} = Wire tension corresponding to wire dead load acting simultaneously with 70% of the seismic loading according to Section 3.1.7.12 "Seismic Forces on Wire Bus (Strain Bus or Cable Bus)," at an ambient temperature determined by the owner.

[b]The combination of SC loads with extreme events listed in this table should be determined by the owner.

[c]Refer to ASCE 7-22, Section C2.4.1, for an explanation of the 0.6 dead load factor; these load combinations are also intended for foundation design.

also prevents the likelihood of converting these structures into other functions or uses. Thus, structure designers can calculate the dead load with reasonable accuracy.

In the case of short-circuit loads, many theories and experiments have proven that the magnitude of this load event is typically much less than what has been predicted using the peak short-circuit force applied statically. The short-circuit loads are also short in duration.

Table 3-18 (USD) and Table 3-19 (ASD) show suggested design load cases, combinations, and minimum load factors to be used for substation structures. The load factors in Tables 3-18 (USD) and 3-19 (ASD) should be used with all material-based design specifications in conjunction with the nominal resistance and safety factors specified in this MOP. Operational loads as appropriate should be included with load combinations in Table 3-18 (apply a load factor of 1.0 to the operating loads) and Table 3-19 (apply a load factor of 0.75 to the operating loads). Load cases that include ice (I_{100} and I_{50}), the effect of icing on the wire dead load and concurrent wind load (W_{WI-100} and W_{WI-50}), are companion load components applied together in the load case.

The general format of the load equations is the combination of load factors times the calculated loads such as 1.1 × dead load. Load factors are used to account for the uncertainties in the estimation of the loads. A load factor of 1.0 does not imply that the structure is not structurally reliable. Rather, it represents a good understanding of the uncertainty in load determination. The load factors given in Tables 3-18 and 3-19 account for the likelihood of the event, in combination with the other events, for the load combination being considered.

Note that when short-circuit loading is combined with other extreme loads in the recommended load combinations, a load factor of less than unity is used. One would expect that the likelihood of a short-circuit event occurring would increase at times when other extreme loading events occur. However, the algebraic combination of the full magnitude of peak short-circuit loading applied statically along with other extreme events may be unnecessarily conservative. For this combination to occur, the peak design load for a particular extreme loading event would need to occur in a direction that would be additive to the short-circuit loading deflection direction. In addition, the peak deflection from an extreme event and peak short-circuit load deflection would need to occur at the same instant. Recall that the peak short-circuit conductor deflection likely occurs in a fraction of a second. It is unlikely that this combination of events would occur simultaneously. Thus, a reduced load factor was applied to the short-circuit load when combined with other extreme events to account for this low likelihood.

A particular structure may not be subjected to all individual load components listed in the load combination equations. The owner should determine whether a load or load combination is appropriate to meet the performance and reliability objectives. The combining of short-circuit loads with other loads (e.g., wind, ice, and earthquake) should be evaluated, and the owner should determine the level of the short-circuit load used in combination with other loads. This MOP does not imply that only the load combinations in Tables 3-18 and 3-19 need be used for designing substation structures. Variations of these or other load cases may be required to account for design conditions such as wind direction or short-circuit fault location.

Some of the load factors in Table 3-18 have been reduced from the prior edition of this MOP. The reduction in extreme wind load factor from 1.2, in the previous edition of this MOP, to 1.0 is a result of changing from wind speed maps with MRI from 50 years to maps with a longer MRI of 300 years. In the previous edition of this MOP, a 1.25 load factor was applied to E in Table 3-18 load combinations based on the assumption of 2% damping for substation structures. This edition of MOP 113 adopts the elastic seismic demands and corresponding seismic response factors on the basis of 5% damping for achieving consistency with relevant references such as

ASCE 7-22, resulting in a load factor of 1.0 on E for the USD method. When damping values other than 5% that are available in the ASCE 7 Hazard Tool website are desired, such as 2% for substation structures, IEEE 693-2018 may be used to develop the damped design spectral accelerations.

Table 3-19 is included to provide load demand provisions for allowable strength design (ASD). The load factors provided in Tables 3-18 and 3-19 for short-circuit loading are based on ASCE 113 committee experience and consensus.

3.4 ALTERNATE DESIGN LOADS AND LOAD FACTORS

Design loads, cases, load combinations, and load factors other than those recommended in this chapter should be substantiated by experience, research results, or test data.

3.5 SERVICEABILITY CONSIDERATIONS

Serviceability is the limit state in which the function, maintainability, durability, and appearance of substation structures and electrical equipment are preserved under normal usage. Serviceability should ensure that operational disruptions to the substation during normal, everyday use are rare.

Unacceptable structural response that is detrimental to serviceability includes the following:

1. Local damage (e.g., yielding, buckling, excessive deformation, or concrete cracking) that may require excessive maintenance or lead to corrosion;
2. Deflection or rotation that may affect structural appearance, structural function, or the operation of electrical equipment;
3. Excessive vibration created by wind;
4. Excessive temperature that may affect the strength of aluminum and copper; and
5. Impact load created during an earthquake, when relative displacement between adjacent electrical equipment connected by a flexible conductor exceeds the conductor slack provided.

Specifying limiting deflections and service loads helps control excessive deflection and vibration. Limiting the operating temperature helps control excessive temperatures. In the past, these guidelines have provided satisfactory structural performance, except for some cases of a VIV of hollow tubular members. These wind-induced vibrations are discussed in Sections 3.1.10 and 6.10.2.

For structures supporting electrical equipment, the manufacturer's recommendations for service loads should be followed. Alternatively, the owner should consult or specify the anticipated loading requirement to the manufacturer.

For dead-end structures, rigid bus structures, and shielding masts, serviceability criteria, in addition to deflection limits specified in Chapter 4, should consist of the following:

1. Minimum phase-to-ground clearance for conductors and bus systems should be maintained during high wind conditions. Flexible conductors can have significant horizontal movements in high wind conditions;

2. Minimum vertical electrical clearances should be maintained for conductors, bus systems, and overhead ground wires during ice conditions. Overhead ground wires are particularly susceptible to large vertical sags in ice conditions;
3. Minimum vertical clearances should be maintained under maximum operating temperature conditions for conductors and bus systems;
4. High-temperature operations may lead to a reduction in aluminum material strength;
5. Coating (e.g., galvanizing or paint) of structural components should be designed to protect the component from corrosion;
6. Provisions should be made to allow for the expansion and contraction of conductors, overhead ground wires, and rigid bus conductors from varying temperatures. The upward vertical movements of the conductors should not be restrained by taps to equipment or structures. See Section 3.1.4 for additional information.

REFERENCES

ACI (American Concrete Institute). 2019. *Building code requirements for structural concrete (with commentary).* ACI CODE-318-19(22). Detroit: ACI.

AISC (American Institute of Steel Construction). 2016. *Seismic provisions for structural steel buildings.* AISC 341-16. Chicago: AISC.

ASCE. 2020. *Guidelines for electrical transmission line structural loading*, MOP 74. 4th edition. Reston, VA: ASCE.

ASCE. 2022. *Minimum design loads and associated criteria for buildings and other structures.* ASCE 7. Reston, VA: ASCE.

Bayar, D. C. 1986. "Drag coefficients of latticed towers." *J. Struct. Div.* 112 (2): 9–167.

BEAIRA (British Electrical and Allied Industries Research Association). 1935. "Wind pressure on latticed towers—Tests on models, report (F/T 84)." *J. Inst. Electr. Eng.* 77: 189–196.

CIGRE (International Council on Large Electric Systems). 1996. *The mechanical effects of short-circuit currents in open air substations (rigid and flexible busbars).* Brochure 105. Paris: CIGRE.

Davenport, A. G. 1979. "Gust response factors for transmission line loading." In *Proc., 5th Int. Conf. of Wind Engineering*, 866–909. Oxford, UK: Pergamon Press.

FEMA (Federal Emergency Management Agency). 2020. "NEHRP recommended seismic provisions for new buildings and other structures." In Vol. 1 of *Part 1 Provisions, Part 2 Commentary.* FEMA P-2082/1. Washington, DC: Building Seismic Safety Council of the National Institute of Building Sciences.

ICC (International Code Council). 2018. *International building code.* Washington, DC: ICC.

IEC (International Electrotechnical Commission). 2011. *Short-circuit currents—Calculation of effects. Part 1: Definitions and calculation methods.* 60865-1 ed.3.0. Geneva: IEC.

IEEE (Institute of Electrical and Electronics Engineers). 2000. *Standard performance characteristics and dimensions for outdoor apparatus bushings.* IEEE C57.19.01. Piscataway, NJ: IEEE.

IEEE. 2008. *Guide for design of substation rigid-bus structures.* IEEE 605. Piscataway, NJ: IEEE.

IEEE. 2011. *Standard requirements for AC high-voltage air switches rated above 1000 V.* IEEE C37.30.1. Piscataway, NJ: IEEE.

IEEE. 2012. *Guide for application of power apparatus bushings.* IEEE C57.19.100. Piscataway, NJ: IEEE.

IEEE. 2018a. *Fall protection of the utility industry.* IEEE 1307. Piscataway, NJ: IEEE.

IEEE. 2018b. *Recommended practice for the design of buswork located in seismically active areas.* IEEE 1527. Piscataway, NJ: IEEE.

IEEE. 2018c. *Recommended practice for seismic design of substations.* IEEE 693-2018. Piscataway, NJ: IEEE.

IEEE. 2018d. *Standard capacitance current switching requirements for high voltage circuit breakers.* IEEE C37.04. Piscataway, NJ: IEEE.

IEEE. 2023. *National electrical safety code.* ANSI C2. Piscataway, NJ: IEEE.

Norris, C. H., and J. B. Wilbur. 1960. *Elementary structural analysis*. 2nd ed. New York: McGraw-Hill.

OSHA (Occupational Safety and Health Administration). 2020. *OSHA safety and health standards for construction*. 29 CFR 1926, Subpart M. Washington, DC: OSHA.

PUC (Public Utilities Commission). 2017. *State of California rule for overhead electric line construction*. General Order 95. Sacramento, CA: PUC.

USDA (US Department of Agriculture). 2001. *Design guide for rural substations*. RUS Bulletin 1724E-300. Washington, DC: USDA.

USGS (US Geological Survey). 2018. *Earthquake hazards*. Washington, DC: USGS.

CHAPTER 4

DEFLECTION CRITERIA (FOR OPERATIONAL LOADING)

Deflection and rotation of substation structures and members can affect the mechanical operation of supported electrical equipment, reduce electrical clearances, and cause unpredicted stress in structures, insulators, connectors, and rigid bus conductors. For these reasons, structural deflections should be limited to magnitudes that are not detrimental to the mechanical and electrical operation of the substation. The deflection limits provided in this chapter are maximum recommended values. The selection of the deflection limits applicable to a specific application is the responsibility of the owner. The owner is responsible for verifying with equipment manufacturers that deflection limits in this chapter will result in acceptable equipment operation.

Loading criteria for deflection limits are recommended in Section 3.1.11, "Loading Criteria for Deflection Limitations."

4.1 STRUCTURE CLASSIFICATIONS AND DEFLECTION LIMITS

This section specifies the structural deflection limits for three classes of typical substation structures: A, B, and C. The sensitivity of equipment to deflection of supporting structures varies considerably. Disconnect switches, with complex mechanical operating components, are susceptible to binding if the structure distorts from the installed geometry. Structures supporting stranded conductors or overhead line dead-ends can typically withstand structure deflections without causing any detrimental effect on-line operation. Therefore, substation structures are classified for the purpose of applying deflection limits that reflect the relative sensitivity (structural or operability) of supported equipment. It should be noted that for long span beams and tall columns, it may be necessary to evaluate the total gross deflections to ensure that the expected electrical clearances are maintained.

4.1.1 Deflection Analysis and Criteria

The following sections define the horizontal and vertical member spans.

4.1.1.1 Horizontal Members. For determination of maximum deflections, the span of a horizontal member is the clear distance between structural connections and the vertical supporting members, or for cantilever members, it is the distance from the point of investigation to the vertical supporting member (Figure 4-1). For simplicity, the horizontal span may be taken as centerline to centerline of the vertical supporting members.

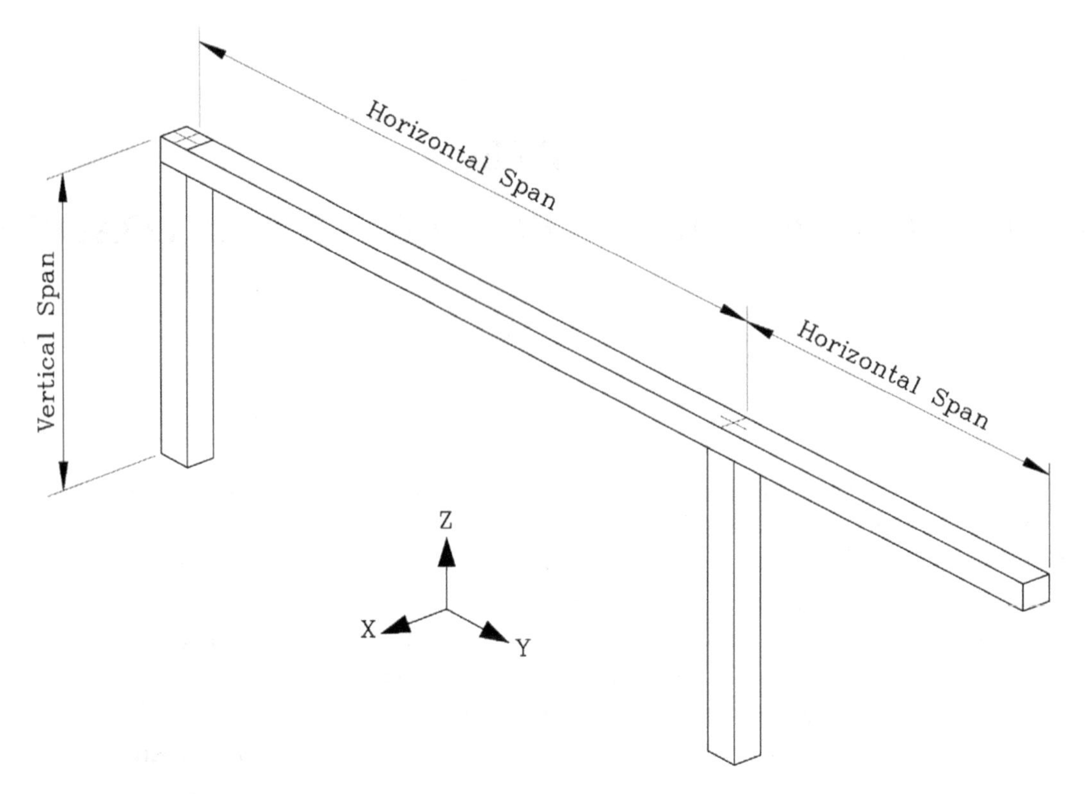

Figure 4-1. Span definitions.

For horizontal members, the deflection is the maximum net displacement, horizontal or vertical, of the member relative to the member connection points. Net displacement is discussed in detail in Section 4.2.4. Deflection analysis typically does not include the foundation displacement or rotation (Section 4.2.3).

4.1.1.2 Vertical Members. For determination of maximum deflections, the span of a vertical member is the vertical distance from the foundation support to the point of investigation on the structure. The deflection to be limited is the gross horizontal displacement of the member relative to the foundation support. Gross displacement is discussed in detail in Section 4.2.4.

4.1.2 Class A Structures

Class A structures support equipment with mechanical devices such as operating rods or control linkages where structure deflection could impair or prevent proper operation. Examples are group-operated switches, vertical reach switches, ground switches, circuit-breaker supports, and circuit-interrupting devices, also referred to as Class A equipment. Equipment manufacturers should be consulted to determine whether any specific structure deflection limits are required for their equipment.

4.1.2.1. Deflection Limits of Horizontal Members in Class A Structures. Horizontal deflection of horizontal members should not exceed 1/200 of the member span (Figure 4-2). Vertical deflection of horizontal members should not exceed 1/200 of the member span (Figure 4-3).

4.1.2.2 Deflection Limits of Vertical Members in Class A Structures. Horizontal deflection of vertical members should not exceed 1/100 of the height of the point of investigation above the foundation (Figure 4-4).

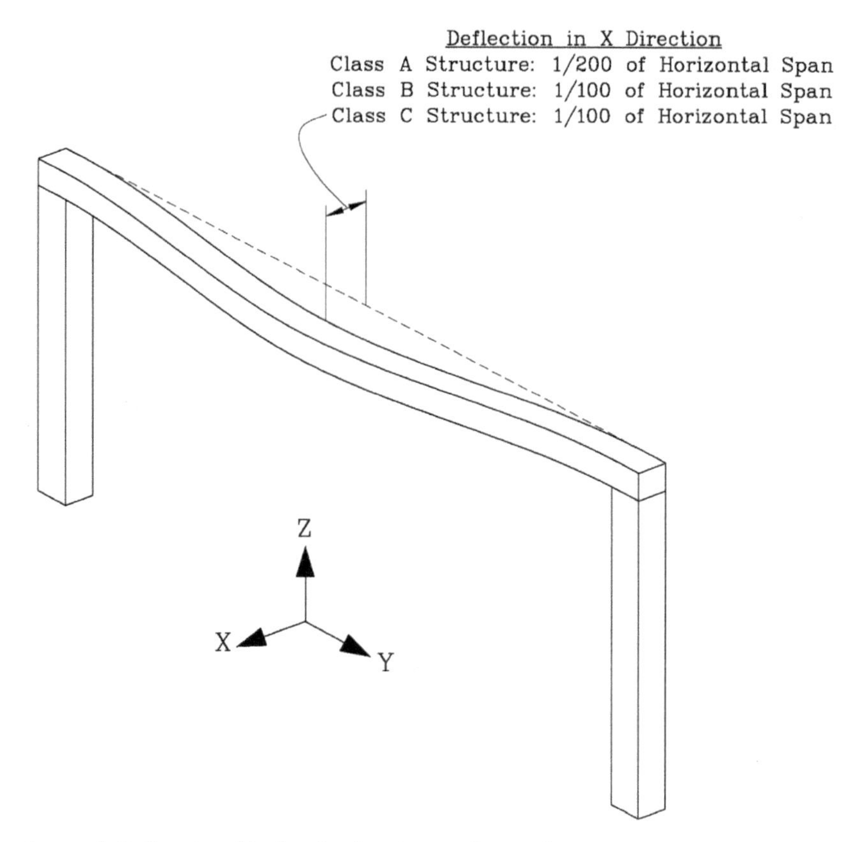

Figure 4-2. Horizontal deflection limits for horizontal members.

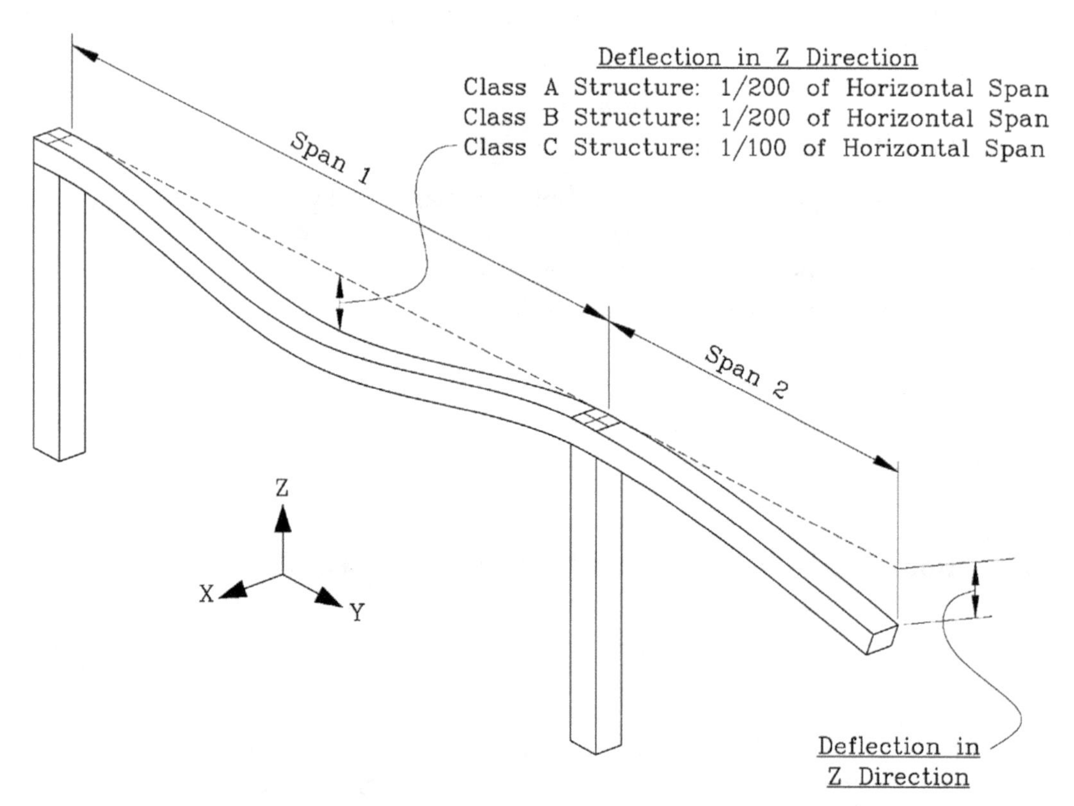

Figure 4-3. Vertical deflection limits for horizontal members.

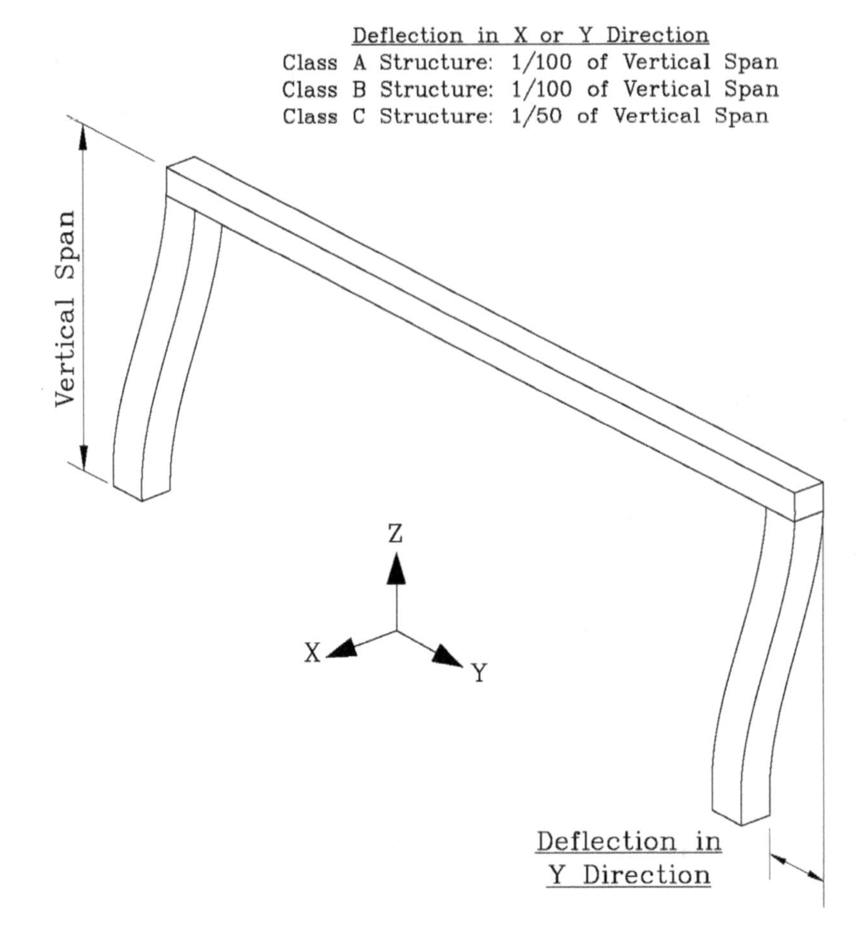

Figure 4-4. Horizontal deflection limits for vertical members.
Note: Member deflection limits apply to the X- or Y-direction; here, only the Y-direction is shown for the purposes of clarity.

4.1.3 Class B Structures

Class B structures support equipment without mechanical devices such as operating rods or control linkages, but where excessive deflection could result in compromised phase-to-phase or phase-to-ground clearances or unpredicted stresses in equipment, fittings, or bus conductors. Examples are support structures for rigid bus conductors, surge arresters, metering devices [such as current transformers (CT's), potential transformers (PT's), and coupling capacitor voltage transformers (CCVT's)], station power transformers, hook stick switches or fuses, and wave traps, also referred to as Class B equipment. Equipment manufacturers should be consulted to determine whether any specific structure deflection limits are required for their equipment.

4.1.3.1 Deflection Limits of Horizontal Members in Class B Structures. Horizontal deflection of horizontal members should not exceed 1/100 of the member span (Figure 4-2). Vertical deflection of horizontal members should not exceed 1/200 of the member span (Figure 4-3).

4.1.3.2 Deflection Limits of Vertical Members in Class B Structures. Horizontal deflection of vertical members should not exceed 1/100 of the height of the point of investigation above the foundation (Figure 4-4).

4.1.4 Class C Structures

Class C structures support equipment relatively insensitive to deflection or are stand-alone structures that do not support any equipment. Examples are support structures for flexible (stranded conductor) bus, shielding masts, and dead-end structures for incoming transmission lines. Deflection limits for these structures are intended to limit P-delta stresses, wind-induced vibrations, and visual impact.

4.1.4.1 Deflection Limits of Horizontal Members in Class C Structures. Horizontal deflection of horizontal members should not exceed 1/100 of the member span (Figure 4-2). Vertical deflection of horizontal members (Figure 4-3) should not exceed 1/100 of the member span.

4.1.4.2 Deflection Limits of Vertical Members in Class C Structures. Horizontal deflection of vertical members should not exceed 1/50 of the height of the point of investigation above the foundation (Figure 4-4).

4.2 SPECIAL CONSIDERATIONS FOR DEFLECTION ANALYSIS

4.2.1 Multiple-Use Structures

Structures can be designed to support several pieces of equipment that require different structure classifications. When evaluating deflection of a multiple-use structure, the deflection limits applicable to any point on the structure are determined by the classification of the structure from that location upward. If there is Class A equipment at or above the location being evaluated, then the analysis of that location is governed by Class A structure limits. If there is only Class B equipment and C structure components at or above the location being evaluated, then the analysis of that location is governed by Class B structure limits. If there are only Class C structure components at or above the location being evaluated, then the analysis of that location is governed by Class C structure limits. As an example, Figure 4-5 shows a line dead-end structure (Class C structure) that also supports a switch (Class A equipment) at a lower elevation. The switch platform and the vertical members from the foundation to the switch platform should meet Class A structure deflection criteria. Members associated with the line dead-end and vertical members from the switch elevation to the line dead-end should meet Class C structure criteria.

4.2.2 Rotational Limitation

Some equipment and rigid bus designs may be sensitive to the rotation of supporting members in addition to the deflection of the member. Special care should be taken when evaluating tall structures because the effect of structure rotations is magnified. Where an analysis is performed of the rigid bus and support system, the sensitivity of the system to accommodate rotation should be evaluated and limits determined if necessary. Rotational limits, if required, should be established by the owner.

4.2.3 Anchorage and Member Connection Restraints

The fixity of the substation structure base plate and foundation can affect the deflection of the structure. If the base plate is not sufficiently rigid, the flexure of the base plate caused by the overturning moment can cause additional column rotation. Depending on the loading,

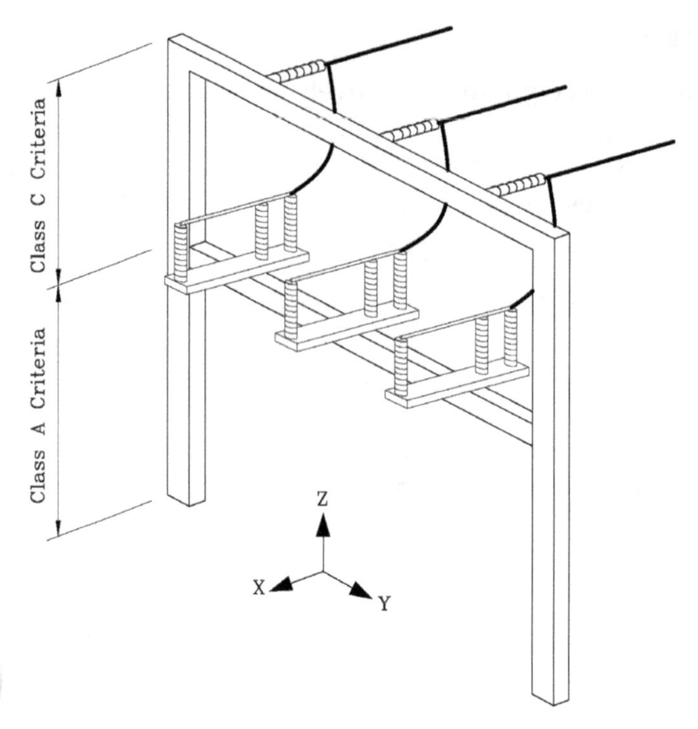

Figure 4-5. Multiple-use structures.

foundation design, and soil conditions, the foundation can deflect and rotate. This will result in additional structure deflection and should be considered when designing substation structure base plates and foundations. A more rigorous soil–structure analysis would be required to account for foundation movement.

Member connection conditions will also influence the horizontal and vertical deflections. Analytical modeling should consider the end conditions, which range from pinned to fixed, and which best represent the final in-service condition. See Section 5.3, "Structure Model," for other structure modeling considerations.

4.2.4 Gross versus Net Deflections

It is important to understand the difference between gross deflection versus net deflection. Gross deflection is the total displacement of a point from its initial position relative to the top of the structure foundation (Figure 4-6). Gross deflection should be considered when checking the electrical clearances of supported equipment. Net deflection is the displacement of a point relative to its point of connection, whether that is a vertical or horizontal structure member. Net deflection for horizontal structure members is currently the substation structure industry standard when evaluating deflection. Net deflection allows for analysis of platform-type structures, such as switch stands, to only limit deflections that would cause problems with switch operation and not result in unnecessarily large columns (see examples in Appendix A.5).

4.2.5 Shielding Masts and Other Tall, Slender Structures

In certain cases, the structure type, design loads, and lower deflection limits for Class C structures can result in a flexible (low stiffness) structure. These structures can be subject to potentially damaging wind-induced oscillations, often referred to as vortex-induced vibrations (VIV). Such structures can be susceptible to fatigue cracking and failure. In addition to the

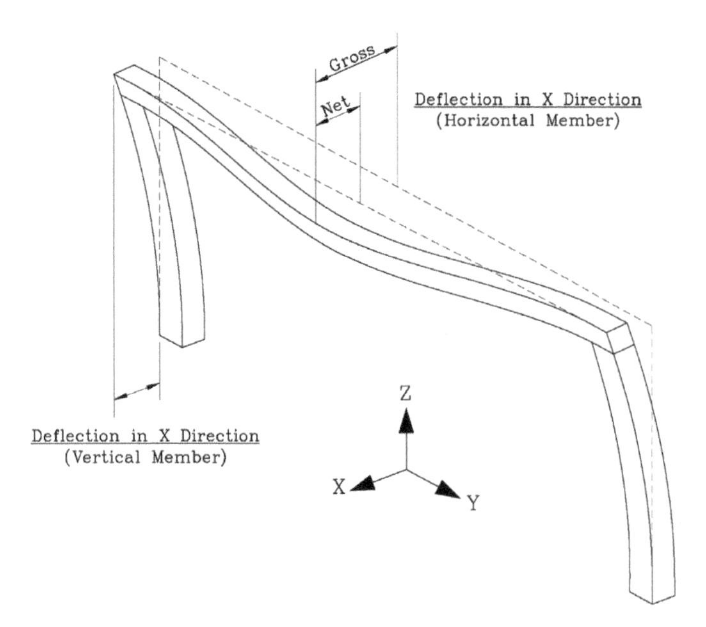

Figure 4-6. Net versus gross deflection.

specified static deflection limits, consideration should be given to the use of dampening devices or other techniques to minimize potential for damage. Methods include the use of internal cables or covered chains, external spoilers, beam tie-downs, or other means of interrupting the oscillations. Chapter 6 provides additional information on structural member vibrations.

4.2.6 Rigid Bus Vertical Deflection Criteria

See IEEE 605-2008 (IEEE 2008), Clause 12.1, for appearance-based rigid bus deflection limits.

4.3 SUMMARY

Table 4-1 summarizes the structure classes and associated deflection limits.

Table 4-1. Summary of Structure Deflection Limits.

Maximum structure deflection as a ratio of span length[a]				
		Structure class		
Member type	Deflection Direction	Class A	Class B	Class C
Horizontal[b]	Vertical	1/200	1/200	1/100
Horizontal[b]	Horizontal	1/200	1/100	1/100
Vertical[c]	Horizontal	1/100	1/100	1/50

[a]For loading criteria for deflection limits, see Section 3.1.11, "Loading Criteria for Deflection Limitations," in Chapter 3, "Loading Criteria for Substation Structures."

[b]See Section 4.1.1.1.

[c]See Section 4.1.1.2.

REFERENCE

IEEE (Institute of Electrical and Electronics Engineers). 2008. *Guide for bus design in air insulated substations.* IEEE 605-2008. New York: IEEE.

CHAPTER 5
METHOD OF ANALYSIS

5.1 OVERVIEW

The design of substation structures requires a knowledge of the equipment being supported by the structure, its operation, and electrical and safety codes.

Analysis, as used herein, is defined as the mathematical formulation of the behavior of a structure under load. The solution yields the calculated displacements, support reactions, and internal forces or stresses. The analysis of a structure begins by developing a model of the structural configuration, connection characteristics, support boundary conditions, material properties, and loading cases.

5.2 STRESS CRITERION VERSUS DEFLECTION CRITERION

An analysis of a structure generates both stresses and deflections, which may be used to determine the adequacy of the structure.

A structure should provide sufficient capability to support its own weight and any combination of loads as specified in Chapter 3, "Loading Criteria for Substation Structures." The internal stresses under these loading cases should be at or below limits specified in Chapter 6, "Design," or verified by physical tests.

A structure designed solely for strength may have excessive deflections. Excessive deflection of a structure may or may not in itself be detrimental, but the effects on substation components that are supported by the deflecting structure are frequently quite significant. Deflection may cause visual displeasure to the viewer or preclude the equipment from performing its intended purpose. A design controlled by deflection is said to be based on serviceability. Some substation structures have rigorous deflection limitations, whereas others are not so restricted. The substation structure deflection criteria are specified in Chapter 4, "Deflection Criteria (For Operational Loading)." These criteria are based on the function of the structure. Deflection may control the design of substation structures.

5.3 THE STRUCTURE MODEL

The *analysis model* is a set of mathematical equations to predict the behavior of the structure. The accuracy of the analysis is only as good as the model. Thus, it is important to construct an adequate mathematical model to simulate the true behavior of the structure.

In developing a model of a structure, assumptions are made concerning the individual elements, their geometrical and mechanical properties, and the connections, loads, and foundation supports. These are discussed in the following paragraphs.

5.3.1 Individual Members and Connections

A structure may consist of only a few members, as in the case of a steel pole, or a group of members, as in the case of a lattice tower. Members may be as simple as a two-dimensional truss element or as complex as a plate in bending in a finite-element analysis. The choice of the element type depends on the structural geometric configuration and the desired load flow into each element. The choice of the element also defines the type of connection in the analysis model.

5.3.2 Truss Model

A truss has a geometric configuration consisting of elements forming triangles. The elements in a truss are assumed to be two force members providing load resistance by only axial forces. The accuracy of the analysis using a truss model depends on the connections being detailed to connect the neutral axes of the truss elements and the end connections to simulate a pinned joint.

A truss model used in an analysis assumes that there is only axial load transmitted by and to the members. Care should be taken in the connection design to ensure that these assumptions remain valid. Eccentricities in the connections will induce moments in the member, which must be considered to accurately represent the state of stress in the member.

A truss model implies that all nodes are pinned. This assumption may sometimes need to be verified for main leg members, which in reality are continuous elements throughout. For this reason, depending on the geometry of the structure, some shear forces could be transferred to the main leg members, thereby inducing bending moments that may need to be checked.

Subtle differences can significantly alter the behavior of a structure. For example, the structure shown in Figure 5-1a could be analyzed as a plane truss and would not have significant bending moments. However, the structure in Figure 5-1b may be governed by bending stresses as a result of the shear acting over the unbraced lower section of the leg.

5.3.3 Frame Model

The elements in a frame model provide resistance to load by forces and bending moments. If the frame model is three-dimensional (3D), moments may be produced in three orthogonal directions. The connections in a frame model may be configured to resist or not resist moment (including torsion). The configuration of the connection is accounted for in a frame model by

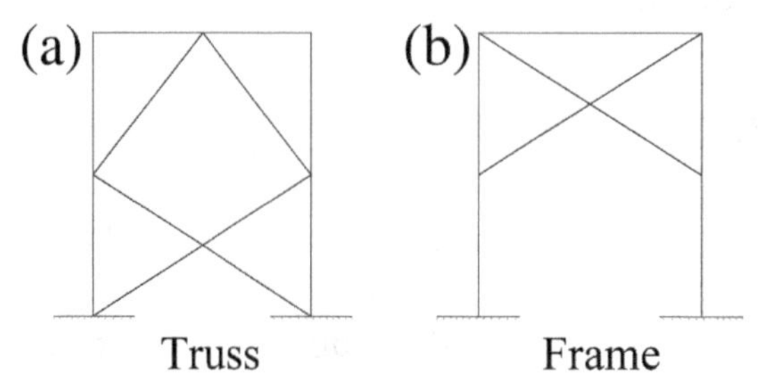

Figure 5-1. Plane truss and plane frame comparison.

defining moment or force restraints. If moment or torsion is to be transferred, the connection should be designed as a rigid connection between the elements of the cross section resisting the moment or torsion, for example, in a wide flange, the flanges resist moment.

5.3.4 Finite-Element Model

More complex elements are used in finite-element analysis. The most common are plane stress or plane strain (membrane forces), plate bending or shell elements, and solid 3D elements. Any or all these elements may be combined with each other or included in a structure having truss or frame elements. Care should be taken when using finite-element analysis because the displacement functions used in formulating the element's performance are inherently approximations. When used correctly, finite-element analysis can calculate accurate stresses and displacements. Poor choice of elements or their mesh density may result in errors of such magnitude as to render the results useless. An excellent treatise on the cautions of using finite-element analysis can be found in MacNeal and Harder (1985).

5.3.5 Loads and Support Conditions

The structural model determines how the loads are applied and the types of support conditions. The joints of a truss element can only translate, whereas the joints of a frame element can translate and rotate. These are the degrees of freedom. A membrane and solid element have degrees of freedom in translation; a plate in bending has rotational degrees of freedom about axes within the plane of the element. The degrees of freedom define the limits of the loads and support conditions. A truss model can only have translational joint loads; a frame model can have both translational and rotational joint loads. A force applied in the span of a truss element is not allowed in the model because it cannot be transferred to the joint without inducing bending in the element, which violates the assumptions defining a two-force member. The effect of these forces should be checked using an analysis of the member, considering the appropriate bending stiffness of the member end conditions. A frame can have loads applied in its span because the flexural strength of the element allows the span load to be transferred to the joints and thus into the rest of the frame.

The allowed loads on finite elements are predetermined by the formulation of the code of the computer program being used and should be understood by the engineer. The allowed joint loads correspond to the degrees of freedom (e.g., a translational degree of freedom allows a translation load).

The support conditions define boundary restraints of the structure. If the support is rigid, it is a known displacement field whose value is zero. A nonzero value defines a support movement. The allowed specified deflections are limited by the degrees of freedom of the joint providing the boundary condition or support for the structure. Not all degrees of freedom at a support need to be specified. Those not specified are unknown and are calculated in the analysis.

It is the engineer's responsibility to ensure that the final physical structural configuration and the model agree. If the analysis uses a truss model, the connections should be designed to provide truss action only or to have the line of action of the internal forces pass through a common point. If eccentricities are not accounted for in the analysis, they should be minimized in the physical design.

This Manual of Practice (MOP) defines the allowable deflections in substation structures. The structure model will provide the calculated deflection. If the model is assumed stiffer than the actual structure, the calculated deflections will be less than the actual deflections, and vice versa. Thus, the model should accurately represent the stiffness of the actual structure for a valid check of the serviceability requirements.

5.4 STATIC ANALYSIS METHOD

5.4.1 Approximate Analysis

Many approximate analysis techniques are documented in structural analysis texts. Some approximate analyses are done on the basis of the experience of the analyst, wherein assumptions are simplified and the stresses and deflections are calculated. The design is considered adequate if all strength and serviceability criteria are satisfied. An approximate analysis is not recommended for complex structural configurations but has value in the preliminary design stage or as an independent cursory check of a computer analysis.

5.4.2 First-Order Elastic Analysis

In a first-order elastic analysis, equilibrium is formulated on the undeformed geometry. Such an analysis is performed using one of the classical structural analysis techniques such as statics, slope deflection, moment distribution, and matrix methods using the stiffness approach. The analysis assumes linear material behavior as an element is loaded and unloaded irrespective of the magnitude of deflection, load, or stress. The calculations in a first-order elastic analysis use the undeformed shape of the structure. Therefore, any amplification in moment or any other force as a result of the deformation of the structure is not considered.

In the case of internal bending moment, this amplification is commonly referred to as the *P-delta amplification.* White and Hajjar (1991) point out that there are two P-delta effects. The first, which is commonly referred to as *P-D*, is caused by joint translation, commonly referred to as *side sway.* The other, which is commonly referred to as *P-d*, is induced by the internal curvature of the bending member irrespective of the translation of the joints. ANSI/AISC 360-22 (AISC 2022) has Figure C-C2.1 showing these two different types of *P*-delta.

A third *P*-delta effect, caused by material non-linearity, occurs if the member design allows for partial yielding. Some design equations account for the *P*-delta effect on stresses derived from a first-order analysis. For example, ANSI/AISC 360-22 includes an alternate method that utilizes "amplification factors" to approximate the *P*-delta effects. Some of the requirements in commonly used documents are summarized in Section 5.4.5.

5.4.3 Second-Order Elastic Analysis

In a second-order elastic analysis or a geometric nonlinear analysis, equilibrium is formulated on the deformed configuration of the structure. This analysis is commonly thought of as including the amplification of moment caused by the axial load times the deformation of the compression member, e.g., the *P*-delta amplification in bending moment. Geometric nonlinear effects can show up in any structure. A lattice tower modeled as a 3D truss has geometric nonlinear effects as the tower moves under load.

Including the geometric nonlinear effects is not as simple as reformulating the analysis using the deflected shape. As the structure deflects, internal forces are produced and need to be accounted for in the analysis of the deformed structure. Cook (1974) gives details of how to perform an analysis including the geometric nonlinear effects. Even so, this method includes only the effects of joint translation and not the internal curvature of the member. Inclusion of internal curvature in the analysis would require subdividing the individual members.

In general, a lattice tower modeled as a truss model has negligible geometric nonlinear effects. A pole structure or cable structure may exhibit significant geometric nonlinear effects.

Some changes to design equations can account for the geometric nonlinear effects by reducing the allowable stress or increasing the force in the member. These changes should not be included if an analysis includes all geometric nonlinear effects.

As computing power has increased in modern times, accounting for second-order effects directly in the analysis has become more routine. Many software packages are capable of iterating the analysis on the deformed shape of the structure until equilibrium is reached and are also capable of subdividing individual members to capture the second-order effects of the internal curvature of the member.

An analysis including the geometric nonlinear effects is most important in determining a realistic value for deflection. This analysis method should be considered for analyzing displacement-sensitive flexible substation (Class A) structures.

5.4.4 First-Order Inelastic Analysis

Material yield or nonlinear member performance is accounted for in a first-order inelastic analysis. A plastic analysis of a rigid moment–resistant frame is a classic example of a first-order inelastic analysis (Beedle 1958). This technique is based on a structure being able to carry load beyond the elastic limit until it reaches its ultimate load through plastic deformation. It is applicable to indeterminate rigid-jointed frames, continuous beams, and in general structures stressed primarily in bending.

A determinate beam or frame under a given set of loads has one point at which the moment is maximum. If this load is increased, this moment increases and ultimately forms a plastic hinge. When the bending stress block reaches yield, a collapse mechanism is formed.

The behavior of an indeterminate (hyperstatic) structure is different from that of a determinate (isostatic) beam. It will not collapse until the number of plastic hinges equals 1 plus the degree of indeterminacy. The concept of plastic analysis and design uses this reserve strength to produce more economical structures. The engineer uses plastic analysis to determine the load that will produce a plastic hinge mechanism of the structure and the internal moments used in plastic design. ANSI/AISC 360-22 recognizes plastic analysis as an acceptable analysis technique. However, this MOP does not recommend the use of plastic analysis.

Mueller et al. (1991) proposed a method to account for nonlinear member performance in the analysis of transmission towers. In this method, the individual member performance is defined by a nonlinear member load versus a member deflection curve.

5.4.5 Analysis Requirements in Commonly Used Documents

The following sections provide a brief summary of the analysis methods required for the design documents commonly utilized for substation structure design.

5.4.5.1 Analysis Requirements in ANSI/AISC 360-22. ANSI/AISC 360-22, *Specification for Structural Steel Buildings*, contains stability requirements in Chapter C. The stability requirements are intended to account for the following:

1. Effects of initial geometric imperfections, commonly referred to as *P-Delta*.
2. Secondary effects caused by structure drift or member deflection, commonly referred to as $P\text{-}\delta$.
3. Account for the material non-linearity caused by residual stresses.

The current preferred method of analysis is the Direct Analysis Method. The *Direct Analysis Method* is a second-order elastic method. Methods other than the Direct Analysis Method are located in Appendix 8 of ANSI/AISC 360-22. The primary practical advantage of the Direct Analysis Method is that it allows for the use of member effective length factor, or *K*-factors, equal to 1.0 for all members rather than adjusting *k*-factors for individual members.

Item 1 is addressed by either modeling the structure with an initial geometric imperfection or by applying a lateral load that is a fraction of the structure weight. This lateral load is meant to mimic the effects of the geometric imperfection. These lateral loads, referred to as notional loads, are often much smaller than the lateral loads applied to substation structures under extreme wind or seismic events, and often need not be applied in combination with other lateral loads. Thus, these loads would often not be included in the controlling load combination. Refer to ANSI/AISC 360-22, Chapter C, "Direct Analysis Method of Design," for more information.

Item 2 is addressed by repeating the analysis using the deformed shape until the structure ceases further appreciable deflection. This is typically accomplished by using computer software. The software can efficiently iterate the analysis on the deformed shape.

Item 3 is addressed by reducing the stiffness of the structural members, which leads to larger deflections, thus increasing the stresses in an iterative analysis.

If computer software is used to implement the direct analysis method, it is extremely important to understand how the particular software package accounts for Items 1 to 3 along with values used as defaults. Some software packages default to include these items in the analysis. Other software packages rely on the user to specify these options.

It is important to note that stiffness reduction used in the Direct Analysis Method to account for member residual stresses (Item 3) can greatly affect the design of substation structures.

According to ANSI/AISC 360-22: Chapter C, Section C2.3, "Adjustments to Stiffness,"

The analysis of the structure to determine the required strengths of components shall use reduced stiffnesses, as follows:

> A factor of 0.80 shall be applied to all stiffnesses that are considered to contribute to the stability of the structure. It is permissible to apply this reduction factor to all stiffnesses in the structure.

According to the commentary of ANSI/AISC 360-22,

> The use of reduced stiffness only pertains to analyses for strength and stability limit states. It does not apply to analyses for other stiffness-based conditions and criteria, such as for drift, deflection, vibration, and period determination.

Therefore, substation structures designed in accordance with ANSI/AISC 360-22 typically require two analyses: one analysis for strength using the Direct Design Method, which accounts for Items 1, 2, and 3, and another analysis for deflection using only Items 1 and 2.

Because substation structures are generally deflection-controlled, if the software accounts for Item 3 by default and the user limits the structure deflection on the basis of deflection value calculated including the stiffness reduction, the final structure design may be more robust than needed.

5.4.5.2 ASCE 48-19, *Design of Steel Transmission Pole Structures.* ASCE 48-19 (ASCE 2019) requires the engineer to use geometrically nonlinear elastic stress analysis methods. If deviating from the design equations in the standard and pushing the structure beyond the elastic state and into the inelastic or collapse state, material nonlinearity may need to be addressed as well.

5.4.5.3 ASCE 10-15, *Design of Latticed Steel Transmission Structures.* ASCE 10-15 (ASCE 2015) does not contain specification provisions regarding the type of analysis that needs to be performed. Rather, it instructs the user to use "established engineering principles" to determine member forces. If a tower is sufficiently rigid, it may be reasonable to assume that a lattice tower does not have significant geometric or material nonlinear effects. However, the engineer should consider whether the structure in question will experience large-enough deflections to warrant an analysis including these effects.

5.5 DYNAMIC ANALYSIS METHOD

Dynamic analysis is a complex topic not easily summarized in a design guide. Performing a dynamic analysis requires a thorough understanding of the background material. There are many informative textbooks and design guides on the subject, and it is recommended that engineers educate themselves on the topic prior to performing a dynamic analysis. The following is a rudimentary discussion of dynamic analyses.

5.5.1 Steady-State Analysis

In a steady-state analysis, the load $\{R\}$ is of the form

$$\{R\} = \sin(\omega t)\{F\} \tag{5-1}$$

where ω is the circular frequency, and t is the time, and the equilibrium equation is as follows:

$$[M]\{\ddot{y}\} + [K]\{y\} = \{R\} \tag{5-2}$$

where

$\{F\}$ = Loads
$[M]$ = Mass
$\{y\}$ = Displacement
$\{\ddot{y}\}$ = Acceleration
$[K]$ = Structural stiffness

The steady state assumes zero damping and requires the loading frequency of all loads to be the same.

5.5.2 Eigenvalue Analysis: Natural Frequencies and Normal Modes

In an eigenvalue analysis, free vibration is considered, and the structure is not subjected to external forces or support motion. Equation (5-3) describes this free vibration state in the following way:

$$[M]\{\ddot{y}\} + [K]\{y\} = 0 \tag{5-3}$$

A solution is sought in the form of

$$\{y\} = \{a\}\sin(\omega t) \tag{5-4}$$

where a_i is the amplitude of motion of the ith degree of freedom. Integrating Equation (5-4) into Equation (5-3) yields the following:

$$[[K] - \omega^2[M]]\{a\} = 0 \tag{5-5}$$

The nontrivial solution of Equation (5-5) results in normal or natural modes of free undamped motion. The shapes are normal mode shapes or modal shapes, each of which has its natural frequency, ω.

5.5.3 Response Spectrum Analysis

A *response spectrum* is a plot of the maximum response (e.g., displacement, velocity, acceleration) to a specified load function for all possible single-degree-of-freedom systems. Typically, the abscissa of the spectrum is the natural frequency or period of the structural system, and the ordinate is the maximum response. For structures designed according to this MOP, refer to Chapter 3, "Loading Criteria for Substation Structures," Section 3.1.7.3, "Design Response Spectra," for the recommended response spectrum.

5.6 RECOMMENDATION FOR AN ANALYSIS METHOD

5.6.1 Static Analysis

For some structures, such as simple bus supports, an approximate analysis (as in Section 5.4.1) may be adequate.

When Ultimate Strength Design (USD) (defined in Chapter 6, "Design") is to be used for structural design, a second-order elastic analysis (as in Section 5.4.3) should be used to analyze the strength and deflection of the structure because the equations used in USD depend on the P-delta moments included in the analysis.

5.6.2 Dynamic Analysis

Structures supporting electrical equipment and the electrical equipment mounted on them should be analyzed together to model accurately the mass distribution and stiffness characteristics of the structure and the electrical equipment. This model may be used to perform a modal (eigenvalue) analysis to determine whether the structure and equipment are rigid or flexible. A *dynamic analysis* is required to determine the magnitude and distribution of seismic forces on the structure when structural irregularities such as geometry, stiffness, or mass distribution exist. Chapter 3, "Loading Criteria for Substation Structures," Section 3.1.7.6.3, "Dynamic Analysis Procedure," provides some examples of such structures and outlines the requirements for this dynamic analysis.

For simple beam or column structures, the fundamental natural frequency can be approximated from the equations shown in Table 5-1.

The natural frequency of the equipment–support structure system can be altered when its base plate is supported by anchor bolts with leveling nuts in Chapter 8, "Connections to Foundations." This possibility should be considered when evaluating the seismic behavior of the equipment–support structure system.

A static analysis may be used for seismic design of rigid structures in substations. See Chapter 3, "Loading Criteria for Substation Structures," Section 3.1.7.6.2.3, "Rigid Substation Structures," for the applicable seismic load.

5.6.2.1 Seismic Analysis. Refer to Chapter 3, "Loading Criteria for Substation Structures," Section 3.1.7.6, "Seismic Analysis," for more on seismic analysis.

Table 5-1. Fundamental Frequency (cycles/s) Equations for Simple Beam or Column Configurations.

With M	With m	With M and m
Cantilever beam or column		
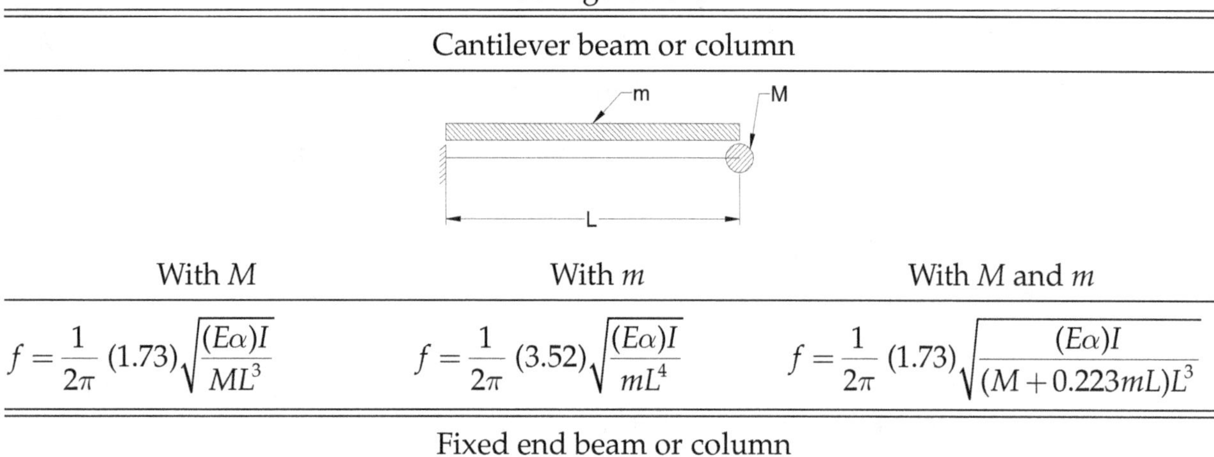		
$f=\frac{1}{2\pi}(1.73)\sqrt{\frac{(E\alpha)I}{ML^3}}$	$f=\frac{1}{2\pi}(3.52)\sqrt{\frac{(E\alpha)I}{mL^4}}$	$f=\frac{1}{2\pi}(1.73)\sqrt{\frac{(E\alpha)I}{(M+0.223mL)L^3}}$
Fixed end beam or column		
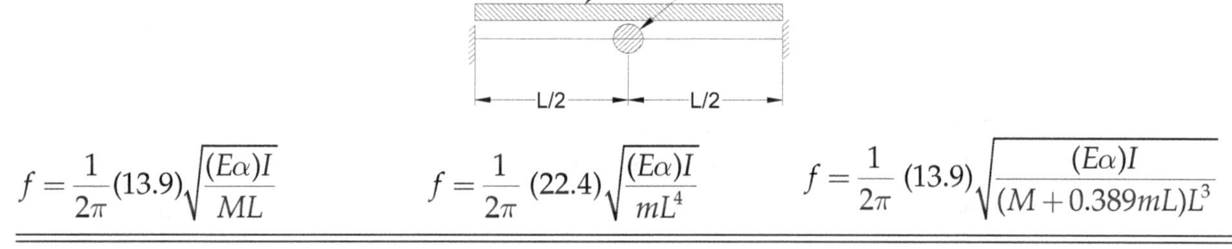		
$f=\frac{1}{2\pi}(13.9)\sqrt{\frac{(E\alpha)I}{ML}}$	$f=\frac{1}{2\pi}(22.4)\sqrt{\frac{(E\alpha)I}{mL^4}}$	$f=\frac{1}{2\pi}(13.9)\sqrt{\frac{(E\alpha)I}{(M+0.389mL)L^3}}$
Pinned end beam or column		
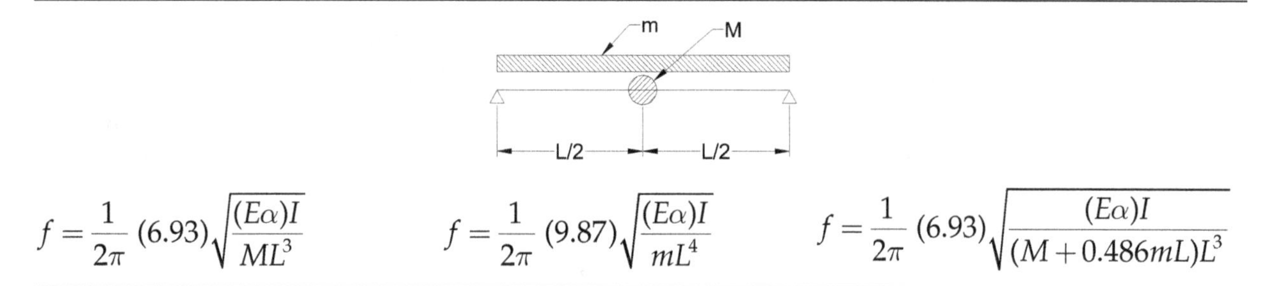		
$f=\frac{1}{2\pi}(6.93)\sqrt{\frac{(E\alpha)I}{ML^3}}$	$f=\frac{1}{2\pi}(9.87)\sqrt{\frac{(E\alpha)I}{mL^4}}$	$f=\frac{1}{2\pi}(6.93)\sqrt{\frac{(E\alpha)I}{(M+0.486mL)L^3}}$

Notes: E = modulus of elasticity [lbf/in.^{2} (kPa)]; $\alpha = g$ when E is in lbf/in.^{2}, where g = acceleration caused by gravity (386 in./s^2); $\alpha = 1$ when using SI units; I = moment of inertia [in.^{4} (mm^4)]; M = concentrated mass [lbm (kg·m)]; m = distributed mass [lbm/in. (kg m/mm)]; L = span of the beam or height of the column [in. (mm)].

5.7 ANALYSIS OF SHORT-CIRCUIT EVENTS

This section is not intended to address the design of bus or insulators. The user is referred to IEEE 605 for the design of these items. This section is intended to address the analysis of the system of rigid conductors and supporting insulators and structures for the effects of short-circuit events so as to determine the appropriate forces to apply in the design of substation structures. Engineers should familiarize themselves with the following reference documents:

- CIGRE Brochures 105 (CIGRE 1996) and 214 (CIGRE 2002): *The Mechanical Effects of Short-Circuit Currents in Open Air Substations (Parts 1 and 2)*

- IEEE 605 (IEEE 2008): *IEEE Guide for Bus Design in Air Insulated Substations*
- IEC 60865-1 (IEC 2011): *Short Circuit Current—Calculation of Effects*

This section presents background information that will help guide the engineer in analyzing the structure for short-circuit loading and the system's response to it.

Refer to Chapter 3, "Loading Criteria for Substation Structures," Section 3.1.8, "Short-Circuit (Fault) Loads," for additional information on short-circuit load development for substation structures.

5.7.1 Rigid Bus Analysis Methods

Rigid Bus systems should be analyzed using one of the following methods:

1. Simplified static analysis using the peak short-circuit force determined from Chapter 3, "Loading Criteria for Substation Structures," Section 3.1.8.1, "Simplified Static Short-Circuit (Fault) Force on Rigid Conductors." This can be accomplished using one of the following analysis methods:
   a. Simplified calculations treating the conductor as a beam supported at insulator locations.
   b. 3D model of bus arrangement.
2. Dynamic Time-History Analysis using time-dependent forces:
   a. Using 3D variation of magnetic fields and resulting forces.
   b. Using the forces determined from Section 3.1.8.1, "Simplified Static Short-Circuit (Fault) Force on Rigid Conductors," which include the assumption of parallel infinitely long conductors.

      Note: This may be used where the parallel, infinitely long assumption adequately captures the structure loads that would result from including 3D forces.

Regardless of the analysis method used, the dynamic and static load events should produce deflections and stresses in bus conductors, insulators, and support structures that are lower than the allowable values for each component.

The following section contains a discussion of the advantages and disadvantages of the methods available for analyzing short-circuit loading.

5.7.2 Rigid Bus Analysis Methods Discussion

The fault force waveform discussed in Chapter 3, "Loading Criteria for Substation Structures," Section 3.1.8, "Short-Circuit (Fault) Loads," is decaying and oscillating at the system frequency (60 Hz in the United States). A conductor in a high-voltage substation may have a natural frequency in the 2 to 10 Hz range. Because of the large range between frequencies, inertia may not have been overcome even before the peak fault force has already passed. Depending on their structural properties, if a fault is interrupted in the 2 to 6 cycle range, the rigid bus conductors and supports may or may not have appreciable structural response to the fault. If the primary power control system fails and the backup system is required to operate, the rigid bus conductors and support system will likely have sufficient time to structurally respond to the fault. It is possible that the inertia of the rigid bus conductors and supports can be overcome, creating deflections and forces. However, it is unlikely that the deflections and stresses experienced will be as large as those predicted from a static analysis with the peak fault force.

5.7.2.1 Short-Circuit Loading Simplified Static Analysis. Using a short-circuit force variation including time and 3D magnetic field effects to perform a dynamic time-history analysis is the most precise method available for determining the effect of short-circuit loading on a bus system. However, an analysis to this level is computationally demanding, and may not be warranted for all cases, particularly for preliminary design and general arrangement layout.

A significant amount of research has been performed in an attempt to quantify the structural response to fault forces. Full-scale tests have been performed, and dozens of papers have been written on the subject. Numerous researchers have attempted to develop a simplified analysis that more closely reflects the true structural deflections and stresses. As a result of the large variations in bus systems used in substations, adequately capturing these dynamic effects in modification factors is quite difficult. Some of the common variations are bus size, insulator/equipment type and size, structural support type, fitting type and resulting joint fixity, use of a damping conductor or discrete dampers, use of A-taps, and use of jumpers attached to a rigid bus. These variations and assumptions lead to differences in the resulting mass, stiffness, and damping of the system and thus alter the dynamic response. In addition, the short-circuit force variation with time from the equations in Chapter 3, "Loading Criteria for Substation Structures," Section 3.1.8.2, "Simplified Static Short-Circuit (Fault) Force (SCF) on Strain Bus," includes the simplifying assumption of parallel, infinitely long conductors, which can be a significant assumption. Due to these variations and simplifying assumptions, there is currently no method in the United States to account for the dynamic structural response using a static short-circuit analysis.

Simplified calculations treating the conductor as a beam supported at insulator locations typically involve the use of beam tables to calculate the maximum conductor moments and support reactions under static loading for a single or continuous beam with equal spans. This has the advantage of being a very rapid analysis method but has multiple disadvantages:

- It cannot account for the fault force dynamics or the dynamic response of the conductors.
- It cannot account for the stiffness of the supporting structures, insulators, or equipment.
- It cannot easily be used for more complex arrangements such as those with changes in bus direction of A-frames.

It is sometimes necessary to model an entire rigid bus system when an electrically or structurally complex arrangement is used. When rigid A-frames are used to connect two bus sections that are turned 90 degrees to each other but separated vertically, a 3D model may be needed to determine the stresses in the bus and the supporting structures. The bus conductor, insulators, and support structures for the substation should be modeled in a structural computer program. This approach removes some of the inaccuracies of the traditional simplified "beam table" method. However, it still does not account for the fault force dynamics or the dynamic response of the system.

5.7.2.2 Dynamic Time-History Model. One can precisely determine the applied forces in the system and the displacements that the system will undergo by employing a dynamic time-history model (DTHM) of the actual arrangement in question, which uses the mass, stiffness, and damping of the structural system to calculate the structural response.

This more rigorous analysis would be especially beneficial in cases where a substation is being upgraded to a higher-required fault current, and the results of a simplified analysis show that significant upgrades are required. A more rigorous analysis would also be beneficial when a necessary design constraint cannot be satisfied with a simplified analysis, such as providing a span long enough to provide the desired drive access or maintaining a required electrical clearance.

As computing power has become more readily available today, this approach has become more appealing. However, rigorous analysis has not become standard practice in the United States. The complexity involved has not been justified by the potential cost savings. In addition, the analysis can be quite sensitive to the input values. To obtain results of any value, accurate values for the static and dynamic properties of the system need to be determined, which may not always be possible. This would be especially difficult for utilities that procure items such as insulators or fittings from multiple manufacturers. In such a case, the physical/structural design may often be performed well in advance of procurement, making a determination of appropriate properties difficult. Another difficult situation could arise when analyzing an arrangement with a connection into a disconnect switch. A *switch* is a complex apparatus, and determining accurate values for its dynamic characteristics may be quite difficult.

When employing this DTHM type of analysis, the methods used should first be tested by performing a trial analysis of a similar case for which full-scale test results are available so that the accuracy of the analysis methods can be verified.

5.7.3 Short-Circuit Analysis Considerations

5.7.3.1 Joint Fixity. When analyzing a bus system, particular attention should be paid to the determination of joint fixities. Assumptions of releases to forces and moments can have a significant effect on the load distribution within the system. The appropriate joint fixity is dependent on the restraint provided by the fittings used to attach the conductors to the insulators. The restraint mechanisms, dimensions, stiffnesses, tolerances, and so on can all significantly vary between manufacturers, and between individual fittings selected. Therefore, the fittings used in a given application should be examined closely to determine appropriate fixities.

Refer to Chapter 6, "Design," Section 6.9.3, "Fittings and Couplers," for information on the types of fittings commonly used in rigid bus systems.

5.7.3.2 Arrangements with A-Frames or Jumper Transitions. A-frames (also referred to as A-taps) are often used to connect a high bus conductors to low bus conductors. An A-frame consists of conductors supported by the low bus conductor extending up diagonally to support the high bus conductor. A-frames are commonly used in substations, as they often reduce the amount of space required and the total number of structures needed. This is because an additional high bus structure is not needed on the end of a bus run to support the high bus. Arrangements with A-frames have generally performed well historically. However, arrangements with this type of support transmit a large load from the high bus to the low bus because of the large moment arm of the forces on the high bus. The extent of this load and its impact varies based on the rigidity provided by the fitting on top of the insulator positioned beneath the A-frame. If this fitting is sufficiently rigid, this load can generate a significant moment in the insulator beneath the A-frame. If this fitting is not sufficiently rigid, this load can induce a substantial moment in the low bus conductor.

These moments may be eliminated by using a flexible cable connection between the high and low bus, as shown in Figure 5-2. A flexible connection enables a more simplified analysis of high and low rigid bus conductor individually. It should be noted that this approach can introduce additional analysis concerns that are difficult to quantify. Bundled jumpers introduce possible pinch effects within the bundle and the point loads at the jumper attachment locations

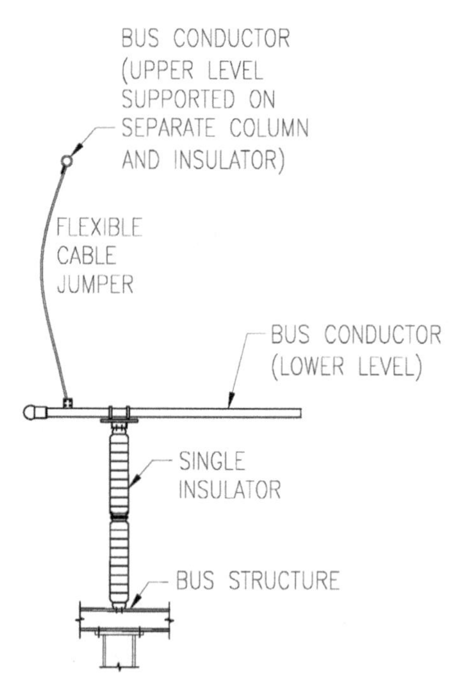

Figure 5-2. Flexible jumper between upper and lower bus.

are difficult to predict because of the longitudinal force caused by the conductor angle as well as the 3D forces in the jumpers caused by the adjacent phases.

If the choice is made to use A-frames, rather than flexible cables as suggested previously, it is critical to model the entire system of rigid bus conductors, insulators, and structural supports, so that the force from the high bus transmitted to the supporting low bus can be accurately quantified and addressed in the design. The short-circuit force, when applicable, should be applied simultaneously to the high bus and low bus. If a single insulator below an A-frame is inadequate to resist the applied forces and moments, common solutions include the use of double insulators or delta-configured insulators, as shown in Figure 5-3.

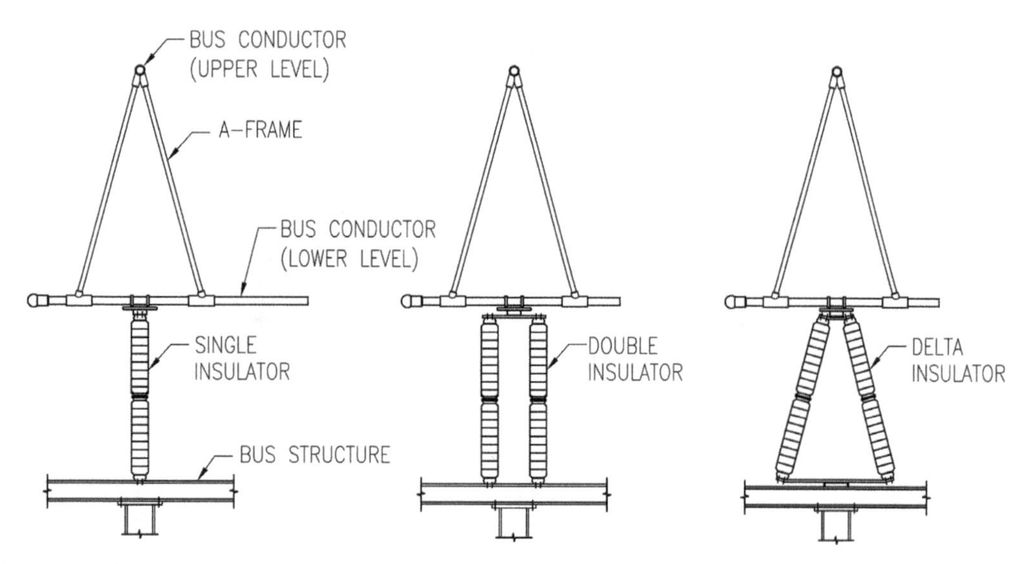

Figure 5-3. Insulator arrangement options.

Double and delta insulator configurations are also useful where space constraints do not allow for additional structural supports, but the capacity of a single insulator is inadequate.

REFERENCES

AISC (American Institute of Steel Construction). 2022. *Specification for structural steel buildings.* ANSI/AISC 360-22. Chicago: AISC.

ASCE. 2015. *Design of latticed steel transmission structures.* ASCE 10-15. Reston, VA: ASCE.

ASCE. 2019. *Design of steel transmission pole structures.* ASCE 48-19. Reston, VA: ASCE.

Beedle, L. S. 1958. *Plastic design of steel frames.* New York: Wiley.

CIGRE (International Council on Large Electric Systems). 1996. *The mechanical effects of short-circuit currents in open air substations (rigid and flexible busbars).* Brochure 105. Paris: CIGRE.

CIGRE. 2002. *The mechanical effects of short-circuit currents in open air substations (Part II).* Brochure 214. Paris: CIGRE.

Cook, R. D. 1974. *Concepts and applications of finite element analysis.* New York: Wiley.

IEC (International Electrotechnical Commission). 2011. *Short-circuit currents—Calculation of effects. Part 1: Definitions and calculation methods.* IEC 60865-1. 3.0 edn. Geneva: IEC.

IEEE (Institute of Electrical and Electronics Engineers). 2008. *Guide for design of substation rigid-bus structures.* IEEE 605. Piscataway, NJ: IEEE.

MacNeal, R. H., and R. C. Harder. 1985. "A proposed standard set of problems to test finite element accuracy." *Finite Elem. Anal. Des.* 1 (1): 3–20.

Mueller, W. H., M. Ostendorp, and L. Kempner. 1991. *LIMIT: A nonlinear three-dimensional truss analysis program.* Tech. Research Rep. Portland, OR: Bonneville Power Administration.

White, D. W., and J. F. Hajjar. 1991. "Application of second order elastic analysis in LRFD: Research to practice." *Eng. J.* 28 (4): 133–148.

CHAPTER 6
DESIGN

6.1 GENERAL DESIGN PRINCIPLES

Specific guidelines for member design and fabrication are not included in this Manual of Practice (MOP). This MOP refers to other documents for design guidelines and notes any exceptions to the referenced documents.

Load factors, load combinations, load cases, and deflection criteria specified in Chapter 3, "Loading Criteria for Substation Structures," and Chapter 4, "Deflection Criteria (For Operational Loading)" should be used with referenced design codes or documents.

There is no intention to exclude any material or section types. If the material or section type is not addressed in this MOP, the engineer should use the appropriate design code or reference document.

6.2 DESIGN METHODS

Allowable Strength Design (ASD) is a method of proportioning structural members such that elastically computed stresses produced in the members by service loads do not exceed specified allowable stresses. ASD is also called *working stress design*. In 1989 ASD was called *allowable stress design*. The base reactions developed through ASD are used in conjunction with foundation soil interaction design.

Ultimate Strength Design (USD) is the method of proportioning structural members such that the computed forces produced in the members by the factored loads do not exceed the member design strength. USD is also called *Load and Resistance Factor Design* (LRFD). USD is recommended for substation structures. The term LRFD will be used in this MOP interchangeably with USD.

Plastic analysis or design is not recommended for substation structures.

Structures that support conductors and overhead ground wires that extend outside the boundaries of the substation should also meet or exceed the load and strength requirements of the *National Electrical Safety Code* (NESC) (IEEE 2023) and local regulatory codes. For structures that are required to meet NESC 2023 load criteria, it is recommended that the USD method be used because NESC 2023 specifies load factors and material strength factors.

6.3 STEEL STRUCTURES

6.3.1 Ultimate Strength Design

6.3.1.1 Lattice Angle Structures. ASCE 10-15 (ASCE 2015) should be used for the design of lattice structures. This standard uses factored design loads, linear material properties, and first- or second-order elastic analysis. This standard does not use strength reduction factors.

ASCE 10-15 has adjusted column equations for angles, which account for the effect of eccentricities in connections that are commonly used in lattice angle structures. The effects of flexural–torsional and torsional buckling are also included. For these reasons, ASCE 10-15 is recommended for designing lattice substation structures constructed using angle sections.

Steel lattice structures using angle sections can also be designed in accordance with ANSI/AISC 360-22. LRFD should be used with factored design loads, linear material properties, LRFD member capacity reduction factors, and second-order elastic analysis.

LRFD states that compression members should be loaded through the centroidal axis. When the loading is not through this axis, LRFD requires that the combined stress equation for bending and axial loads be used and that flexural–torsional and torsional buckling be considered for these members.

6.3.1.2 Standard Structural Shapes Other Than Angles. Examples of standard structural shapes other than angles are wide flanges, channels, HSS tubes, pipes, and tee sections. LRFD should be used for design of these member shapes with factored design loads, linear material properties, LRFD member capacity reduction factors, and second-order elastic analysis.

To account for uniformly tapered open member shapes, an equivalent slenderness ratio can be calculated in the same manner, as discussed in Section 6.3.1.3, "Hollow Tubular Member Shapes."

6.3.1.3 Hollow Tubular Member Shapes. Custom-fabricated hollow tubular member shapes include 4-, 6-, 8-, and 12-sided polygonal sections and circular sections. These members should be designed and fabricated in accordance with ASCE 48-19 (ASCE 2019a).

The local buckling equations for polygonal sections in ASCE 48-19 identify the stress level for collapse of the section, rather than the stress level at which local buckling is initiated.

The local buckling equations for 8- and 12-sided members are based on bending tests. In these tests, one of the flats is initially loaded in uniform compression and becomes inelastic before the adjacent flats because the adjacent flats are at a lower, non-uniform stress level. If the section does not collapse, further increases in compressive stress are distributed to adjacent flats, giving the section additional post buckling strength. If these sections are loaded with a uniform compressive stress over the entire cross section, there would not be a redistribution of stresses to adjacent flats because they are all loaded equally. Accordingly, the additional post buckling strength does not exist, and the local buckling equations in ASCE 48-19 may overpredict the collapse strength of 8- or 12-sided members loaded in predominantly axial compression. For this reason, it is recommended that for 6-, 8-, and 12-sided polygonal sections, the permitted compressive stress (F_a) caused by axial force and bending moment ($P/A + Mc/I$) should be based on the limits in ASCE 48-19 for rectangular shapes with $f_a > 1$ kip/in.^{2}, where f_a is the compressive stress caused by axial loads.

ASCE 48-19 uses factored design loads, linear material properties, and a second-order geometric nonlinear analysis method (*P*-delta effect is included). ASCE 48-19 also allows for uniformly tapered members.

ASCE 48-19 is the recommended method for design because the effective length factor, K, amplification factors to account for P-delta effects, and factors to account for tapered members do not have to be determined with the required geometrically nonlinear (P-delta) elastic stress analysis of the tapered members. ANSI/AISC 360-22 (AISC 2022) can also be used for the design of hollow structural shapes.

6.3.1.4 Local Buckling of Irregular Polygonal Shapes. The local buckling assumptions in ASCE 48-19 cover regular polygonal shapes of 4 or more sides where all flats are of the same width. For flats that bisect the neutral bending axis, the local buckling strength k factor can be increased over the k factor for flats in uniform compression. For structures that use an irregular polygonal shape and have high w/t ratios for the two flats that are on the neutral axis, the buckling stress of the two long flats can be calculated using Equation (6-1) (Bleich 1952):

$$\sigma_{cr} = \frac{\pi^2 E}{12(1-\upsilon^2)}\left(\frac{t}{w}\right)^2 k \quad \text{but less than } F_y \tag{6-1}$$

where

σ_{cr} = Critical buckling allowable stress,
E = Young's modulus,
F_y = Yield strength stress,
υ = Poisson's ratio,
t = Plate thickness,
w = Plate width in compression, and
k = Local buckling factor as defined subsequently.

For flats that have tension on one corner and compression on the other corner, the width w can be taken as the distance from the corner in compression to the point along the flat that has zero compression. In this case, k is 7.7. For flats where both corners are in compression, k is 4.0, as in ASCE 48-19, and the width w can be taken as the length of the flat between the actual inside bend radii or four times the thickness, as specified in ASCE 48-19.

The stress at each corner of the polygon must be calculated for each load case and for the entire length of the member because the buckling strength is load case and section position–dependent.

6.4 CONCRETE STRUCTURES

Concrete structures are designed to accommodate cracking behavior. In a corrosive environment, water may be absorbed into the open cracks and corrode the reinforcing steel. Typical substation concrete structures should provide enough concrete cover to protect the reinforcing steel. For structural members subjected to sustained flexure loading, such as dead-end structures, it may be desirable to allow no tensile stress along the member cross section under everyday loading conditions. This zero-tension criterion will prevent cracks from staying open under normal situations and will preclude the reinforcing steel from corroding. This zero-tension condition can be resolved with prestressed concrete members or controlling cracking according to ACI CODE-318-19 (ACI 2019), ACI PRC-224-01 (ACI 2001), and ACI CODE-350.3-20 (ACI 2020).

6.4.1 Reinforced Concrete Structures

Reinforced concrete structures should be designed and constructed in accordance with ACI CODE-318-19. This code uses the USD method with factored design loads, linear material properties, and second-order elastic analysis. Member strength reduction factors should be used as specified in ACI CODE-318-19. Fiberglass reinforcing bars (ACI 2015) may be needed to prevent eddy heating near the strong magnetic fields in air core reactors.

6.4.2 Prestressed Concrete Structures

Prestressed concrete structures should be designed and constructed in accordance with ACI CODE-318-19. This code uses the USD method with factored design loads, linear material properties, and second-order elastic analysis. Member strength reduction factors should be used as specified in ACI CODE-318-19. The following documents may be referenced for design aids and examples: *PCI MNL-120 Design Handbook—Precast and Prestressed Concrete*, 8th edition (PCI 2017) and PTI *Post-Tensioning Manual*, 6th edition (PTI 2006).

6.4.3 Prestressed Concrete Poles

The prestressed concrete pole–type structures, either static cast or spun cast, should be designed and constructed in accordance with ASCE 123-12 (ASCE 2012), *Prestressed Concrete Transmission Pole Structures: Recommended Practice for Design and Installation*. This guideline uses the USD method and, in general, follows all ACI and PCI recommendations.

6.5 ALUMINUM STRUCTURES

The Aluminum Association's *Aluminum Design Manual* (AA 2020) for aluminum structures is recommended for use in this MOP for aluminum design. Additional information on aluminum structure design is available in ASCE (1972), Mooers (2006), and Kissell and Ferry (1995).

6.5.1 Typical Substation Alloys and Tempers

The following materials are alloys used for substation structures:

- Alloy 6061-T6: This is a moderate-high strength, heat-treated alloy that has good machining and welding properties. It is commonly available as a plate, bar, or extrusions.
- Alloy 6063-T5: This is a moderate-strength, heat-treated alloy that has good machining and welding properties. It is commonly available as a plate, bar, or extrusions.
- Alloy 7075-T54: This is a high-strength, heated-treated alloy that has good machining properties, but has poor welding properties using conventional processes. It is available as a plate, bar, or extrusions, but it is not as commonly available as the other alloys that are presented.

Aluminum alloy strength ratings are specified in the Aluminum Design Manual 2020. Pipe Schedules: Pipes that are used for substation structures are typically Schedule 40. Occasionally, Schedule 80 pipes are used to obtain greater strength and reduced deflection.

All aluminum alloys can be anodized to resist crevice corrosion and stress corrosion cracking. The weldable alloys have reduced yield and ultimate strengths in the heat-affected zone (HAZ) near the welds.

6.5.2 Applications to Substation Structures

The use of aluminum for structural supports has a history in applications such as light poles, substation structures, and highway sign structures. Aluminum structures are typically used in substations of voltages 230 kV and below and can be used for metering supports, high and low bus supports, CCVT structures, disconnect switch supports, reactor supports, and lattice A-frame pull-off structures.

Aluminum has a higher strength-to-weight ratio in comparison with steel and thus can lead to the formation of lighter structures. Additional advantageous properties of aluminum are that it is nonferrous, can tolerate high magnetic fields, and can (for most alloys) be cast, extruded, and machined.

6.5.3 Use Limitation with Aluminum Substation Structures

There are some limitations to the use of aluminum for specific structural applications in substations that should be considered by the engineer.

Typically, for structures over 25 ft (7.62 m) high, the aluminum structure tends to lose efficiency from a design and cost perspective.

A larger member cross section may be needed to maintain the required deflection criteria because of a smaller Young's Modulus than steel and recommended fatigue stress threshold by code. Fatigue behavior is a function of size because the larger the member, the more likely a flaw will be present, and fatigue cracks tend to initiate at defects.

Welding, where possible, effectively eliminates any tempering of the aluminum alloy and reverts the metal to an annealed condition. This is typically associated with a reduction in strength. Where welding is required, the reduction of strength may dictate the use of larger members, the use of post weld heat treatment, or the consideration of a bolted connection.

Aluminum is not suitable when exposed to extremely high temperatures because of the relatively low melting point.

6.5.4 Aluminum Connections

6.5.4.1 Bolted Connections. High-strength aluminum alloys 2024 and 7075 are less corrosion-resistant than 6061, but they are acceptable for most applications. Sometimes, higher-strength alloys are anodized for additional protection. Where 2024 and 7075 are specified by the designer, a 0.2 mil (0.00508 mm) thick anodized coating is recommended. This recommendation also applies where they will be exposed to moisture. The 2024-T4 aluminum alloy is typically used for bolts and fasteners in accordance with ASTM F468 (ASTM 2023) and ASTM F467-13(2018) (ASTM 2018b) with Alclad 2024-T4 flat washers. Bolted connections may be as prescribed by the Aluminum Association's *Aluminum Design Manual* (AA 2020). Stainless-steel bolts have also been used at a safe distance from salt water.

6.5.4.2 Weldments. Typically, only base plates, cap plates, gusset plates, beam seats, grounding pads, and equipment support attachment plates are permanently welded in place. An effective method of welding a base plate to a pipe column is performed by using a computer-controlled machined depression of approximately 1/16 in. (1.59 mm) deep pattern to accept insertion of the base of the column and attendant gusset plate prior to welding. It is imperative that shop-applied and, occasionally, field-applied welds are installed by prequalified welders using compatible filler material. The Aluminum Association's *Aluminum Design Manual* (AA 2020) requires a reduction in allowable stresses within 1 in. (25.4 mm) of welds. Plate thickness is sometimes governed by weld heat dissipation consideration (Section 6.10.5.3).

6.5.5 Aluminum Design Resources

6.5.5.1 Ultimate Strength Design. Aluminum structures should be designed and fabricated in accordance with the Aluminum Association's *Aluminum Design Manual* (AA 2020), using the specified resistance and safety factors.

6.5.5.2 Allowable Strength Design According to IEEE 693. To design aluminum structures in accordance with the ASD requirements of IEEE 693 (IEEE 2018a), the structure safety factors (Ω) in the *Aluminum Design Manual* (ADM) (2020) should be increased by a factor of 1.13.

The IEEE 693 refers the user to the latest edition of the ADM using the bridge-type structure safety factors. The Aluminum Association's *Aluminum Design Manual* (AA 2015) ADM (2015) had both bridge-type structure safety factors and building-type structure safety factors. The more recent ADM (2020) removed the Ω bridge-type structure safety factors that were approximately 13% higher than the building-type structure safety factors listed in the Aluminum Association's *Aluminum Design Manual* ADM (2015). The 1.13 factor is used to give equivalent reliability to aluminum structures in accordance with IEEE 693, where the use of bridge-type structure safety factors was specified but is no longer available in ADM 2020.

6.6 WOOD STRUCTURES

6.6.1 Ultimate Strength Design

Wood structures and poles should be designed and constructed in accordance with ASCE MOP 141 *Recommended Practice for the Design and Use of Wood Pole Structures for Electrical Transmission Lines* (ASCE 2019) and the *National Electrical Safety Code* (IEEE 2023). ANSI O5.1-2017 (ANSI 2017) can be used for wood pole stresses with the NESC 2023 defined 0.65 strength factor. Additional design information can be found in ANSI/AWC NDS-2015 (AWC 2015).

6.6.2 Allowable Strength Design

Wood structures and poles should be designed and constructed using ANSI/AWC NDS-2015.

6.7 SEISMIC DESIGN GUIDELINES

Each substation installation should be evaluated on the basis of its relative criticality to the owner's power system. Installations or specific equipment defined as critical or essential are those that are vital to power delivery and cannot be bypassed in the system or are undesirable to lose because of economic effects. Equipment that can be bypassed for short-term emergency operations is considered nonessential.

Displacements caused by the seismic events among components of different seismic response potential should not impair the performance of the mounted equipment, cause secondary induced stress, reduce the required electrical clearance, or cause other safety hazards. The seismic qualification report or the equipment manufacturer should be consulted for seismic displacement requirements specific to the particular equipment. Seismic displacement criteria and limits required to meet the intended seismic performance level may be different from the deflection limits recommended in Chapter 4, "Deflection Criteria (For Operational Loading)," for other loading conditions from Chapter 3, "Loading Criteria for Substation Structures."

Connections between equipment and components and their effects on one another require specific attention. Rigid electrical bus connections between equipment that restricts seismic-induced displacements may cause equipment damage.

All components should be designed to withstand stresses caused by the seismic loading in Section 3.1.7, "Seismic Loads."

For lattice dead-end structures that do not support electrical equipment, the wind, ice, and wire tension loads usually control the design, when compared to loading combinations containing seismic loads. One situation where this may not be the case is when the dead-end structure supports the full transmission line wire tensions and the wind load effects are smaller than the seismic load effects. There has not been a documented case of failure for these types of structures because of seismic inertial loads.

Additional recommendations for seismic design of rigid bus systems are provided in Section 6.9.

6.7.1 Structures That Support Electrical Equipment Qualified for IEEE 693

Design of structures that support seismically qualified electrical equipment should satisfy the requirements of IEEE 693 (IEEE 2018a). Equipment qualified to the requirements of IEEE 693 is often mounted on dedicated supports. A dedicated support is a structure designed exclusively to support only a single piece of substation equipment. Dedicated supports may be seismically qualified in conjunction with a specific piece of equipment by the prescribed methodology in IEEE 693 for the supported equipment. The dedicated support can be a pedestal supporting a cantilever-type piece of equipment, such as a surge arrester or a switch structure that includes the switch frame and columns that support a disconnect switch.

Equipment may also be mounted on intermediate supports, which are structural members or sub-assemblies located in between the equipment and a primary substation structure, such as a dead-end or switch/bus structure. An example of a structural member intermediate support is a beam that the equipment, such as a reactor, is mounted on a steel rack, box-type structure. See Section 3.1.7.10.2, "Equipment to Structure Anchorage Design Forces," for equipment to structure anchorage design forces.

6.7.2 Structures Not Covered by IEEE 693

Structures not included in IEEE 693 (IEEE 2018a) are dead-end, rigid bus, strain bus, cable bus structures, lightning masts, and shielding masts. These structures, also defined by IEEE 693 as primary substation structures and intermediate support, should be designed to withstand seismic loads in Section 3.1.7, "Seismic Loads," and load combinations in Section 3.3, "Load Factors and Combinations." The seismic design loads for nonstructural components such as rigid bus work, appurtenances, and equipment that are not subject to seismic qualifications according to IEEE 693 are provided in Section 3.1.7.11, "Seismic Demand on Other Components." The stresses in steel members and connections should be in accordance with Section 6.3.1. Connections and anchorages should be designed with overstrength factors Ω_0 or Ω_{PL} as recommended in Sections 3.1.7.4, 3.1.7.8.2, 3.1.7.10, and 8.3.

6.8 BASE PLATE DESIGN

This section provides a method to determine the plate thickness for a base plate on leveling nuts. These design methods may not apply to flange plates used in column connections.

It is conservative to use this procedure for base plates mounted directly on concrete. AISC Design Guide 1 (AISC 2006) can be used as a reference to design base plates on concrete. Although the design of anchor rods is discussed in Chapter 8, "Connections to Foundations," it is important to know that the number of anchor rods will affect the determination of the base plate thickness. In general, a greater number of small rods will allow the use of thinner base plates than a lower number of larger rods. However, when the total installed cost of the foundation and anchor rods is considered, a slightly thicker base plate with fewer rods may prove to be the most economical choice because of reduced construction costs. It should be noted that the bending of less stiff thinner plates may affect load distribution to the anchor rods on leveling nuts (Chapter 8, "Connections to Foundations") and also increase the deformations of the structure. The use of thicker plates, however, may result in welding and galvanizing issues caused by excessive restraint and higher thermal stresses occurring during galvanizing, which may result in weld cracks. The trade-off between rigidity and the effects of using thicker plates should be considered when sizing base plates. Thinner base plates may have insufficient stiffness and redistribute bending stresses that could cause premature failure in the shaft wall. ASCE 48-19 also contains methods for base plate design.

Figure 6-1 shows some base plate connections that may be used in substation structures. The effective length of the bend line (b_{eff}) of the suggested bending planes 1-1, 2-2, and 3-3 depends on the size and shape of any galvanizing drain holes in the base plate for Figure 6-1a, c, d. In some instances, the b_{eff} will be reduced if the column is inserted into a hole, same size and shape of the column, in the base plate. In this instance, the column can extend half way through the base plate thickness with fillet welds on the inside and outside to attach the column to the base plate.

Several factors should be considered when specifying the center opening size in a base plate. Larger center openings may provide increased access for welding. Center openings

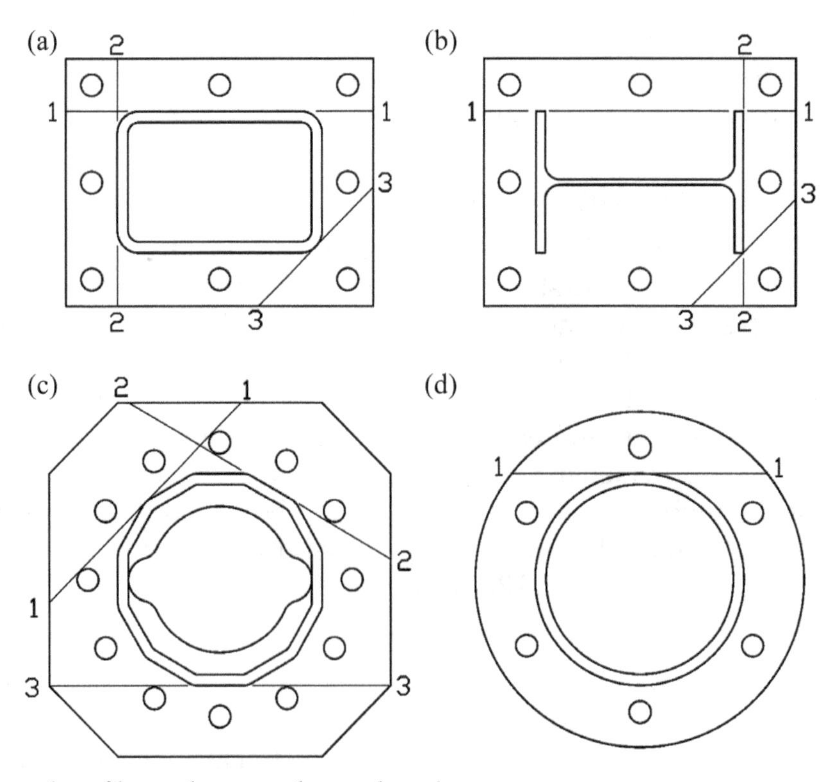

Figure 6-1. Examples of base plates used on substation structures.

reduce the mass of base plates and may reduce the immersion time required in a galvanizing kettle. Larger center openings may also reduce the thermal stresses that occur during the hot-dip galvanizing process as well as facilitate the removal of ash inside a tubular section upon removal from the kettle. Smaller center openings will, however, increase the strength and stiffness of a base plate. Some methods for determining the thickness of base plates depend on beam action and assume that the stiffness of the plate continues beyond the tube wall toward the interior of the tubular section. The reduction in stiffness as a result of the presence of a center opening of a base plate in a T-joint (Section 6.10.5.2) should be considered when determining the thickness. Base plate design methods should be validated by experimental or analytical investigations.

6.8.1 Determination of Anchor Rod Loads

The anchor rod setting plan is determined by the geometry of the column, the loads imposed on the column, and the proper clearance between the nuts and the column. Assuming that the base plate behaves as an infinitely rigid body, the load in anchor rod i (BL_i) can be calculated by using the following equation:

$$BL_i = \left(\frac{P}{A_{BC}} + \frac{M_x y_i}{I_{BC_x}} + \frac{M_y x_i}{I_{BC_y}} \right) A_i \tag{6-2}$$

where

BL = Anchor rod load,
P = Total vertical load at the base of the column,
M_x = Base moment about the x-axis,
M_y = Base moment about the y-axis,
x_i, y_i = x and y distances of anchor rod i from reference axes,
A_i = Net area of anchor rod i,
$A_{BC} = \sum_{i=1}^{n} A_i$ (A_{BC} is the total anchor rod cage area),
$I_{BC_x} = \sum_{i=1}^{n}(A_i y_i^2 + I_i)$ (I_{BCx} is the total anchor rod cage inertia about the x-axis),
$I_{BC_y} = \sum_{i=1}^{n}(A_i x_i^2 + I_i)$ (I_{BCy} is the total anchor rod cage inertia about the y-axis),
n = Total number of anchor rods, and
I_i = Moment of inertia of anchor rod i.

Because I_i is often small, it may be omitted when calculating I_{BCx} and I_{BCy}.
Figure 6-2 illustrates the application of Equation (6-2) to determine anchor rod loads.

6.8.2 Determination of Base Plate Thickness

A common design procedure for base plates assumes that anchor rod loads produce uniform bending stress (F_b) along the effective portion of bend lines located at the face of the column. Each bend line is characterized by the following:

k = Number of anchor rod load BL_i's contributing moment along the bend line,

c_i = Shortest distance from the center of each anchor rod (i) to the bend line,

b_{eff} = Length of the bend line (depending on the shape of the column, the shape of the base plate, and k), and

Figure 6-1 suggests some possible bend lines in various types of base plates. The most difficult task in the analysis of the base plate is the determination of the proper effective bend

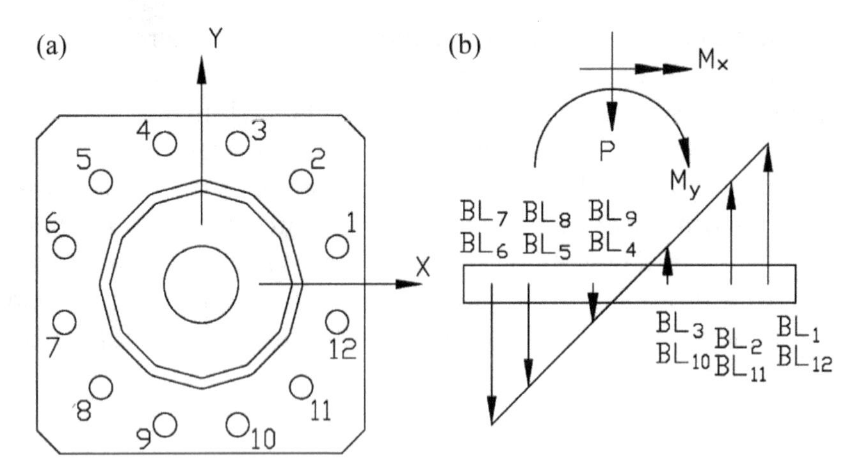

Figure 6-2. Rigid plate free body diagram.

lines (b_{eff}) used to calculate the bending stress. Many times, b_{eff} should be limited to ensure that the bend line will be effectively loaded. One method to calculate b_{eff} assumes that the anchor rod reactions are resisted at a bend line that is tangential to the column. The effective length of this bend line (b_{eff}) is assumed to be limited by the distance between the projected length of the first and last bolt acting on the bend line plus the sum of the perpendicular distances from these bolts to the bend line. Manufacturers have used similar methods for many years and have had successful verifications of this approach through full-scale testing. ASCE 48-19, Appendix F, can also be used to design base plates for tubular steel pole columns.

The base plate bending stress F_{PL} for the assumed bend line can be calculated by using Equation (6-3):

$$F_{PL} = \left(\frac{6}{b_{eff}t^2}\right)\left(BL_1c_1 + BL_2c_2 + \cdots + BL_kc_k\right). \tag{6-3}$$

where t is the base plate thickness, and BL_i is the effective anchor rod load for each anchor rod that causes a moment on the assumed bend line b_{eff}. The base plate thickness is determined by keeping F_{PL} below the yield stress F_y for anchor rod loads corresponding to the factored loads. To determine t_{min}, Equation (6-3) can be solved for t and rewritten as Equation (6-4):

$$t_{min} = \sqrt{\left(\frac{6}{b_{eff}\left(F_y\right)}\right)\left(BL_1c_1 + BL_2c_2 + \cdots + BL_kc_k\right)} \tag{6-4}$$

6.8.3 Anchor Rod Holes in Base Plates

To accommodate construction tolerance in the placement of cast in place anchor rods in the foundations, the recommended size of the anchor rod holes in the base plate is listed in Table 6-1. The table covers a wide range of anchor rod diameters, from very small equipment supports with four anchor rods to large dead-end structures with multiple large-diameter anchor rods. Steel templates with holes 1/16 in. (1.6 mm) larger than the anchor rod may be used at the top and bottom of the anchor rod group to ensure that the base plate will fit the cast-in-place anchor rods during construction. The bottom anchor rod group template should have a center hole large enough to allow concrete to flow through during foundation construction. Anchor rod installation tolerance is covered in Chapter 8, "Connections to Foundations." The use of hardened washers above and below the oversized holes in the base plate is recommended.

Table 6-1. Base Plate Anchor Rod Hole Diameters.

Anchor rod diameter	Hole diameter	Anchor rod diameter	Hole diameter
0.625 in.	0.8125 in.	1.625 in.	1.9375 in.
0.750	0.9375	1.750	2.0625
0.875	1.0625	1.875	2.1875
1.000	1.2500	2.000	2.3125
1.125	1.4375	2.250	2.5625
1.250	1.5625	2.500	2.8125
1.375	1.6875	2.750	3.0625
1.500	1.8125	3.000	3.3125

Note: 1 in. = 25.4 mm.

6.8.4 Base and Flange Plate Design for Deflection-Sensitive Structures

Substation structures that support equipment or rigid bus that are sensitive to deflection, or may be controlled by the deflection limits recommended in Chapter 4, "Deflection Criteria (for Operational Loading)," can have relatively large tubular or rolled shape members to limit deflection for the loading from Section 3.1.11. These members could have relatively low stresses at the base or flange plate. The designer should consider designing the base plate and anchor rods based on a percentage of the moment capacity of the members in the connection. The percent of the moment capacity is at the discretion of the designer and is intended to allow the flange or base plate to be more rigid than if they were sized solely for the actual forces and moments at the connection. The base or flange plate thickness must be checked for stress for all appropriate load cases from Chapter 3, "Loading Criteria for Substation Structures."

6.9 RIGID BUS DESIGN

Rigid bus design should be approached as a system requiring both an electrical engineer and a designer. The electrical engineer should be responsible for selecting the electrical parameters such as the minimum size bus required for ampacity (current-carrying capacity), insulators, hardware, and electrical clearances and determining the short-circuit fault current. The designer should be responsible for selecting support locations and structural analysis and design of rigid bus conductors, insulators, and support structures. The designer should also examine the bus arrangement and determine boundary conditions for bus fittings and determine whether a simple static or complex dynamic analysis is required.

IEEE 605 (IEEE 2008) contains numerous considerations and additional information in the design of rigid bus systems outside the scope of this section that the engineer should consider.

6.9.1 Bus Layout Configuration

A rigid bus is usually configured in two perpendicular directions at two elevations with either A-frame, rigid pipe, or flexible drops connecting the high and low bus spans. Flexible drops are preferred from a structural standpoint because they uncouple the structural response of the high bus and low bus spans. Refer to IEEE 605-2008 for bus layout configuration considerations.

6.9.2 Rigid Bus Materials and Shapes

6.9.2.1 Materials

6.9.2.1.1 Copper. Copper can often be found in legacy arrangements. Copper sees limited use in new installations as a rigid bus in high current situations and in box-type distribution structures where angles can be directly bolted to each other without connection hardware and spans are typically very short. The limited use of copper is attributed to aluminum being more economical. Copper comes in many alloys and tempers. Design values should be obtained from the manufacturer of the copper shapes used in a specific application.

6.9.2.1.2 Aluminum. Typical aluminum alloys and tempers used in a rigid bus design are shown in Table 6-2. Other aluminum alloys and tempers may be used. See the Aluminum Design Manual 2020 (AA 2020), IEEE 60-2008, ASTM B241/B241M (ASTM 2022b) and ASTM B317/B317M-07(2015)e1 (ASTM 2015) for additional material properties.

6.9.2.2 Aluminum Shapes

6.9.2.2.1 Tubular. Seamless aluminum pipe in Schedule 40 or Schedule 80 is the most common shape used in substation rigid bus constructions. The usual diameters range from 4 (101.6 mm) to 8 in. (203.2 mm) for bus spans and 2 (50.8 mm) to 4 in. (101.6 mm) for A-frame constructions.

Round tubular shapes are considerably more rigid than other structural shapes of the same ampacity. These shapes are efficient structurally and electrically, and their larger diameter helps minimize corona at higher voltages.

Circular cross sections are prone to Aeolian vibration caused by vortex shedding at moderate wind speeds. See Section 6.10.2, "Vortex-Induced Oscillation and Vibration," for information regarding rigid bus vibrations.

6.9.2.2.2 Angle. Angle shapes in copper or aluminum are sometimes used in box-type structures where spans are small and the simplicity of direct bolted connections is desired.

6.9.2.2.3 Flat Bar. Flat bars are used in the same situations as angles. They are also used in multiple bar configurations where large currents are transmitted.

6.9.2.2.4 Integral Web. Integral web sections have a large cross-sectional area and multiple locations for attachment of jumpers to the cross section that make them useful for large current-carrying applications. They also have a large section modulus in one direction that is usually oriented horizontally to counter the largest bending moments.

6.9.3 Fittings and Couplers

Typical fittings used to attach conductors to insulators can, in general, be classified as expansion-, fixed-, and slip-type fittings.

Table 6-2. Aluminum Alloy Tempers and ASTM Standards.

Alloy and temper	ASTM standard
6061-T6	B241
6063-T6	B241
6101-T6	B317

6.9.3.1 Expansion Fittings. Expansion fittings support the conductor while providing flexible resistance or free movement of the conductor caused by thermal expansion and contraction. This fitting type is often assumed to have moment releases about three primary axes and force release parallel to the conductor.

6.9.3.2 Fixed Fittings and Slip Fittings. Fixed fittings and slip fittings connect the conductor to the insulators by welding the conductor to the fitting or clamping or swaging the conductor. These may or may not provide enough rotational stiffness to be considered a fixed joint for structural modeling purposes.

As stated in Section 5.7.3.1, the individual fitting should be investigated closely to determine the restraint provided.

6.9.3.3 Bolted-type Fittings. Bolted-type bus support fittings permit easier installation than welded-type fittings and do not require a reduction in stress level because of welding. However, welded-type fittings provide a better path for current flow. The same oxide coating that provides aluminum its anti-corrosion benefits (refer to Section 6.5.1) can hamper the conductivity of bolted aluminum connections. Where required, contact preparations (cleaning and joint compounds) or plated terminals can reduce or eliminate the detrimental conductivity effects of the oxide.

6.9.3.4 Couplers. Internal welded bus couplers consist of a short tube inside the rigid bus conductors, which permits the connection of individual bus lengths by welding. The design stress level should be reduced within 1 in. (25.4 mm) of the weld. Whenever possible, the rigid bus coupler should be located at a low-stress location along the bus. If a coupler can obtain 100% capacity of the rigid bus, the rigid bus splice can be located anywhere along the bus. Swaged couplers and fittings are also available.

Currently, couplers are selected from catalogs that may not give dimensional data or alloy type. The designer should contact the manufacturer to obtain the alloy and dimensional properties of the selected coupler.

6.9.4 Insulators

6.9.4.1 Porcelain Insulators. Porcelain station post insulators are assemblages typically consisting of end fittings, a bonding medium, and a porcelain body. Although all components of the insulator can be sources of failure, the focus of most investigations is the porcelain body because of the relative modest cantilever strength. However, the cemented joint can be a failure mode in situations dominated by torsional or tensile loading. For most insulators, the torsion and tensile ratings are typically limited by the cement strength rather than the porcelain strength. Porcelain insulators typically have strengths that are 1 to 3 standard deviations above the rated strength. The porcelain insulator is coated with glaze, approximately 3 mil (0.0762 mm) thick, to obtain the rated cantilever strength. Glaze has a higher coefficient of thermal expansion and preloads porcelain in compression.

Porcelain typically has a modulus of elasticity ranging from 10,000 ksi (68.9 GPa) to 15,000 ksi (103 GPa) and ultimate compressive strength ranging from 4 ksi (27.6 MPa) to 15 ksi (103.4 MPa). Contact the insulator manufacturer if the porcelain properties are not known.

Manufacturers provide ratings for tension, compression, torsion, and bending strength. Compression strength is usually much higher than the tensile strength. Typically, porcelain station post insulators have relatively good axial compression characteristics in comparison with cantilever strength.

For bending strength, manufacturers specify a cantilever rating. The cantilever rating is the maximum horizontal load that can be applied at the top of the insulator with the insulator base fixed. The rated cantilever strength usually represents a reliable breaking strength of the insulator. Manufacturers typically recommend multiplying the cantilever rating by a strength reduction factor of 0.4 for allowable loads. It may be necessary for the designer to consider the combined stress at the base of a station post insulator from shear and torsion forces applied at the top of the insulator by the rigid bus. Interaction equations for combining axial stress, bending stress, and torsional stress may be available from the manufacturer of the post insulator.

6.9.4.2 Composite. The materials that provide the structural strength of composite insulators are much different from porcelain, and their behavior and properties are also quite different. Composite station posts derive their strength from a fiberglass core that is covered by elastomeric weather sheds. Composite insulators are typically lighter and more flexible than porcelain posts of similar electrical ratings.

Composite insulators are not as brittle as porcelain insulators, and their strength depends on the duration of loading. As loading is increased on a composite insulator, the glass fibers progressively break, leading to loss of stiffness, and eventually rupture, whereas porcelain behaves nearly linearly up to fracture. Composite insulators have higher strength capacities under short-term loading compared to sustained loading. Solid rod (solid core) as well as hollow core composite station posts are available, but the use of a solid rod is more widespread compared to hollow core, although hollow core composite insulators are frequently used in apparatuses such as transformer and circuit breaker bushings, instrument transformers, and cable terminations. At higher basic impulse levels (BIL), solid rod station posts become excessively flexible, which may limit their use; hollow core insulators have higher stiffnesses and are more suitable for higher-voltage applications. Composite insulators can be designed with glass fibers that are oriented in a manner that most efficiently resists the applied load effects, whereas porcelain is isotropic.

6.9.5 Bus System Design

Refer to IEEE 605 (IEEE 2008) for the forces on the bus and insulators and the design of the bus system.

6.9.6 Rigid Bus Seismic Considerations

The following seismic issues should be addressed:

- Expansion fittings may create impact loads because the thermal expansion–fitting gap is too small to allow for relative displacements at the top of insulators. Bus conductors may also pull out of the expansion fittings during an earthquake.
- Cast bus fittings may be brittle and lack sufficient strength to resist seismic loads. Forged fittings should be considered.
- Depending on boundary conditions at the top of the insulator and the direction of the seismic loads, moments may be developed at the top of the insulator.
- If a segment of conductor is supported with fixed and slip fittings and the ground acceleration is parallel to the conductor, the insulators and supports connected to the fixed fittings require sufficient strength to resist the seismic force created by the entire mass of the conductor segment.

- Additional porcelain strength is obtained by increasing the cross-sectional area. High-strength insulators have a corresponding increase in weight, which for seismic loads may offset the increase in strength.
- Flexible connections between rigid bus conductors and electrical equipment should be considered to reduce the transfer of seismic forces. Information on flexible bus connections can be found in IEEE 1527 (IEEE 2018b) and IEEE 605 (IEEE 2008).
- Catenary-hung flexible conductors (strain bus or jumper), used in place of rigid bus conductors, can generate dynamic loads during an earthquake. IEEE 1527 (IEEE 2018b) provides guidelines for the design of flexible bus.

6.10 SPECIAL CONSIDERATIONS

When designing a substation structure, the engineer may encounter some special situations that are unique to utility structures. This section provides some guidance in these areas.

6.10.1 Precautions Regarding the Magnetic Fields of Air Core Reactors

Air core reactors will intrinsically generate a magnetic field in operation. The effects of the magnetic field are normally considered by the manufacturer, but engineers should be aware of the magnetic field and its potential effects when designing and laying-out substations.

The effects of the magnetic field are dependent on the following primary factors:

- Distance: The strength of the magnetic field drops off exponentially with the distance from the reactor.
- Current Magnitude: The magnetic field strength around the reactors is also a function of the current magnitude flowing through the reactor at a given moment.
- Current Frequency: The frequency of the current can also play a significant role in determining acceptable magnetic clearances from air core reactors. For reactors designed to carry predominantly direct current, the recommended magnetic clearances are usually lower than those for filter reactors designed to carry harmonic currents at frequencies much higher than the usual power frequencies.
- Application: Duty cycles, current flow, and ambient conditions can impact temperature rises and are therefore a consideration with regard to acceptable heating.
- Materials: The material properties of any nearby object will impact the acceptability of exposure to a magnetic field. Among the considerations are its conductivity, permeability, and strength at elevated temperatures.
- Geometry: The orientation of the part within the magnetic field and the presence of closed loops can impact the extent of heating.

Historically, the approximate "rule-of-thumb" guidelines for the effects of the magnetic field were typically expressed as distance "D" that corresponded to the diameter of the reactor. Today, reactor manufacturers are able to provide precise information on magnetic clearances on the basis of accurate calculations of magnetic field strengths anywhere around the reactor(s). This magnetic clearance information is typically provided on reactor outline drawings as two contours: the distance to small metallic parts and the distance to closed loops.

With the two aforementioned contours and the reactor centerline spacing (also typically given on the reactor outline drawing), the use of a reactor within a substation can be successfully implemented by observing the following:

- Force: The magnetic field between adjacent air core reactors may interact and generate forces. These forces can be significant when reactors carry very high currents caused by power system short-circuit faults. Thus, the minimum recommended centerline spacing specified by the reactor manufacturer should be followed when laying out the substation (this information is provided on the reactor outline drawing).
- Eddy Heating: Metallic objects within the reactor magnetic field will generate eddy heating. Normally, the reactor manufacturer will ensure that the reactor and its provided support structure are constructed with materials that are designed to withstand the projected eddy heating.

Those laying out substations should consider the following guidelines with regard to eddy heating:

- Small aluminum and steel objects should not be within the clearance to small metallic parts not forming closed loops provided on reactor outline drawings.
- Austenitic stainless-steel anchorage should be specified if the anchorage is within the clearance to small metallic parts.
- Heating effects on large aluminum or steel objects should be checked with the reactor manufacturer.

The reactor manufacturer should be contacted if it is not possible to achieve the clearance guidelines mentioned above. Often, lower clearances are possible after the manufacturer takes into account the full details of the orientation and materials used for the object in question.

Objects that form a closed electrical loop may experience induced currents because of magnetic coupling to the reactor stray magnetic field, and these circulating currents may result in heating. Normally, the reactor manufacturer will ensure that the reactor and its provided support structure have geometries that eliminate closed loops in the vicinity of the magnetic field. Those laying out substations should also ensure the following:

- Closed loops are not within the clearance to parts forming closed loops provided on reactor outline drawings.
- Closed loops can be formed in the rebar in concrete structures. Possible mitigation strategies for a rebar within the closed loop clearance include the use of isolation material to break the closed loop in the rebar or to use fiberglass rebar (refer to ACI PRC-440.1-15 for guidance).
- Closed loops can be formed by chain link fencing. There are several ways of limiting fencing closed loops and the reactor manufacturer should be contacted for details.
- Closed loops can be formed by lattice support structures and should be avoided within the closed loop clearance for the reactor.
- Closed loops can be formed by grounding connections and should be avoided within the closed loop clearance for the reactor.

If it is not possible to achieve the clearances listed above, contact the reactor manufacturer. Often, lower clearances are possible after the manufacturer takes into account the full details of the orientation and materials used for the object in question.

Signal wiring for system control, protection, voice or data communication, and any electronic apparatus should be properly shielded or moved away from the magnetic field of the reactor. The reactor manufacturer may be contacted to provide the expected magnetic field at any location to ensure that the magnetic field contributed by the reactor is less than the maximum limit allowed by the signal circuitry.

Beyond the potential of unwanted heating, there may be safety guidelines that dictate that the reactor either be elevated or fenced off to maintain compliance. The reactor manufacturer may be contacted to provide the expected magnetic field at any location to ensure that the magnetic field contributed by the reactor is less than the maximum safety limit. Recommended safe limits for magnetic field strength are often provided by local legislation or by international standards such as IEEE C95.1 (IEEE 2019).

6.10.2 Vortex-Induced Oscillation and Vibration

Wind-driven vortex-induced oscillation (VIO) occurs when vortices are shed in a periodic and alternative fashion from the sides of a member. As each vortex is shed, the member is pulled in that direction, resulting in alternative forces perpendicular to the direction of the oncoming wind. Under sustained wind conditions, and with amenable structural characteristics, the periodic and alternating forcing can result in oscillations of excessive amplitudes. Vortex-induced oscillation requires relatively steady winds to produce oscillations of excessive amplitudes. The strength of the vortices being shed by the member is affected by the turbulence in the wind, where the vortices are more organized in smoother (i.e., low-turbulence) wind conditions and thereby have the potential of producing oscillations with higher amplitudes. Normally, the amplitude of these oscillations is small, but it may be greatly increased when the frequency of the vortex-shedding oscillations is close to one of the natural frequencies of the structure or component or when the structural damping is very low. When this occurs, the structure or component is susceptible to fatigue failure. Fatigue failure typically occurs at locations with the greatest rigidity, which are typically member connections.

Long, slender structures or components can have a natural frequency that can be excited by winds between 10 mph (16 kph) and 35 mph (56 kph). These wind speeds can be expected almost anywhere; however, only isolated cases of vortex excitation of substation structures have been reported. The majority of known cases consist of 345 kV and higher dead-end structures and tall slender lightning masts. In most of these cases, this excitation occurred when the structures were still unloaded during the construction phase. These observations are possibly a result of conductors, insulators, and ground wires increasing the total damping of the structure-wire system.

Analytical procedures for calculating the response of a structure to VIO have not yet been made practical for design use for the variety of structures applicable to substations. There are three approaches to controlling the amplitude of VIO for strain bus structures and lightning masts:

- Increase the stiffness of the member. This can be accomplished by using larger sections, bracing between members, or adding guy wires.
- Increase the damping of the member. This can be accomplished by hanging a chain inside a vertical member that has a weight approximately equal to 5% of the member weight.
- Add spiral strakes along the length of a member. These strakes suppress the periodicity of the vortex formation and reduce the correlation between the aerodynamic forces acting along the length of the member. The pitch, spacing, number, and length of spoilers along the member's length are derived from wind tunnel tests and are contained in Scruton (1963) and ASCE (1961).

6.10.3 Galvanizing Steel Considerations

Members and connections should be designed and detailed to allow for proper drainage and venting during the galvanizing process. The American Galvanizer's Association has two pamphlets *Recommended Details for Galvanizing Structures* (AGA 2012) and *The Design of Products to be Hot-Dip Galvanized After Fabrication* (AGA 2018), which show proper design and detailing practices. ASTM A123/A123M (ASTM 2017) and ASCE 48-19 should also be consulted.

Tubular structures present special challenges for galvanizing, and failure to address proper design and detailing for galvanizing practices may lead to structural failures because of internal corrosion. In addition to the drainage and venting requirements in the references listed, the engineer should avoid the design of complex built-up members with interior stiffeners that impede the free flow of molten zinc and make interior inspection difficult or impossible (ASTM A385/A385M) (ASTM 2022a). Preferably, all members should be able to be galvanized in one single dip. If it is not possible to do a single dip, the engineer should ensure that the fabricator and galvanizer are qualified to use the double-dip process without flux becoming entrapped beneath a layer of zinc in the double-dipped area.

6.10.4 Painted or Metallized Steel Considerations

Tubular steel members to be painted or metallized should have their interiors sealed because there is usually no effective method to apply the surface coating to the interior of the member or to inspect it once it is in place.

6.10.5 Member Connection Design

The design of the connections between members is an important part of the overall structure design. The external forces on the structure should be transferred through the members by axial, shear, torsion, and moment forces into the foundation with properly designed connections that mirror the assumptions made in the theoretical model. If a member is assumed to be fixed against rotation, the connection should be rigid enough to provide rotational stiffness.

6.10.5.1 Bolted Connections in Steel. When bolts are used to connect members, they should be placed in shear or in tension. Placing the bolt shaft in bending should be avoided. The section modulus of the threaded shaft is small and this can result in high bending stresses. If tension and shear are combined in a connection, interaction equations should be followed in the selection of the bolts. ANSI/AISC 360-22 and ASCE 48-19 can be used for the design.

Care shall be taken when using unsymmetrical bolt patterns on base plate and flange plate connections for tubular shafts. The unsymmetrical bolt patterns may result in higher stress concentrations at the shaft wall compared to symmetrical patterns. The same effect occurs with bolt patterns that have a large center-to-center spacing or that have a large spacing between the pole shaft and the flange bolts or anchor rods. These effects may be undesirable depending on the magnitude and type of loading, in which case the use of unsymmetrical bolt patterns should be based on comprehensive testing or analytical models for the type of loading anticipated.

6.10.5.2 Welded Connections in Steel. The proper design and fabrication of welded connections require considerations for material selection, welding procedure qualifications, detailing, and inspection. Members are usually connected by combinations of flat plates welded together. These flat plates are sized for weak axis bending after finding the appropriate bending plane and plate bending moment. Blodgett (1966), Section 6.6.2, limits the effective plate-bending plane for weak axis plate bending caused by a point load to a value of 12 times the plate thickness. AISC 360-22 can also be used for welded connection design.

Base plate and flange plate welded connections are critical joints for substation structures. The most common application of these connections is for base plates supported by anchor rods and splices between tubular sections, respectively. The design and fabrication of these connections have been based on years of field experience combined with experimental and analytical investigations without the benefit of a comprehensive uniform standard. Flange and base plate connection usually consist of comparatively thin wall tubular sections connected to thicker transverse plates often with limited or no access to the interior of the tubular section. T-joints are the most commonly used joint used to connect transverse plates to tubular sections. A T-joint has the base of the section butted up against a transverse plate. The section wall tension and compression stresses are transferred by applying normal (through-thickness) stresses to the transverse plate. Complete penetration T-joints are typically used for flange and base plate connections on large tubular columns shafts. A reinforcing fillet weld is used around the perimeter of the shaft to reduce through-thickness stresses. An interior weld is typically used if there is enough room to get inside the shaft to prevent acid entrapment during the galvanizing process. Smaller columns with lighter loads are typically connected to base plates with fillet welds or partial penetration welds on the basis of load requirements.

AWS D1.1/D1.1M (AWS 2020) classifies connections into three general categories: nontubular, tubular, and cyclically loaded. The tubular joint provisions of AWS D1.1/D1.1M were adopted from the welding practices and experiences with offshore platforms, which have unique tubular connections, unlike flange and base plate connections used for substation structures. In addition, substation structures are typically designed for static loading conditions. As a result, the provisions of AWS D1.1/D1.1M for nontubular joints are recommended for flange and base plate connections for substation structures.

6.10.5.3 Welded Connections in Aluminum. Care should be taken when designing welded connections for aluminum alloy structures. The heat-affected zone (HAZ) adjacent to the weld can have lower allowable stresses. The HAZ adjacent to the weld could revert to the stress allowable of the untempered aluminum alloy. If at all possible, the welded joints should be heat-treated after welding to restore the strength properties lost in the welding process. Additional information on welding aluminum is available in AWS D1.2/D1.2M (AWS 2014) and the Aluminum Association's *Aluminum Design Manual* (AA 2020).

6.10.5.4 Concrete Structure Connections. Substation structures can be made of precast or cast-in-place concrete elements. Bolts are typically used when members other than concrete are to be connected to a concrete member or when two precast concrete members are to be connected to each other. If a new member is to be installed on an existing concrete structure using concrete anchors, the anchor manufacturer's specifications and limitations should be followed along with ACI CODE-318-19, Chapter 17 for anchorage design.

When connecting two or more concrete members that are not one continuous placement of concrete, the concrete elements can be connected by welding embedded angles or plates. The embedded angles and plates should be designed to develop the connection loads through proper concrete embedment. Care should be used when welding is done on these embedded steel angles to prevent the spalling, or splintering, of concrete.

6.10.5.5 Connections in Wood Structures. The connections designed for wood structures should follow the guidelines in ASCE MOP 141 *Recommended Practice for the Design and Use of Wood Pole Structures For Electrical Transmission Lines* (ASCE 2019b). Additional design information can be found in ANSI/AWC NDS-2015 *National Design Specification (NDS) for Wood Construction.*

6.10.6 Weathering Steel Structures

ASTM A242/A242M (ASTM 2018a), ASTM A588/A588M (ASTM 2019), and ASTM A871/A871M (ASTM 2020) state that atmospheric, corrosion-resistant, high-strength, low-alloy steels can be used uncoated in most environments. As the bare steel is exposed to the normal environment, a tightly adherent oxide layer forms on the surface, which protects the steel from further corrosion. Proper design, detailing, fabrication, erection, and maintenance practices for the application of such steels should be followed to achieve the benefits of the enhanced atmospheric corrosion resistance property.

The oxides from the weathering steel structures may be deposited on underhung insulators and other equipment. This is likely caused by water runoff. There is no direct evidence that this staining will provoke an electrical flashover. However, the stain can roughen the insulator surface and elevate the level of contamination accumulation. Water on the stained surface also tends to remain longer than on unstained insulators. These, combined with other meteorological factors, such as coastal salt spray and abundant humidity, may create a favorable condition for electrical flashover. It may be advisable to increase the leakage distance for the insulators to prevent such an incidence.

The design and detailing of weathering steel connections should be done in accordance with ASCE 48-19. Large edge distances and bolt spacing can cause joint failure over long periods of time when the two connecting surfaces become filled with products of corrosion and expand. This process is commonly called *pack out* and can lead to connection bolt failure when two weathering steel surfaces are in contact as in a bolted lap splice.

6.10.7 Guyed Substation Structures

Guyed structures are infrequently used in substations. Substation yards typically have limited space for positioning guys. A geometric nonlinear analysis should be used for such structures to accurately simulate the forces in the members and guy (Chapter 5, "Method of Analysis"). Particular care should be taken when checking for overall structure buckling. Local plate buckling may cause a problem on hollow polygonal compression members. ASCE 91 (ASCE 1997) provides guidance for the design of this structure type.

6.10.8 Aluminum with Dissimilar Materials

Aluminum has the tendency to corrode when in contact with dissimilar materials, such as steel, wood, or concrete. This corrosion may be caused by galvanic currents between the anodic aluminum and a cathodic material in the presence of an electrolyte whereby the anodic aluminum corrodes, or by a chemical reaction such as direct oxidation in a fresh concrete or mortar alkaline solution. The moisture in wood or the treatment salts (chemicals) may act as an electrolyte in accelerating corrosion of the aluminum. Guidance for aluminum in contact with other materials is provided in the *Aluminum Design Manual* (AA 2020).

6.10.8.1 Steel in Contact with Aluminum. Aluminum surfaces to be placed in contact with steel should be given one coat of a zinc chromate primer complying with *Federal Specification*

TT-P-645B (USDC 2006) or the equivalent, or one coat of a suitable nonhardening joint compound that can exclude moisture from the joint during prolonged service. Additional protection can be obtained by applying the joint compound in addition to the zinc chromate primer. The zinc chromate primer should be allowed to dry before the parts are assembled.

Aluminum surfaces to be placed in contact with aluminized, hot-dip galvanized, or electrogalvanized steel may not need to be painted in low-to-moderate humidity conditions. This is attributed to the low galvanic potential between aluminum and the materials listed previously.

In damp, moist, or polluted environments (areas with a lot of rain, coastal environments, areas with industrial pollution, etc.), these measures may not be sufficient, and an alternative isolation barrier between dissimilar metals may be warranted. An additional reference is the *Aluminum Design Manual* (AA 2020).

6.10.8.2 Wood in Contact with Aluminum. Aluminum surfaces to be placed in contact with wood should be given a heavy coat of an alkali-resistant bituminous paint before installation. The paint should be applied in the same condition in which it is received from the manufacturer without adding any thinner.

6.10.8.3 Concrete in Contact with Aluminum. Aluminum should not come into contact with uncured concrete or concrete exposed to moisture. Aluminum reacts with the alkaline constituents of the cement and generates hydrogen gas. The hydrogen gas will cause expansion of the mortar and reduce the concrete's compressive strength and may be a cause of corrosion of the aluminum.

Aluminum base plates should be separated from concrete foundations by mounting the base plate on galvanized steel anchor rods and leveling nuts above the concrete, by placing a galvanized steel plate between the aluminum plate and the concrete, or by other dielectric isolation. Where aluminum is used in conjunction with steel, the aluminum surface should be treated as specified in Section 6.10.8.1.

REFERENCES

AA (Aluminum Association). 2015. *Aluminum design manual, including specification for aluminum structures.* Arlington, VA: AA.

AA. 2020. *Aluminum design manual, including specification for aluminum structures.* Arlington, VA: AA.

ACI (American Concrete Institute). 2001. *Control of cracking in concrete structures.* ACI PRC-224-01. Detroit: ACI.

ACI. 2015. *Guide for the design and construction of concrete reinforced with fiber-reinforced polymer bars.* ACI PRC-440.1-15. Detroit: ACI.

ACI. 2019. *Building code requirements for structural concrete (with commentary).* ACI CODE-318-19. Detroit: ACI.

ACI. 2020. *Code requirements for environmental engineering concrete structures and commentary.* ACI CODE-350.3-20. Detroit: ACI.

AGA (American Galvanizers Association). 2012. *Recommended details for galvanized structures.* Aurora, CO: AGA.

AGA. 2018. *The design of products to be hot dip galvanized after fabrication.* Aurora, CO: AGA.

AISC (American Institute of Steel Construction). 2006. *Design guide 1: Base plate and anchor rod design.* 2nd ed. Chicago: AISC.

AISC. 2022. *Specification for structural steel buildings.* ANSI/AISC 360-22. Chicago: AISC.

ANSI (American National Standards Institute). 2017. *Wood poles—Specifications and dimensions.* ANSI O5.1-2017. New York: ANSI.

ASCE. 1961. "Wind forces on structures." *Transactions* 126 (3269): 1124–1197.

ASCE. 1972. "Guide for the design of aluminum transmission towers." *J. Struct. Div* 98 (ST12): 2785–2801.

ASCE. 1997. *Design of guyed electrical transmission structures*. ASCE 91. New York: ASCE.

ASCE. 2012. *Prestressed concrete transmission pole structures: Recommended practice for design and installation*, MOP 123. Reston, VA: ASCE.

ASCE. 2015. *Design of latticed steel transmission structures*. ASCE 10-15. Reston, VA: ASCE.

ASCE. 2019a. *Design of steel transmission pole structures*. ASCE 48-19. Reston, VA: ASCE.

ASCE. 2019b. *Wood pole structures for electrical transmission lines: recommended practice for design and use*, MOP 141. Reston, VA: ASCE.

ASTM International. 2015. *Standard specification for aluminum-alloy extruded bar, rod, tube, pipe, structural profiles, and profiles for electrical purposes (bus conductor)*. ASTM B317/B317M-07(2015)e1. West Conshohocken, PA: ASTM.

ASTM. 2017. *Standard specification for zinc coating (hot-dip galvanized) on iron and steel products*. ASTM A123/A123M. West Conshohocken, PA: ASTM.

ASTM. 2018a. *Standard specification for high-strength low-alloy structural steel*. ASTM A242/A242M. West Conshohocken, PA: ASTM.

ASTM. 2018b. *Standard specification for nonferrous nuts for general use*. ASTM F467-13(2018). West Conshohocken, PA: ASTM.

ASTM. 2019. *Standard specification for high-strength low-alloy structural steel with 50 ksi [345 MPa] minimum yield point to 4 inch [100 mm] thick*. ASTM A588/A588M. West Conshohocken, PA: ASTM.

ASTM. 2020. *Standard specification for high strength low-alloy structural steel plate with atmospheric corrosion resistance*. ASTM A871/A871M. West Conshohocken, PA: ASTM.

ASTM. 2022a. *Standard practice for providing high-quality zinc coatings (hot-dip)*. ASTM A385/A385M. West Conshohocken, PA: ASTM.

ASTM. 2022b. *Standard specification for aluminum and aluminum-alloy seamless pipe and seamless extruded tube*. ASTM B241/B241M. West Conshohocken, PA: ASTM.

ASTM. 2023. *Standard specification for nonferrous bolts, hex cap screws, socket head cap screws, and studs for general use*. ASTM F468. West Conshohocken, PA: ASTM.

AWC (American Wood Council). 2015. *National design specification for wood construction*. Leesburg, VA: AWC.

AWS (American Welding Society). 2014. *Structural welding code—Aluminum*. AWS D1.2/D1.2M. Miami: AWS.

AWS. 2020. *Structural welding code—Steel*. AWS D1.1/D1.1M. Miami: AWS.

Bleich, F. 1952. *Buckling strength of metal structures*, 1st ed. New York: McGraw-Hill.

Blodgett, O. W. 1966. *Design of welded structures*. Cleveland: James F. Lincoln Arc Welding Foundation.

IEEE (Institute of Electrical and Electronics Engineers). 2008. *Guide for design of substation rigid-bus structures*. IEEE 605-2008. Piscataway, NJ: IEEE.

IEEE. 2018a. *Recommended practice for seismic design of substations*. IEEE 693. Piscataway, NJ: IEEE.

IEEE. 2018b. *Recommended practice for the design of flexible buswork located in seismically active areas*. IEEE 1527. Piscataway, NJ: IEEE.

IEEE. 2019. *IEEE standard for safety levels with respect to human exposure to electric, magnetic, and electromagnetic fields, 0 Hz to 300 GHz*. IEEE C95.1. Piscataway, NJ: IEEE.

IEEE. 2023. *National electrical safety code*. ANSI C2-2023. Piscataway, NJ: IEEE.

Kissell, J. R., and R. L. Ferry. 1995. *Aluminum structures: A guide to their specifications and design*. New York: Wiley.

Mooers, J. D. 2006. "Aluminum substation structures." In *Proc., Electrical Transmission Line and Substation Structures Conf.*, 137–148. Reston, VA: ASCE.

PCI (Precast/Prestressed Concrete Institute). 2017. *Design handbook, precast and prestressed concrete*. 8th ed. MNL-120. Chicago: PCI.

PTI (Post-Tensioning Institute). 2006. *Post-tensioning manual*. 6th ed. Farmington Hills, MI: PTI.

Scruton, C. 1963. "On the wind-excited oscillations of stacks, towers, and masts." In *Proc., 1st Int. Conf. on Wind Effects on Buildings and Structures*, National Physical Laboratory, Aerodynamics Division, June 26–28, 798–837. Teddington, Middlesex, UK: H.M. Stationary Office.

USDC (US Department of Commerce). 2006. *Federal specification TT-P-645B: Primer, paint, zinc-molybdate, alkyd type*. Washington, DC: USDC.

CHAPTER 7
FOUNDATIONS

A substation consists of interconnected electrical equipment and elements. The complete structural system should be designed considering the allowable structural deflections, rotations, and movement to maintain safe and reliable operation of the equipment and the interconnected parts. Foundations should be designed to safely transfer the vertical and lateral loads from the electrical equipment and structures to the supporting soil. As part of the complete load path, foundations are a critical component to ensure that the equipment and their supporting structures are safe, stable, and can function properly.

This chapter identifies various types of foundations that are commonly used to support substation equipment and structures, discusses geotechnical considerations for the design of substation foundations, and provides design considerations that may be unique to construction within a new or existing substation. It is the responsibility of the engineer to determine the applicable design loads and design requirements for specific sites. Loads should be combined and loading factors applied in accordance with this chapter and Chapter 3, "Loading Criteria for Substation Structures," Section 3.3, "Load Factors and Combinations." A site-specific custom foundation design could avoid additional costs resulting from the use of an overdesigned foundation and is usually necessary where standard equipment foundations may not be applied.

This chapter is focused on the design of new, cast-in-place concrete foundations. Substation structure foundations may also be of precast concrete type, which is beyond the scope of this document. The information presented in this chapter may be applicable for analyzing existing foundations or when retrofitting an existing foundation with new structures or equipment. Requirements for the analysis of existing foundations should be determined by the engineer and based on sound engineering principles and relevant experience. This Manual of Practice (MOP) is not a substitute for engineering competency, nor is it to be considered as a rigid set of rules.

The Naval Facilities Engineering Systems Command (NAVFAC) document, *Foundations and Earth Structures*, NAVFAC DM 7.02 (USDN NAVFAC 1982), is referenced in this chapter. NAVFAC DM 7.02, although an old document, contains a wide breadth of information and considerations for the design of both shallow and deep foundations. It is important to note that NAVFAC DM 7.02 is not a code and should be used only as a reference.

7.1 FOUNDATION TYPES

7.1.1 Shallow Foundations

Shallow foundations may be used where there is a suitable bearing stratum near the surface, where there are no highly compressible layers below, and where calculated settlements are acceptable. Where the bearing stratum at the ground surface is underlain by weaker and more compressible materials, the use of deep foundations or piles should be considered (NAVFAC DM 7.02). Common types of shallow foundations include spread footings, combined footings, mat foundations, and grade beams, as shown in Figure 7-1.

Various types of substation structures are supported by shallow foundations. Shallow foundations are able to distribute loads effectively through bearing pressure transferred to the soil. For the majority of structures, the design of shallow foundations is controlled by limiting settlements (NAVFAC DM 7.02).

7.1.1.1 Spread Footings. A *spread footing foundation* is a single pedestal supported on a single enlarged, reinforced concrete footing. Pedestals and footings may be of a variety of shapes such as circular, square, rectangular, and trapezoidal. Pedestals are typically centered within the footing but may be offset where existing features such as existing facilities or property lines require this.

7.1.1.2 Combined Footings. A *combined footing* is a foundation for more than one piece of equipment or structural support. Combined footing foundations may be a practical type of foundation when the structure or equipment to be supported has multiple closely spaced columns or supports.

7.1.1.3 Mat Foundations. A *mat, pad,* or *slab foundation* is a shallow reinforced concrete footing that has no pedestals that are placed at or near grade level. Mat foundations are capable of being sized to transfer large loads to the soil. Equipment that may require continuous support, such as transformers, shunt reactors, control enclosures, and switchgears, are often supported on mat foundations. For the purposes of this MOP, mat, pad, and slab foundation are used interchangeably. Anchorage requirements and designing foundations to a minimum depth for frost protection typically results in a mat foundation thickness that may be assumed to be rigid in comparison with the surrounding soil. For certain foundations, such as thin slabs, the engineer may consider the soil–structure interaction. Refer to Section 7.3.5 for additional information.

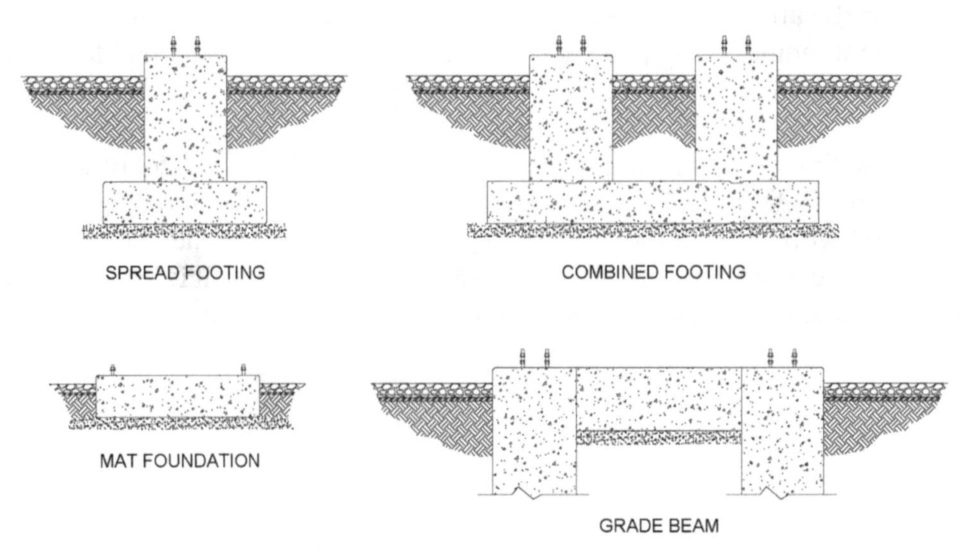

Figure 7-1. Various types of shallow foundations.

7.1.1.4 Grade Beams. A *grade beam* is a shallow foundation that consists of a reinforced concrete beam used to transmit the load from a bearing wall, or other similar continuous loads, to the foundations. Grade beams are designed to support the bending moments from the continuous loads and may be used with a variety of foundation types such as drilled shafts, piles, or spread footings. Grade beams are commonly used to support control enclosures, firewalls, sound barriers, or other structures requiring continuous support.

7.1.2 Deep Foundations

The term *deep foundation*, in general, refers to a foundation with a depth-to-width ratio (*D/B*) exceeding 5 (NAVFAC DM 7.02). Deep foundations are used in a variety of applications. Such applications may be where the upper soil stratum is weak or compressible, underwater, in cases where existing structures or foundations are in close proximity to the location of the new foundation, or to provide additional uplift and lateral capacity where loads may be too high to be transferred in a practical fashion by a shallow foundation.

Various types of substation equipment and foundations are supported on deep foundations. Choosing the appropriate type of deep foundation is dependent on the subsurface conditions and local experience and practice (IEEE 691) (IEEE 2001). The performance of deep foundations is highly dependent on the installation procedure, quality of workmanship, and any changes to the design that may have been made in the field. Because of these factors, the inspection of deep foundations by a qualified geotechnical engineer or technician during installation is recommended (NAVFAC DM 7.02).

7.1.2.1 Drilled Shafts. *Drilled shafts* are a common type of deep foundation (also called *piers*, *drilled piers*, and *drilled caissons* for supporting substation structures, as shown in Figure 7-2. Drilled shafts are reinforced concrete foundation elements, with or without enlarged end-bearing areas, extending to the depth of suitable supporting soils (ACI PRC-336.3) (ACI 2014). A variety of diameters and depths are available. Drilled shaft foundations can be sized to support the applied loads and anchoring requirements of the structure or equipment.

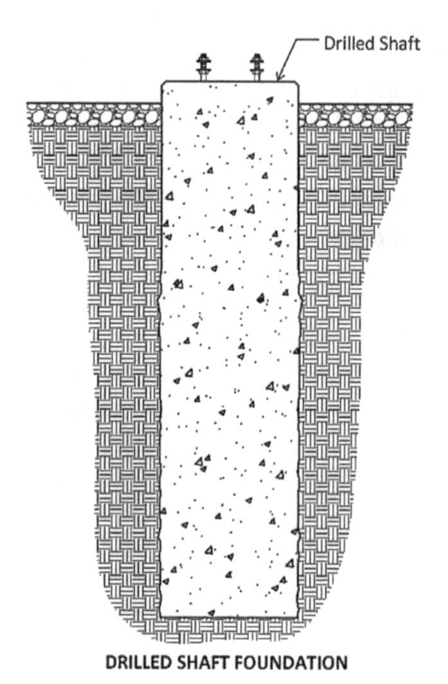

Figure 7-2. Drilled shaft foundation.

The performance of drilled shafts is highly dependent on soil conditions and installation methods. Casing or slurry construction methods may be required if the foundation extends beyond the water table or in collapsible soils. Concrete should be placed continuously and in a manner to avoid cold joints or other discontinuities. Additional costs for shoring excavations should be expected where drilled shaft construction requires casing. Rock excavation will also increase the cost of drilled shaft construction when shafts extend into rock that cannot be excavated with conventional drilling equipment.

Drilled shaft foundations transmit axial loads to the soil through end-bearing, skin friction between the sides of the concrete foundation and the soil, or a combination of both processes if permitted according to geotechnical recommendations. Overturning moments and lateral loads are transferred, in general, as pressure applied to the confining soils. The design of drilled shafts is often controlled by limiting the rotation and horizontal displacement at the top of the foundation. See Section 7.6.6 for additional commentary regarding allowable deflection and rotation of drilled shaft and deep foundations.

7.1.2.2 Piles. *Pile foundations* are a type of deep foundation capable of transferring axial and shear loads to the soil. A variety of pile types are commonly used. These types may include precast concrete, cast-in-place concrete, concrete-filled steel pipe, steel H-piles, helical piles, timber piles, and others.

Similar to drilled shafts, the design and performance of piles is highly dependent on the soil conditions and installation methods. Some types of piles require design to be performed by the manufacturer, as proprietary methods and tools are used for design and analysis. The use of piles is most practical when soil conditions are not wet, sandy, or predominately cohesionless. See Section 7.6.3 for more information regarding the spacing of deep foundations and group effects.

7.1.3 Direct Embedment

Direct embedded foundations are poles or columns that are installed directly into an open excavation with backfill placed in the void space around the pole or column. The backfill material may be cohesionless material such as gravel or gravel sand mixes. Cohesive materials such as silts and clays, concrete, or a controlled low-strength material (CLSM) may be used.

Direct embedded foundations are typically more economical than installing a reinforced concrete foundation. However, direct embedded poles or columns should be limited to structures with lower overturning moments and may allow more rotation and horizontal displacement than desired for the supported equipment or structure. Direct embedded foundations for poles or columns and other structures within a substation are typically limited to structures that are not deflection sensitive and do not carry critical electrical components such as lightning masts.

7.1.4 Helical Screw Anchor Piles

Helical screw anchor piles are a manufactured deep foundation system consisting of varying sizes of tubular hollow steel shaft sections with one or more helical plates welded around it. The pile shaft transfers a structure's loads into the pile. The piles are either connected to the structure by welded plates or are enclosed within a concrete cap. These systems are typically proprietary and utilize software to establish the empirically derived relationship between the installed capacity and the torsional resistance encountered during installation.

The advantages of helical screw anchor piles include the following: (1) ease of installation, (2) smaller equipment required than other deep foundation options, (3) can be installed in soft soils and/or high water table conditions, and (4) little vibration during installation.

7.2 GEOTECHNICAL SUBSURFACE EXPLORATION

7.2.1 General

Some form of subsurface exploration is recommended for all foundation installations. Temporary installations and small, lightly loaded distribution-voltage structures and foundations may require minimal information. However, major substation expansions and new substations may require a comprehensive subsurface exploration and report including recommendations for the design and construction methods of various types of foundations. The soil resistivity may be necessary to provide adequate data for structure grounding design. Coordination is required between the engineers responsible for structural engineering and geotechnical engineering to ascertain whether the investigation and report are in conformance with project-specific requirements.

7.2.2 Existing Geological Data

Existing subsurface and geotechnical data are available from a variety of sources. The US Geological Survey (USGS) may be able to provide information, including the distribution of surface and ground waters, soil conditions and classes, type of rock, and other geological formations. The Natural Resources Conservation Service (NRCS), a division of the US Department of Agriculture, may have additional soil survey information available. Such sources of public information provide limited information and are intended to be used for reference only as they may not be representative of the site-specific conditions (see IEEE 691) (IEEE 2001).

7.2.3 Site-Specific Subsurface Exploration

A site-specific subsurface exploration and laboratory testing program is recommended for new substations, major substation expansions, in areas of poor subsurface conditions, or otherwise as determined by the engineer. Information gathered from such an investigation may be used to determine the most suitable and economical foundation designs. Subsurface conditions can vary significantly from location to location, and may also vary significantly within the footprint of a single site. A subsurface investigation provides the information needed to properly design the foundations that support critical electrical infrastructure.

To obtain a subsurface investigation, a qualified geotechnical engineer should be selected to perform the work. Close coordination between the engineer and the testing firm is required to ensure that the testing program is compatible with the design requirements and types of foundations to be designed for the substation site.

If available, the engineer may provide to the testing firm the types of anticipated foundations, preliminary or estimated foundation loads (identified as factored or unfactored), load type and combinations, site grading design, and any limitations including but not limited to settlement, differential settlement, and expansion that are to be considered in the design of the foundations. The engineer should work with the geotechnical testing firm to identify boring locations in areas of critical equipment or at highly loaded foundations such as transformers, dead-ends, and control enclosures. Additional boring locations may be needed to determine whether the profile of the subsurface conditions across the site is accurate. In addition to borings, a geophysical survey can provide further subsurface information. The geotechnical testing program should determine whether the proposed foundation types are suitable and provide recommendations for foundation types and construction methods.

Design parameters provided by the geotechnical testing firm should be in accordance with those required by the software to be used for foundation design. It is important for the engineer to understand that a variety of design software is available and that the input parameters may vary between these design tools. As such, the engineer should also specify to the geotechnical testing firm the design parameters required and the software used for the design of foundations.

7.2.3.1 Seismic Considerations. The subsurface investigation should include an evaluation of the site-specific seismic design parameters and seismic hazards. Based on the project needs, proposed foundation types, and seismic design category, potential geologic and seismic hazards to be identified and evaluated may include slope instability, liquefaction, differential settlement, and surface displacement. The geotechnical report should provide the required seismic design parameters and site-specific ground motion parameters (S_{DS}, S_{D1}). Additional seismic hazards and mitigation methods should be discussed in the report on the basis of soil types and subsurface conditions.

7.2.3.2 Soil Borings. Boring locations should be selected by the engineer and geotechnical testing firm. Coordination between the engineer and the testing firm is essential. The number of borings, depth, and boring locations should provide a representation of the subsurface conditions across the site, with special consideration given to critical locations such as transformers and dead-ends.

Before boring, it is important to verify that no existing underground utilities will be contacted or damaged during drilling operations for the safety of the personnel and for the operation of the equipment. Buried electrical conduits and underground lines are common within substations. Record drawings may not accurately depict the locations of such underground obstructions, or locations may not be shown at all. The engineer, owner, and testing firm should have a plan for drilling and excavation within an existing substation. Work should be performed with extreme care when drilling near energized overhead or underground transmission lines, bus, or other energized parts. See Section 7.6.5 for additional recommendations for excavations within existing substations.

7.2.3.3 Geotechnical Report. The results of the subsurface exploration and testing program should be presented in report format and certified by the responsible professional engineer. The contents of the report should provide the engineer with the information and recommendations necessary to select and design foundations of the appropriate type and size on the basis of subsurface conditions. In addition, a record of the drilling and sampling program, a description of the geology and subsurface conditions, groundwater conditions, seismic hazards, recommended cement type, boring logs, and the results of the testing program should be contained within the report. Allowable bearing capacities should consider any limitations on the settlement, rotation, and deflection of the foundations and supported structures. Refer to Section 7.6.6 for an additional discussion of foundation movement.

Additional information and recommendations should be provided in the geotechnical report beyond recommendations for foundation design. Such recommendations are beyond the scope of this manual and should be coordinated on the basis of project-specific needs. Additional information may include design parameters and recommendations for the design of site grading, retaining walls, pavement, soil stabilization, soil resistivity measurements, or other project-specific needs.

7.2.3.4 Other Considerations. Additional design considerations are provided in Section 7.3. Such considerations can impact the design and performance of foundations and, where applicable, should be addressed by the geotechnical testing firm with recommendations for design or mitigation in the geotechnical report.

7.3 ADDITIONAL DESIGN CONSIDERATIONS

7.3.1 Frost Action

Frost action is caused by the formation of ice lenses in frost-susceptible soils. Freezing water in the void spaces increases the volume of the ice lens, causing the pore pressure of the remaining water to decrease. This tends to result in additional draw of water from deeper depths (US Army Corps of Engineers Engineering Manual 1110-3-138) (USACE 1984), which dries up the soil and makes it susceptible to consolidation and wetting in the frost zone, thus creating fissures.

At a minimum, foundations should extend to a depth below the anticipated frost line to avoid movement and settling caused by frost action. In cases where it is not practical to extend the foundation beyond frost depth, over-excavation is recommended, and the void space should be backfilled and compacted with a suitable structural fill that is free-draining and less susceptible to frost action than the native soils. A clean, medium coarse sand or gravel layer is typically suitable for controlling frost action (USACE 1984). Pressure injection with lime slurry or lime fly-ash slurry may also be a suitable method of control. Movement of foundations caused by frost action can either cause stress in structures and other rigid substation equipment connections or cause switches to bind and become inoperable.

Soils having predominately silt size particles are more likely to be impacted by frost action. Smaller pore sizes may also increase the capillary action, thus increasing the susceptibility of the soils to frost action. The geotechnical report should identify whether soil conditions are susceptible to frost action. A frost penetration map is available in IEEE 691, Figure 38 (IEEE 2001), for the continental United States.

7.3.2 Expansive or Collapsible Soils

Soils that are susceptible to changes in volume from water content may cause movement and differential settlement of foundations (USACE 1984). Seasonal wetting and drying can cause soil movements that lead to long-term damage of foundations and the supported structures.

Foundations placed within an active zone of expansive soils should be designed to resist differential volume changes and prevent damage to the supported structure. To help prevent movement and damage to the supported structure, foundations should be designed to prevent uplift and settlement and resist forces exerted on the foundation as a result of soil volume changes.

Clays with a high plasticity index that experience volume changes because of changes in water content are considered expansive soils. These soils may heave under low applied pressure and collapse under higher pressures (USACE 1984). Expansive soils may induce significant uplift forces on the sides of foundations placed through layers of expansive soils and may cause settlement issues where shallow foundations do not extend beyond expansive layers. Where expansive soils are encountered, the effects may be reduced by embedding the foundation to a depth below the zone of moisture change, suitable for handling uplift and downdrag forces from the expanding material layer. Chemical treatments may also be considered to decrease the potential of volume change in the soil. Similar to the ramifications of frost action, expansive soils can cause movement in the foundations that can induce additional stresses in the structures and rigid equipment attachments. Refer to IEEE 691, Figure 40 (IEEE 2001), for a graph of Plasticity Index to Soil Swell Potential.

Downdrag forces, also called *negative skin friction*, may occur in deep foundations when soil layers adjacent to the foundation settle after installation of the foundation. The friction of the settling soil against the edges of the foundation induces additional axial loads. These downdrag forces may occur when new fill is placed on existing soil, causing the existing soils to consolidate and settle, in a liquefaction event, or in expansive soils. Downdrag forces should not be disregarded and the foundation should be sized to accommodate the additional loads without exceeding allowable end-bearing capacities.

7.3.3 Corrosion

Certain types of soils may attack and corrode concrete and reinforcing steel. Chlorides and sulfates are chemicals that, when in high concentration, can have negative impacts on concrete foundations. The geotechnical laboratory testing program should provide information on the content of such materials in the soil, and recommendations for mitigation in the geotechnical report. The concrete and reinforcing steel can be properly protected by selecting the appropriate type of cement, reducing the water-cement ratio, or applying a bituminous coating to the exterior of the foundation. Coated reinforcing bars may be used to protect the reinforcing steel from corrosion, but they are not common in substation foundations. See Section 7.5 for an additional discussion of durability requirements for concrete foundations.

7.3.4 Seismic Loads and Dynamic Loads

Special care should be exercised when designing foundations in high seismic regions or regions where repeated cyclic motion may occur, such as those from vibrating machinery, vehicular traffic, blasting, or pile driving. Design should be focused on minimizing movement and settlement of the foundation and the supporting soils. Structures and foundations should be designed to support seismic or other dynamic loads that may be applied to the structure or supporting soils. The geotechnical report should indicate any potential risks associated with liquefaction or other similar considerations in regions of seismic activity.

Saturated or partially saturated loose soils may be prone to liquefaction caused by dynamic loading. Liquefaction occurs when pore water pressure in the soil increases and the pore water in the void spaces is pulled to the surface. As water is pulled to the surface, the soil beneath consolidates into void spaces that are now free of water. Liquefied soil loses its stiffness and strength, resulting in a loss of bearing capacity. The geotechnical report should identify whether the soils at the site are prone to liquefaction.

Shallow foundations that do not extend beyond the liquefiable soil layers may tilt or experience settlement. A loss of lateral support can occur in deep foundations, leading to a potential for buckling failure. See Section 7.6.8 for additional seismic foundation design considerations that have been used to protect sensitive or heavy equipment that is subject to high seismic loading, such as transformers. If the liquefiable zone is located close to an open face or lies on a gentle slope, the liquefied soil mass may displace laterally. Such an event is referred to as a *lateral spread*, which can result in large, damaging displacements to foundations and other buried items. Foundations for equipment and supports, which are seismically qualified in accordance with IEEE 693-2018 (IEEE 2018), should be designed for foundation loads given in that standard.

7.3.5 Soil–Structure Interaction

Foundations are typically designed without considering the effects of soil–structure interaction (SSI). The foundation is assumed to be infinitely rigid in comparison with the soil, and loads are transferred to the soil on the basis of contact surface area and traditional stress theories. For light structures in relatively stiff soil, disregarding SSI may be a reasonable approach. The effects of SSI are more prominent for heavy structures supported on soft soils and in high seismic regions.

Soil–structure interaction is the interaction of the foundation element in direct contact with the soil and the soil supporting the foundation. Designing to consider SSI is a process that takes into account the influence of the soil response to the motion of the structure and the influences of the motion of the structure to the response of the soil. Displacement of the soil and foundation are not considered independently when SSI is taken into account.

The results of SSI may lead to an undesired increase in the period of the supported structure and increased deformation of the supporting soils. This may, in some cases, have significant influence on the response of the structure to seismic loading. Consideration of SSI, or disregarding SSI, should be at the discretion of the engineer and should take into account the supported structure, foundation type, and soil conditions at the site.

7.4 LOADING CONSIDERATIONS

Utilities may have unique load combinations and criteria for the design of substation structures and foundations based on regulatory codes, operational requirements, and regional experience. Substation foundations may be subjected to a variety of axial, uplift, shear, and overturning moment loads. It is the responsibility of the engineer to determine the combination of loading that governs the design of the foundation, including all possible loading conditions that the foundation may see throughout its service life.

7.4.1 Load Application

To properly size and design a foundation, it is necessary to understand the load path from the equipment or supporting structures to the foundation. The type of structure or equipment supported, as well as its anchorage to the foundation, will determine the types of loads that are transferred from the structure to the foundation. For example, lattice towers, or other structures with multiple legs and a single bolt connection to the foundation at each leg, may be considered pinned. This connection will transfer axial and shear loads but will not transfer moments from the structure at the connection to the foundation.

Steel columns with base plates and multiple anchor rods in an anchor group may be considered fixed connections. These connections will transfer axial, shear, and moment loads to the foundation in each direction of loading.

Unique cases may be presented where the structural anchoring to the foundation transfers loads as a fixed connection in one direction (moment transfer) but acts as a pinned connection in the other (no moment transfer). The engineer must understand how loads will be transferred from the structure to the foundation through the anchorage. See Chapter 8, "Connections To Foundations," for the design of anchorage to foundations.

7.4.2 Load Combinations

Current foundation design practice for stability, bearing, and soil–structure interaction uses unfactored loads (service level loads) with load combinations from Chapter 3, Table 3-19. Bearing pressure applied to the soil should not exceed the allowable load-bearing pressure provided in the geotechnical report or the presumptive load-bearing capacities at the site. The global stability of the foundation should also be verified using unfactored loads. The foundation should be sized to resist overturning and sliding forces with additional safety factors. Deep foundations, such as drilled shafts or piles, should meet the desired deflection and rotation performance criteria from the applied loads, factored or unfactored, as appropriate, with corresponding design limits.

Strength-based design, using factored loads and the load and resistance factor design [USD/LRFD, a refined Ultimate Strength Design (USD)] method, should be used for designing the concrete, reinforcing steel, and other structural components. The factored load combinations in Chapter 3, Table 3-18, may be used to determine the ultimate loads for structure component design. Factored loads should be used to check one- and two-way shear, reinforcing moment capacity, shear friction in construction joints, shear ties, column interaction, pile section capacities, and all other structural components as applicable on the basis of the applied loading and the type of foundation. Table 7-1 provides the user with guidance on the use of the different load combinations for typical foundation design checks.

Table 7-1. Applicable Load Combinations for Design of Foundations.

Foundation type	Design consideration	Table 3-18 Load combinations	Table 3-19 Load combinations
Spread footing/ mat foundation	Soil capacity		
	• Ultimate bearing pressure	X	
	• Allowable bearing pressure		X
	Foundation stability*		
	• Overturning	X	X
	• Sliding	X	X
	Concrete/steel design		
	• Concrete/reinforcement design	X	
	• Anchorage design	X	
Drilled pier/ pile	Soil Serviceability		
	• Pier/pile top rotation		X
	• Pier/pile top deflection		X
	Soil strength/foundation stability		
	• Foundation movement arrested by soil	X	
	Concrete/steel design		
	• Concrete/reinforcement and steel design	X	
	• Anchorage design	X	

*Foundation Stability may be checked using either ASD or LRFD load combinations depending on the safety factor or resistance factor used.

7.5 DURABILITY OF CONCRETE

Concrete used for foundations should be capable of resisting weathering exposure, chemical attacks, and erosion while still maintaining desired properties. Chapter 19 of ACI CODE-318-19 (ACI 2019) and ACI PRC-201.2-23 (ACI 2023) provide recommendations for selecting minimum compressive strength, maximum water-to-cement ratios, cementitious materials, and air content for concrete mix designs in various exposure categories. Selecting an appropriate mix design will help ensure that a foundation maintains integrity throughout its service life.

Alkali-silica reactions may occur over time and can cause expansion in the aggregate that may lead to spalling, loss of strength, and potential failure of the foundation. Alkali-silica reactions occur between highly alkaline cement pastes and reactive noncrystalline silica. Random cracking and spalling concrete may be indicators of an alkali-silica reaction. If aggregates are determined to be potentially reactive, they may be tested to determine whether special requirements such as pozzolans, slags, or blended cements are needed to help control alkali-silica reactions. Refer to "Effect of Alkali-Silica Reaction/Delayed Ettringite Formation Damage on Behavior of Deeply Embedded Anchor Bolts" (Sungjin et al. 2009).

Freezing and thawing cycles in concrete can cause damage if the water pressure in the pores of the concrete exceeds the tensile strength of the concrete. Water expands as it freezes and successive freeze–thaw cycles can cause expansion, cracking, scaling, and crumbling of concrete, leading to a loss of strength, or exposing reinforcing steel. Air-entrained concrete can help mitigate cracking from freeze–thaw cycles by acting as expansion chambers and relieving pressure build-ups that may cause cracks.

Portland cement typically does not have good resistance to acidic materials. Seawater and oxidized sulfide minerals in adjacent soil are typical sources of sulfates that may cause harm to concrete foundations. Sulfate reactions can cause expansion of the concrete and extensive cracking. Selecting the appropriate type of cement (Type II, V, or II/V) and limiting the permeability of the concrete may help minimize the impact of acidic attack.

7.6 SPECIAL CONSIDERATIONS

In addition to sizing foundations for the applied loads, special consideration should be given to the design of foundations or groups of foundations located within a substation and supporting associated equipment. Many of the special considerations listed in the following sections may be useful when designing foundations for any application. However, some of these considerations are applicable only when designing and installing foundations in an electrical substation or foundations supporting high-voltage electrical equipment.

7.6.1 Operational Loads

Operational loads are defined as any loads associated with the operation of the substation equipment. Operational loads may be applied to the foundation in various directions and should be presented on equipment manufacturer drawings. For example, circuit breakers may transmit operational loads to the foundation during opening and reclosing of the breaker. Although operational loads may be short in duration, they may be significant, and the foundation should be sized considering such loads when they are expected to transfer from the structure or equipment to the foundation.

7.6.2 Construction Loads

Construction loads are loads applied to the structure or foundation during installation of equipment or wires. Equipment such as transformers, as well as other heavy equipment, may require the use of special installation methods for placement and removal on or off of the foundation. The design of the foundation should consider construction loads such as loads from jacking points or cribbing that may bear on the foundation during placement. Equipment manufacturer drawings should provide the locations of jacking points.

Similarly, substation structures should be designed to support loads that may be applied only during construction. Construction loads may include lifting or moving of the equipment or structure, the weight of a worker, wire stringing loads, or other loads that may be higher than the loads that the structure is anticipated to see during its design life. Structures or other elements that are not designed for construction loads should be indicated on the plan drawings. For example, concrete oil containment walls or flooring may be designed for construction loading according to ASCE 37 (ASCE 2002). However, such walls and flooring are typically not designed for the bearing or cribbing loads that may be applied when moving a transformer (or other heavy equipment) to the foundation pad. The heavy hauling and rigging company should take precautions to span these components to prevent damage.

7.6.3 Group Effects

When deep foundations are spaced in close proximity, the loading transferred to the soil from the foundations may impose additional loads on adjacent foundations and the supporting soil.

The vertical and lateral load capacities and displacements of these foundations are influenced by the spacing of the individual foundations and whether they are connected by pile caps and the types of supporting soils (cohesive or cohesionless). For axial loads, the vertical load capacity will depend on whether the loads are transferred to the supporting soil strata in point-bearing (end-bearing) or side resistance (skin friction). The total load capacity of deep foundations installed as a group may be less than the load capacity of an isolated deep foundation multiplied by the number of foundations in the group. Publications such as the NAVFAC (USDN NAVFAC 1982), AASHTO (1978), ACI PRC-336.3 (ACI 2014), FHWA-NHI 06-089 (USDOT FHWA 2006), FHWA-NHI 16-064 (USDOT FHWA 2016), FHWA-NHI 18-024 (USDOT FHWA 2018), and other texts provide methods to account for the group effects of deep foundations. The minimum recommended spacing of deep foundations will vary with the soil type and installation method. The final design should conform to the spacing and any load capacity reduction factors for group effects specified in the geotechnical subsurface report.

During construction, care should be taken when installing closely spaced drilled shaft foundations to prevent the collapse of adjacent open holes. Careful control of construction operations and sequencing, and possibly the use of temporary casing, may be required to avoid unpredictable adverse consequences to drilled shaft performance.

7.6.4 Slopes and Excavations

A reduction in the allowable vertical bearing capacity applies when installing foundations on, or in close proximity to, a slope. Slopes may also decrease the lateral stability. The grade, depth of foundation bearing, and applied loading are all critical factors in determining the suitability of the foundation installed on or near a slope (NAVFAC). Adjacent slopes may be inside or outside of the substation yard. Oil containment pits around transformers, other oil containing facilities, and similar areas that may require excavations adjacent to foundations should also be considered when designing foundations on or directly adjacent to these regions.

7.6.5 Constructability

Constructability plays an important role in the selection and design of foundations. The geotechnical report should provide information regarding any limitations that subsurface conditions may have on the constructability of anticipated foundation types.

When new foundations are required inside an existing substation, existing overhead clearances, access limitations, underground obstructions, existing foundations, and vibration-sensitive substation equipment are some items that may affect constructability of the new foundations.

Overhead obstacles such as structures, bus, or other wires (energized or de-energized) may dictate the use of a shallow foundation rather than a deep foundation. Shallow foundations can usually be installed with lower profile equipment than deep foundations. Deep foundations, such as drilled shafts or piles, may not be practical in some cases where overhead clearance requirements needed for installation cannot be met. Low-profile drill rigs and drilling methods are available, but this specialized equipment may not be available to the contractor performing the work and may increase installation costs.

It is not uncommon to install new structures and foundations within existing substation yards and it is important to identify underground obstructions that may limit the footprint or locations of new foundations. Existing drainage piping, conduits, duct banks, and underground transmission lines should be avoided if possible or relocated prior to the installation of new foundations. Locations of underground obstructions should be identified for the safety of construction personnel and for the operation of existing equipment. It may not be able to easily relocate existing underground utilities or transmission lines because of outage requirements.

The location, size, and depth of new foundations installed near existing foundations or underground utilities may transfer additional loads to the existing foundations (or vice versa). Existing foundations may require stabilization during construction if excavation for the new foundation may undermine their integrity. Similarly, where driven piles are to be installed adjacent to vibration-sensitive substation equipment, a monitoring program may be required to verify that vibrations resulting from the piling driving activity do not exceed allowable frequencies so as to not interrupt the performance of vibration-sensitive equipment.

Because the locations of existing underground facilities are not always known within an existing substation, the contractor can have the utilities marked. When excavating near marked utilities or when utilities have not been marked, the contractor can carefully excavate, by hand, hydrovac, or otherwise, to a prescribed depth prior to the use of larger drilling or excavation equipment. This provides a measure of safety for the construction personnel and minimizes the risk of an unplanned outage if energized wires or conduits are encountered.

The foundation placement and curing time should follow American Concrete Institute (ACI) requirements and the Owner's Construction Specifications concerning formwork removal, structure installation, and construction loading during curing.

7.6.6 Settlement, Rotation, and Deflection

After the installation of a foundation, immediate settlement will occur. After the foundation is loaded to its design capacity and throughout the life of the foundation, additional settlement will take place. Limiting the bearing pressures applied from the foundation to the supporting soil is one way to minimize settlement. The bearing area of foundations should be sized to keep the amount of settlement below the design limit. Greater amounts of settlement and movement may be allowed for those foundations supporting equipment with flexible attachments (e.g., a strain bus structure).

Movement in deep foundations is typically measured by the rotation and horizontal displacement of the top of the foundation. In establishing performance criteria, consideration should be given to how much recoverable, as well as non-recoverable, displacement and rotation can be permitted (IEEE 691) (IEEE 2001). In some cases, permanent rotations may be aesthetically unacceptable. In other cases, this deflection may reduce electrical clearances or increase stress in the structure or rigid attachments. Settlement, rotation, and deflection of the foundation should be limited to levels that are suitable for the performance of the electrical equipment and structure being supported with consideration to economy of design and past performance history.

7.6.7 Uplift

There are many reasons that uplift loading may be applied to a foundation or group of foundations. Uplift loads are any loads that act in the opposite direction of axial compression loads and in a direction that lifts the foundation from the ground. Uplift loads may be present in areas of high seismic loading, caused by operational loads, in multi-legged frame structures, and in substation dead-end structures. Accordingly, for such structures, foundations should be sized and designed to resist uplift loads. Dead-end structures with high wire tensions may transmit high uplift loads to the foundations. H-frame wire-supporting structure foundations may experience uplift when large horizontal line angles are present or where wire attachment elevations on the adjacent structure are significantly higher. A-frame wire-supporting structures are typically used for high tensions, and uplift should always be expected if the structure supports wire tensions from one side only.

7.6.8 Seismic Base Isolation

In the building industry, *seismic base isolation* is a well-known and well-established method to provide seismic protection. Some utilities have applied this technology to protect high-voltage transformers in a substation. Transformers, because of their mass, are subject to very high seismic demands. Minimizing damage to the transformer bushings, tanks, and other components and keeping the transformer in-service after a seismic event provides inherent value to the utility, as shown in Figure 7-3.

Seismic base isolation bearings provide a disconnect between the superstructure above grade and its supporting foundation. In a seismic event, loads are transferred from the soil to the foundation as a result of ground acceleration. Isolation devices included at the base of the superstructure allow movement of the foundation with minimal translation of horizontal loads and movement to the supported equipment, providing freedom from the ground motion and acceleration. Isolation devices should be sized according to site-specific seismic parameters and seismic demand from the supported equipment. Tests have indicated that base isolation can reduce the demands on the supported equipment by up to 50%.

7.6.9 Grounding

Grounding of substation equipment and structures is needed for many reasons, the most important of which is protecting people from contacting energized parts. In addition to

Figure 7-3. Transformer foundation with seismic base isolation.
Source: Courtesy of Seattle City Light.

protecting personnel, grounding provides protection to equipment, enclosures, and structures. Proper grounding also reduces electrical noise. Ground wires or rods provide a path for the flow of electrons from energized parts to the earth.

Substations are designed with a complex ground grid, composed of buried wires (typically copper), ground rods, ground wells, or other means of electrical grounding to provide a grid that is of sufficient size to safely disperse electrical currents to the earth. Energized equipment, static wires, structures, and any structure or object that has the potential to become energized is grounded to this system.

In some cases, structures have been grounded to the concrete encased reinforcing cage in the foundation. Testing has indicated that this is a suitable method for low fault currents in areas where ground rods may be less effective. For substation structures, however, it is recommended to ground structures directly to the ground grid within the substation. Fault currents travel through the path of least resistance from the reinforcing steel to the soil if the reinforcing steel is not grounded effectively to another available electrode. As the fault current follows this path, heat will be generated and can cause any moisture in the concrete to vaporize. As the water expands and turns to steam, the concrete may crack or spall. High fault currents (500 to 2,600 A) that have been tested with concrete-encased electrodes have experienced damage ranging from minor damage to complete destruction. Refer to IEEE 142-2007 (IEEE 2007) and IEEE 80-2013 (IEEE 2013) for additional discussion of concrete-encased electrodes.

7.6.10 National Electrical Safety Council District Loading and Foundation Design

A common area of confusion when designing foundations for wire-supporting structures is how to design for the *National Electrical Safety Code* (NESC) (IEEE 2023) district loading. The NESC 2023 Rule 250B districts (light, medium, and heavy) designate ice thicknesses, wind speeds, and wire temperatures to be applied to the wires and structures. Rule 253 specifies load factors to be applied to these loads. Rule 261 instructs the user to include strength factors on the basis of the element in question and the grade of construction.

The following interpretation of NESC 2023 is the opinion of the members of the ASCE 113 committee and is not necessarily the interpretation of the NESC 2023 committee. Before proceeding, it should be noted that NESC 2023 is a safety code, intended to provide minimum requirements to ensure safety, and is not intended to be a design code.

The application of load factors and strength factors in NESC 2023 Rules 253 and 261 lead the user to interpret the loading criteria as USD, or LRFD. This does not pose a problem for design of steel or concrete elements of foundations, as these are commonly designed using LRFD. But design checks for foundation soil capacity, foundation stability, and foundation serviceability are still predominantly performed using unfactored service loads and Allowable Strength Design (ASD) methodology rather than LRFD.

Foundation soil capacity, foundation stability, and foundation serviceability design can be performed using ASD with the load combinations specified in Chapter 3, Table 3-19, as both LRFD- and ASD-suggested combinations have been provided. However, Rule 250B loads are deterministic loads and are not compatible with the probabilistic load/resistance design concept. A common misconception is that the NESC 2023 load factors can simply be taken as 1.0 to derive unfactored NESC 2023 district loading, which can then be used for ASD foundation design. However, it is the understanding of this committee that the NESC 2023 committee did not intend for the Rule 253 load factors to be manipulated on the basis of design methodology. The loads should remain at ultimate design level without modification.

Using factored NESC 2023 district loads as ultimate loads, an LRFD approach can be accomplished for several typical foundation checks. When obtaining a geotechnical report, recommended ultimate capacities should be obtained to accommodate this design approach, such as a recommended ultimate bearing pressure for design of a dead-end spread footing.

However, it can be difficult to determine whether one should use these large, factored loads for serviceability and stability checks. Considering that NESC 2023 is a safety code and it specifies a strength factor of 1.0 for foundations when subjected to the Rule 250 loads, the apparent intent of the NESC 2023 loading and strength rules is to prevent failure of the foundation. Its intent is not to dictate serviceability requirements. Often, foundations have large excess capacity beyond the serviceability restrictions that are used in foundation designs. Because the NESC 2023 250B district loads are ultimate loads, it would be unnecessarily conservative to apply typical serviceability-level deflection/rotation/settlement limits to foundations when subjected to these loads (e.g., 0.5 in. pier top deflection limit, 0.5 degree rotation limit).

To help eliminate the confusion, Table 7-2 provides the user with guidance on the use of the different load combinations for typical foundation design checks. This table was created with dead-end/wire-supporting structures in mind, as these are the only substation structures to which NESC 2023 is applicable.

Table 7-2. Applicable Load Combinations for Design of Foundations for Line Termination Structures.

Foundation type	Design consideration	IEEE 2023 Rule 250B district loading[a]	Table 3-18 Load combinations	Table 3-19 Load combinations
Spread footing/ mat foundation	Soil capacity			
	• Ultimate bearing pressure	X	X	
	• Allowable bearing pressure			X
	Foundation stability[b]			
	• Overturning	X	X	X
	• Sliding	X	X	X
	Concrete/steel design			
	• Concrete/rein-forcement design	X	X	
	• Anchorage design	X	X	
Drilled pier/ pile	Soil serviceability			
	• Pier/pile top rotation			X
	• Pier/pile top deflection			X
	Soil strength/ foundation stability			
	• Foundation movement arrested by soil	X	X	
	Concrete/steel design			
	• Concrete/rein-forcement and steel design	X	X	
	• Anchorage design	X	X	

[a]IEEE 2023 Rule 250 B loads only apply to structures that support conductors that extend outside the substation. The column for Rule 250B would be used in conjunction with either the column for Chapter 3, Table 3-18 or Table 3-19.

[b]Foundation stability may be checked using either ASD or LRFD load combinations depending on the safety factor or resistance factor used.

REFERENCES

AASHTO (American Association of State Highway and Transportation Officials). 1978. *Manual on foundation investigations*. Washington, DC: AASHTO.

ACI (American Concrete Institute). 2014. *Report on design and construction of drilled piers*. ACI PRC-336.3. Detroit: ACI.

ACI. 2019. *Building code requirements for structural concrete (with commentary)*. ACI CODE-318-19. Detroit: ACI.

ACI. 2023. *Guide to durable concrete.* ACI PRC-201.2-23. Detroit: ACI.

ASCE. 2002. *Design loads on structures during construction.* ASCE 37. New York: ASCE.

IEEE (Institute of Electrical and Electronics Engineers). 2001. *Guide for transmission structure foundation design and testing.* IEEE 691. Piscataway, NJ: IEEE.

IEEE. 2007. *Recommended practice for grounding of industrial and commercial power systems.* IEEE 142-2007. Piscataway, NJ: IEEE.

IEEE. 2013. *Guide for safety in AC substation grounding.* IEEE 80-2013. Piscataway, NJ: IEEE.

IEEE. 2018. *Recommended practice for seismic design of substations.* IEEE 693-2018. Piscataway, NJ: IEEE.

IEEE. 2023. *National electrical safety code.* ANSI C2-2023. Piscataway, NJ: IEEE.

Sungjin, B., O. Bayrak, J. O. Jirsa, and R. E. Klingner. 2009. "Effect of alkali-silica reaction/delayed ettringite formation damage on behavior of deeply embedded anchor bolts." *ACI Struct. J.* 106 (6): 848–857.

USACE (US Army Corps of Engineers). 1984. *Pavement criteria for seasonal frost conditions.* EM 1110-3-138. Washington, DC: USACE.

USDN NAVFAC (US Department of the Navy, Naval Facilities Engineering Command). 1982. *Foundation and earth structures.* NAVFAC DM 7.02. Washington, DC: Government Printing Office.

USDOT FHWA (US Department of Transportation, Federal Highway Administration). 2006. *Soils and foundations reference manual.* Vol. 2. FHWA-NHI 06-089. Washington, DC: Government Printing Office.

USDOT FHWA. 2016. *Design and construction of driven pile foundations.* Vol 1. FHWA-NHI 16-064. Washington, DC: Government Printing Office.

USDOT FHWA. 2018. *Drilled shafts: Construction procedures and design methods.* FHWA-NHI 18-024. Washington, DC: Government Printing Office.

CHAPTER 8
CONNECTIONS TO FOUNDATIONS

To resist loads applied to structures and equipment, it is recommended that all substation structures and equipment be positively anchored to their foundations.

Connections of substation structures and equipment to their foundations should be designed with adequate strength and stiffness to provide stability, meet the serviceability requirements of Chapter 4, "Deflection Criteria (For Operational Loading)," of this Manual of Practice (MOP), and resist the forces defined in Chapter 3, "Loading Criteria For Substation Structures." Anchors should be proportioned to resist the governing combination of axial (tensile and compressive), shear, and bending forces that are applied to the anchor.

The utility may decide not to anchor equipment under 400 lb (181 kg) with a center of mass less than 4 ft (1.2 m) high or equipment whose displacement will not compromise the critical functions of adjacent equipment/systems or personnel safety.

Different types of anchorage are used to connect substation structures to their foundations. The most common means of transferring structure reaction forces to the foundations is by anchor rods. This type of anchorage provides a good transfer of load. Anchor rods (i.e., anchor bolts) may be configured to provide anchorage into concrete by means of a headed end, threaded with a nut on the end, a hook, or a straight length of deformed reinforcing bar. Cast-in-place headed rods or threaded rods with a nut on the end are the recommended anchor rod types as they can be designed to provide a ductile failure mode. Hooked rods have been reported to fail by anchor pullout in a brittle mode and are not recommended by this MOP. Anchor rods confined by the foundation reinforcement may increase the structural capacity of the anchorage system.

Specialized anchorage systems used in the industry also include the use of embedded standard shapes (W, WT, C, etc.), embedded plates with headed studs, or embedded plates attached to standard shapes.

Under certain conditions, anchorage may need to be installed in existing foundations using post-installed anchors such as adhesive anchors, expansion anchors, undercut anchors, and threaded anchors screwed into holes drilled in the concrete.

Stub angles and direct-embedded poles can also be used to transfer loads from the structure to the foundations and are addressed in ASCE 10-15 (ASCE 2015) and ASCE 48-19 (ASCE 2019b), respectively.

The approach for the design of anchor rods in this chapter is based on ultimate strength design loads (also referred to as *factored loads*). All equipment anchorage assemblies, including anchor rods, should be designed for forces determined from a structural analysis or tests, whichever is applicable.

8.1 FOUNDATION TYPES AND ANCHORAGE SYSTEMS

ACI CODE-318-19 (ACI 2019), Chapter 17, is the most common reference used for designing anchorage to concrete. Although other references are available, they are not as prevalent because ACI is the referenced standard in relevant building codes. ACI CODE-318-19, Chapter 17, does not address anchorage to all foundation types typically found in substations. The discussion that follows addresses areas when ACI CODE-318-19, Chapter 17, is relevant and when it is not. For design situations not addressed by ACI CODE-318-19, alternative methods or references are suggested.

ACI CODE-318-19, Chapter 17, is most applicable to anchorage in structural slabs with or without horizontally placed rebar. Anchorage consists of relatively short-headed anchor rods, hooked anchor rods, headed studs, expansion anchors, undercut anchors, or adhesive anchors. The ACI CODE-318-19, Chapter 17, design process fits these anchor types and provides a step-by-step methodology with good correlation to the figures presented and equations derived within that chapter.

Although these anchor types are common, they are not the only types of anchorage systems typically used in substations. The following is a general discussion, and not an all-inclusive list, of foundation types and anchor systems where anchorage design is determined using ACI CODE-318-19, Chapter 17, with modifications or different guidelines.

8.1.1 Spread Footing Foundation

A *spread footing foundation* (also referred to as a *pad and pedestal foundation*) typically has a square or rectangular horizontal footing (pad) with vertically oriented rebar that extends into the pedestal. The vertical rebar is typically enclosed with ties and within close proximity to the anchor rods (Figure 8-1).

The anchor rod tensile failure mode of concrete breakout is similar to that proposed by ACI CODE-318-19, Chapter 17; a postulated failure surface begins at the head of the anchor and is intercepted by the vertical rebar. The vertical rebar is used to transfer the loads from the headed anchor into the foundation. The required development length of the vertical rebar typically determines the length of the headed anchor rod. This follows ACI CODE-318-19, Chapter 17, requirements for anchor reinforcement, except that the vertical rebar is typically larger than the maximum pier reinforcement rebar size allowed in the ACI CODE-318-19, Commentary.

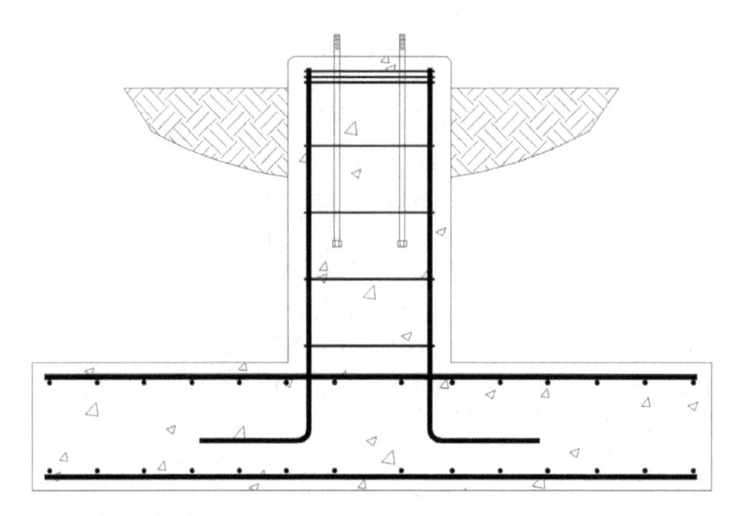

Figure 8-1. Spread footing foundation.

The tension side-face blowout failure mode is similar to that specified in ACI CODE-318-19, Chapter 17, but can be complicated if the design uses the available ties to evaluate this failure state. It is noted that *Anchorage Design for Petrochemical Facilities* (ASCE 2013) states that the transverse reinforcement does not increase the side-face blowout capacity, but that it just increases the magnitude of load-carrying capacity after side-face blowout has occurred (Section 3.5.2.f). The pullout failure mode as specified in ACI CODE-318-19, Chapter 17, is applicable to spread footing foundations.

The shear failure mode of concrete breakout from the anchor rod is similar to that proposed by ACI CODE-318-19, Chapter 17. Fracture begins at the face of the anchor rod, but that is where the similarity to ACI CODE- 318-19, Chapter 17, methodology ends. The distance to the free edge is typically small and a rebar tie is typically encountered. This is not adequately addressed by ACI CODE-318-19, Chapter 17. Also, the concrete pryout failure mode is not representative, as the anchor rods in substation structure foundations are typically long and flexible, and therefore, shear is not rigidly transferred to the anchor head to create shear pryout (Figures 8-2 and 8-3).

The guide *Anchorage Design for Petrochemical Facilities* (ASCE 2013) addresses various methodologies to assess the concrete strength for anchorages described previously. Its methodology is based on a strut–tie system that is found in ACI CODE-318-19 but not explicitly covered in Chapter 17.

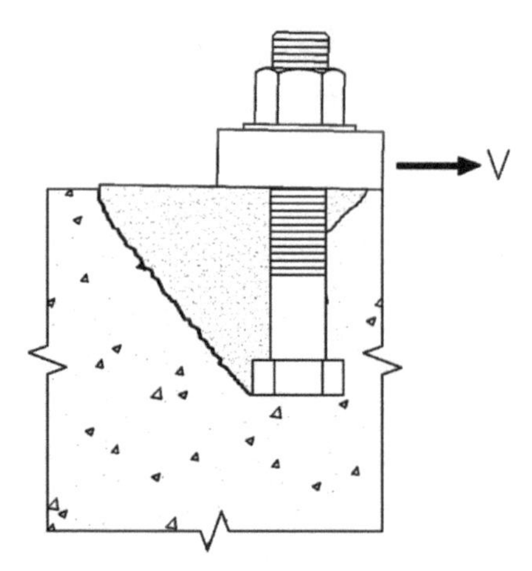

Figure 8-2. Shear pryout with a short anchor rod far from a free edge (not likely to control in substation structures).

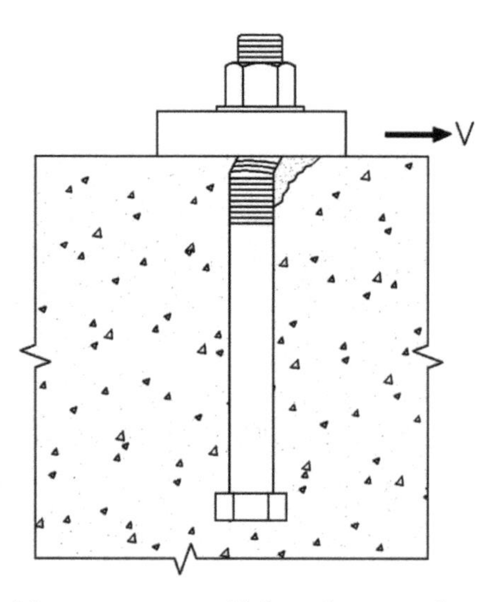

Figure 8-3. Steel failure preceded by concrete spall for a long anchor rod.

8.1.2 Drilled Pier Foundation

Another common foundation type is a circular *drilled pier foundation* (Figure 8-4). These are long cylindrical foundations with vertical reinforcing steel placed around the outside edge of the foundation with ties or spirals used to properly support the vertical steel during concrete placement and provide shear capacity for the pier. Typical anchorage consists of longer-headed anchor rods, headed studs, and threaded reinforcing steel with or without a nut at the embedded end. The anchor rods are typically placed inside the rebar cage in relatively close proximity to the vertical steel. The anchorage to the foundation load transfer mechanism for this type of foundation is the same as noted previously for the pedestal anchorage, except for the geometric complexity of a curved surface onto which failure planes are projected, instead of a flat surface. A Florida Department of Transportation (FDOT)/University of Florida (FDOT 2007) report BD545 RPWO No. 54 provides guidance on concrete breakout area determination for round piers, as shown in Figure 8-5.

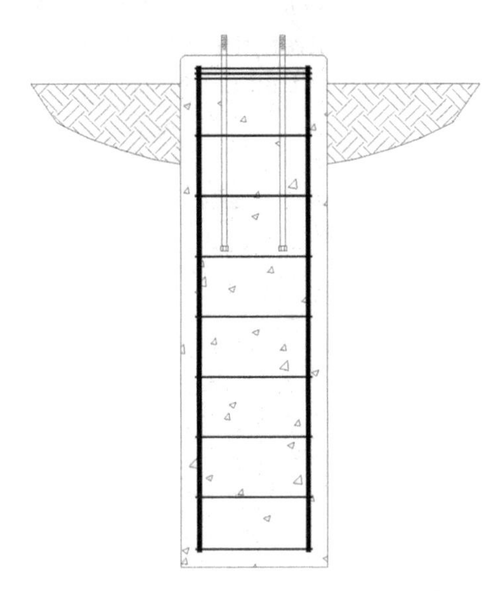

Figure 8-4. Drilled pier foundation.

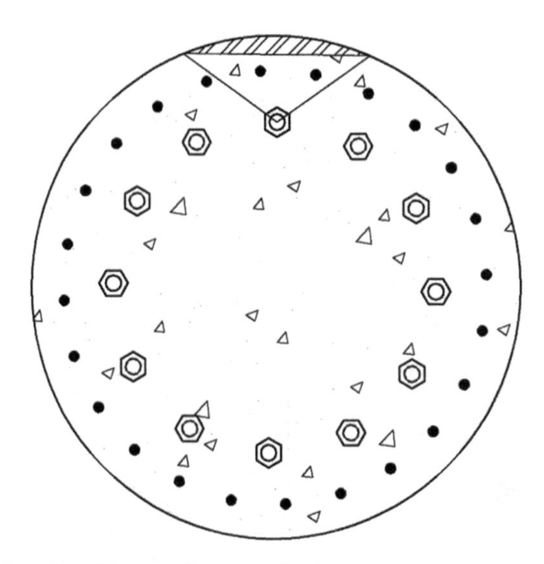

Figure 8-5. Concrete shear breakout area of a round pier.

8.1.3 Anchor Rods Installed without Grout Beneath Base Plates

Base plates of hollow structural sections (HSS) columns are often not supported directly on the concrete foundation or grout creating a condition not addressed in ACI CODE-318-19, Chapter 17. This practice mitigates concerns about corrosion resulting from water trapped inside hollow structural columns, as well as potential damage to column bases resulting from the expansion of any trapped water caused by freezing weather conditions. To level the column, leveling nuts are installed under the base plate, leaving a gap between the bottom of the base plate and the top of the foundation concrete. Such an arrangement subjects the anchor rods to bending. This arrangement and resulting load condition are not addressed in ACI CODE-318-19, Chapter 17. A suggested methodology for addressing this type of anchor system is provided in Section 8.4.2.2.

8.1.4 Embedded Structural Steel

Embedded structural shapes are sometimes used as anchors in seismically active regions. Embedded steel plates with welded studs can be used, but the required capacity cannot always be achieved using this type of system, thus requiring a more robust solution using embedded structural shapes, such as W-sections or channels. Welded studs may be used to anchor the embedded steel to the foundation concrete. ACI CODE-318-19 Chapter 17 does not provide specific guidance for calculating the concrete capacity of anchorage using embedded steel shapes. No well-established methods for designing such systems are currently available. The engineer may consider the resistance for various potential failure mechanisms such as concrete breakout and shear friction provided by reinforcing steel by using ACI CODE-318-19 as a guide. Such systems should be designed as non-ductile, unless a clearly defined ductile yielding mechanism can be identified.

8.2 ANCHOR MATERIALS

Specifying of the anchor material should be in accordance with the appropriate ASTM standard. The ASTM standard provides information that may guide the engineer in understanding the available manufacturer options, appropriate matching hardware, supplemental requirements, and other insights on the design and specification of anchors.

Information on anchor rod material may be available in ASTM F1554/F1554M (ASTM 2020b), ASTM A36/A36M (ASTM 2019a), ASTM A307 (ASTM 2021b), ASTM A193/A193M (ASTM 2023a), and ASTM A354/A354M (ASTM 2017b), with ASTM F1554/F1554M (ASTM 2020b) (Grade 55) being more commonly specified. ASTM F1554/F1554M (ASTM 2020a) is available in Grades 36, 55, and 105, with recommended heavy hex nuts conforming to ASTM A194/A194M (ASTM 2023b) or ASTM A563 (ASTM 2021a) and washers conforming to ASTM F436/F436M (ASTM 2019b). It should be noted that when ASTM F1554/F1554M (ASTM 2020b), Grade 36, is specified, a weldable Grade 55 may be furnished at the supplier's option except when the lower grade anchor rods are specified and designed to be ductile. The standard ASTM F1554/F1554M (ASTM 2020b), Supplement S1, lists requirements for weldable Grade 55 rods. Supplement S4 lists Charpy impact requirements for Grades 55 and 105. If any of these supplement requirements must be satisfied on a specific project, the engineer should list these requirements on the construction documents. Threads on anchor rods should be Unified Coarse Thread Series as specified in the latest issue of ANSI/ASME B1.1 (ANSI/ASME 2019),

Table 8-1. Anchor Material Properties.

Material ASTM standard	Yield strength, min F_y (kip/in.^{2})	Tensile strength F_u (kip/in.^{2})
ASTM A36/A36M	36	58–80
ASTM F1554/F1554M, Grade 36	36	58–80
ASTM F1554/F1554M, Grade 55	55	75–95
ASTM F1554/F1554M, Grade 105	105	125–150
ASTM A615/A615M, Grade 60	60	90 min
ASTM A615/A615M, Grade 75	75	100 min
ASTM A615/A615M, Grade 80	80	105 min
ASTM A706/A706M, Grade 60 (weldable)	60–78	80 min
ASTM A706/, Grade 80 (weldable)	80–98	100 min

Note: 1 kip/in.^{2} = 6.89 MPa.

with Class 2A tolerances. Nut dimensions, threads, and tapping threads oversize after hot-dip galvanizing should conform to requirements specified in ASTM A563 for Class 2B tolerances. ASTM F1554/F1554M (ASTM 2020b) and A36 (ASTM 2019a) steel rods can be supplied with forged heads up to a certain bar diameter on the basis of the manufacturer's capabilities. Dimensions of the forged head should correspond to the heavy-hex bolt head dimensions according to ANSI/ASME B18.2.1 (ANSI/ASME 2012). Testing requirements for forged heads must conform to ASTM F606 (ASTM 2021c). Charpy V-notch tests can be performed on the forged head when the rod material is specified with notch toughness requirements.

Table 8-1 shows the properties of the recommended anchor rod materials. Only one grade of bar should be used for each diameter to prevent the possibility of errors in the field during construction. Anchor rods should not be less than 0.75 in. (1.91 cm) in diameter, unless the design strength is greater than or equal to two times the required strength. Other materials with higher strengths or other desirable properties may be used when circumstances dictate, but the material supplier should be consulted for availability, compatibility, and any special design considerations associated with that material.

For anchor material that could experience extreme cold temperatures, the engineer may consider minimum Charpy V-notch impact strengths. It is up to the individual utility to determine whether the anticipated magnitude of low service temperatures warrants the specification of Charpy V-notch testing. The notch-toughness test temperature for cold weather applications is usually selected to be higher than the lowest anticipated exposure temperature. The *Steel Bridge Design Handbook: Bridge Steels and Their Mechanical Properties* (USDOT FHWA 2015) published by the Federal Highway Administration may be used as a guide to determine the appropriate notch-toughness test temperature for cold weather applications.

Anchor material to be used in areas with high seismic accelerations (ASCE 7-22) (ASCE 2022), Seismic Design Category C and D, Section 11.6 could be specified with minimum Charpy V-notch impact strengths when the Owner deems it is appropriate. Charpy V-notch testing may have an impact on anchor cost and procurement time.

Deformed reinforcing bar material used for anchor rods should be in accordance with ASTM A615/A615M (ASTM 2022a) or ASTM A706/A706M (ASTM 2022b). When high-tensile-strength, deformed reinforcing bars are used, they can be obtained with a minimum Charpy V-notch requirement of 15 ft·lb (20.3 N·m) at –20 °F (–28.9 °C) when tested in the longitudinal direction.

Shear stud connectors can be used to anchor plates to the concrete. Shear studs may be manufactured from ASTM A108 (ASTM 2018b) steel. The shear stud manufacturer should be consulted for available material grades and stud dimensions. Stud welding should conform to AWS D1.1 (AWS 2020).

A protective coating should be applied to steel anchorage used in an outdoor substation to resist corrosion. The ASTM specification corresponding to the anchor rod material should be consulted for permissible coating methods. The anchor rod projection above the top of concrete plus a minimum of 6 in. (15.24 cm) into the concrete foundation should be galvanized. Some galvanizers may prefer to galvanize the entire rod. Rods, nuts, washers, and steel hardware components should be hot-dip galvanized in accordance with ASTM A153 (ASTM 2023c) supplemented by ASTM F2329/F2329M (ASTM 2015) as appropriate, and reinforcing steel in accordance with ASTM A767/A767M (ASTM 2019d). Safeguard products against steel embrittlement in conformance with ASTM A143 (ASTM 2020a). Avoid the use of steel with an ultimate tensile strength greater than 150 ksi, as these steels have been shown to have a potential for hydrogen embrittlement resulting from the pickling process prior to galvanizing. Material that cannot be hot-dip galvanized may be galvanized by mechanically depositing a protective coating, if appropriate, for the application and environmental conditions and if approved by the owner. Steel with tensile strength greater than 150 ksi may be protected by coatings conforming to ASTM F1136/F1136M (ASTM 2019c) or ASTM F2833 (ASTM 2017a). It should be noted that the dimensions of anchor rods that can be mechanically galvanized may be limited and that this process does not produce the alloying layers of the steel and zinc that take place during the hot-dip galvanization process. With proper cleaning processes, mechanical deposition of zinc does not produce hydrogen embrittlement. Anchor rods and nuts must be supplied as an assembly and galvanized by the same process; that is, a hot-dipped rod should not be used with mechanically galvanized nuts, and vice versa.

8.3 ANCHOR ARRANGEMENTS AND GENERAL DESIGN CONSIDERATIONS

The definition of a ductile anchor has two components: ductile failure mode and ductile material. Both items need to be considered in the design of an anchor. A *ductile failure mode* is achieved by ensuring that the steel element governs the design and the concrete strength is not the controlling mode of failure. The use of ductile anchorage material is especially critical in seismic areas. For a specific anchor load, a design with ductile, lower-strength, larger-diameter anchor rods are preferred to that with less ductile, higher-strength, smaller-diameter anchor rods. Accordingly, it is recommended that anchorages for seismic applications use ductile steel elements, as defined in ACI CODE-318-19, Chapter 17. ASTM F1554/F1554M (ASTM 2020b), Grades 36 and 55 (all diameters), and deformed reinforcing bars must comply with ACI CODE-318-19, 17.10.5.3(a)(vi) and are considered ductile steel elements to resist earthquake effects in accordance with ACI CODE-318-19. When ASTM F1554/F1554M (ASTM 2020b), Grade 36 anchors, are specified and designed as ductile anchors in accordance with ACI CODE-318-19, 17.10.5.3(a), substitution with weldable ASTM F1554/F1554M (ASTM 2020b), Grade 55 (with Supplementary Requirements S1) anchors, is not permitted.

The two most commonly used arrangements to anchor a structure to the foundation are base plates supported by anchor rods with leveling nuts (Figure 8-6) and anchor rods with the base plates on concrete or grout (Figure 8-7). Anchor rods should be installed by using templates to ensure that the rods are installed in the proper arrangements in the foundation. The anchor rod installation tolerances listed in AISC 303-22 (AISC 2022a), the *Code of Standard Practice for Steel*

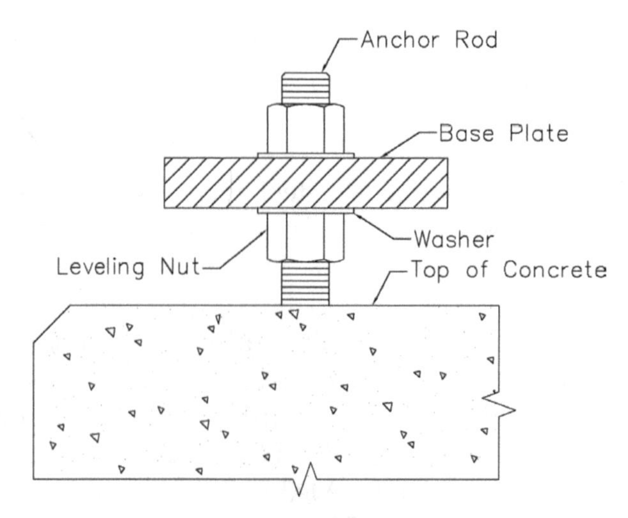

Figure 8-6. A base plate supported by anchor rods with leveling nuts.

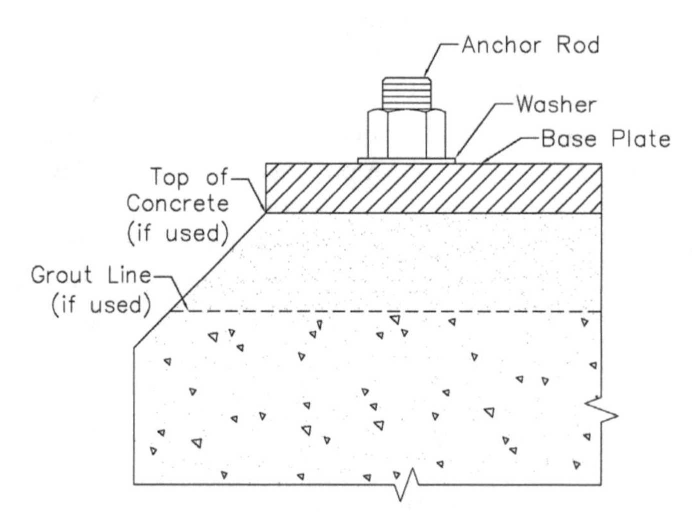

Figure 8-7. Anchor rods with a base plate on concrete or grout.

Buildings and Bridges, and ACI PRC-117.1-14 (ACI 2014), are compatible with the larger hole sizes that are recommended for base plates in the AISC *Steel Construction Manual* (AISC 2022b). The base plate hole sizes for substation structure anchor rods given in Table 6-1 in Chapter 6, "Design," Section 6.8.3, "Anchor Rod Holes in Base Plates," are smaller than those recommended by the AISC *Steel Construction Manual.* It is important to explicitly state the required anchor rod placement tolerances in the project contract documents. If the anchor rods are to be used to transfer the horizontal reaction forces at the substation structure base plates or equipment bases, the hole sizes should not exceed the oversize hole sizes listed in Table 6-1 in Chapter 6, Section 6.8.3. Structures with anchor rods are usually supplied with templates to locate the anchors in the concrete.

Anchors resisting seismic forces should be designed to be ductile in accordance with ACI CODE-318-19, 17.10.5.3(a). However, providing for a ductile failure may not be practical or economical in all situations. In such cases, anchors may be designed in accordance with ACI CODE-318-19, 17.10.5.3(b), (c), or (d). ACI CODE-318-19, 17.10.5.3(d), requires nonductile anchors to be designed for seismic loads amplified by an overstrength factor Ω_0. The overstrength factor to be used for substation structures is provided in Section 3.1.7.4.

The overstrength factor Ω_{PL} to be used for all IEEE 693 (IEEE 2018) qualified substation equipment is provided in Chapter 3, Sections 3.1.7.10.1a, "Design Level Qualifications," and 3.1.7.10.1b, "Performance Level Qualifications." For anchorage design at the Design Level, Ω_{PL} is 1.0 for ductile design and 1.5 for nonductile design. For anchorage design at the Performance Level, Ω_{PL} is 1.0 for both ductile and nonductile designs.

Engineers should exercise caution when dealing with eccentric loads on anchors. For example, when a single anchor is used at the leg of a support structure. The load path to the anchor point and resolution of all applied forces must be performed to prevent undesirable effects such as prying on anchors.

8.3.1 Base Plates Supported by Anchor Rods with Leveling Nuts

The use of leveling nuts to support substation structures has several advantages. This method eliminates the need for close tolerance work on the foundation elevation and trueness of surface while providing flexibility for the structure alignment because of fabrication tolerances and equipment installation. Washers should be installed between the leveling nut and the bottom of the base plate and between the top of the base plate and the top nut to spread the bearing load out from the nuts to the base plate. Leveling nuts also keep the base plate from resting in any standing water on the foundation and prevent trapping of moisture inside HSS or tubular columns. The disadvantage of using a leveling base is that the shear and axial loads must be resisted entirely by the anchor rods and base plates. These combined stresses can require larger anchor rods and thicker base plates for heavily loaded structures. Therefore, in some cases, the leveling base arrangement may not be the best alternative.

If the anchor rods have been designed to support the base plate on the leveling nuts, then it is not necessary to install nonshrink grout under the base plate, and the design procedures in Section 8.4.2.2 may be used. If grout is installed, the design procedures in Section 8.4.2.1 may be used.

The section of anchor rod between the nut above the base plate and the leveling nut under the base plate should be pretensioned. Studies have shown that this pretension improves fatigue strength and assures good load distribution among the anchor rods. The procedure for pretensioning is a turn-of-nut procedure. Table 8-2 indicates the recommended nut turn requirements for various anchor rod grades. Details on the nut tightening procedure may be

Table 8-2. Recommended Pretensioning of Anchor Rods.

	Nut rotation[a,b,c]	
Anchor rod diameter (in.)	F1554, Grade 36	F1554, Grades 55 and 105; A615, Grades 60, 75, and 80; and A706 Grades 60 and 80
$\leq 1\ 1/2$	1/6 turn	1/3 turn
$> 1\ 1/2$	1/12 turn	1/6 turn

[a]Nut rotation is relative to the anchor rod. The tolerance is plus 20 degrees.

[b]Applicable only to UNC threads.

[c]Beveled washer should be used if (1) the nut is not in firm contact with the base plate, and (2) the outer face of the base plate is sloped more than 1:40.

Source: NCHRP Report 469 (USDOT NCHRP 2022).

found in the National Cooperative Highway Research Program (NCHRP) Report 469 (USDOT NCHRP 2022). When it is required that the nuts be prevented from loosening, a jam nut or other suitable device can be used.

8.3.2 Anchor Rods with Base Plates on Concrete or Grout

The use of grout is beneficial in cases where large shear loads are present. The grout should be inspected regularly and replaced when necessary. The grout will typically have a much shorter life compared to the concrete in the foundation. Bolting the structure directly to the foundation requires close tolerances for the top of concrete elevation and a level surface. A bed of nonshrink grout can be used to ensure that uniform bearing is achieved. The nonshrink grout should be freeze–thaw damage resistant, as appropriate, and applied to a properly prepared concrete surface. Provisions should be made to allow drainage of moisture from under the base plate when the equipment support is detailed to require such drainage. The grout may provide a better load distribution beneath the base plate and better shear transfer between the base plate and the foundation and eliminates the bending in the anchor rods. Using friction to transfer shear forces is limited to nonseismic forces and should be used only in situations where the base plate is in compression.

After the concrete and grout below the base plate have reached the design strengths, and after ensuring that the threads on the anchor rods and nuts are clean, install washers and tighten nuts in a star pattern to the full effort of the installer using an ordinary wrench with a moderate length cheater bar. These anchors should not be fully tensioned as a base plate supported on leveling nuts. If the anchors are required to be pretensioned for better fatigue performance, then anchors installed in a sleeve (pipe sleeve) embedded in the concrete should be considered.

8.4 ANCHORS CAST IN PLACE

8.4.1 Types of Anchors

The three types of anchors most commonly used are headed rods, deformed reinforcing bar rods, and hooked rods (Figure 8-8).

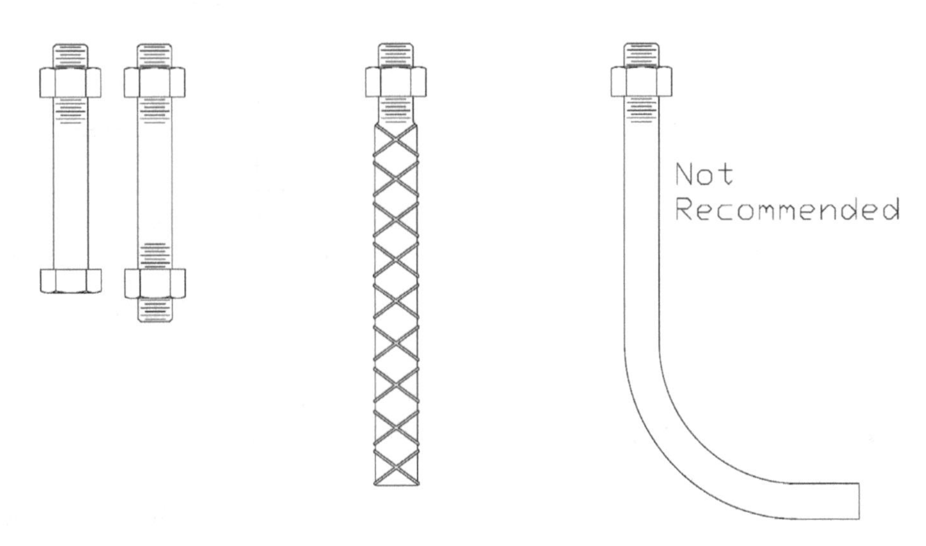

Figure 8-8. Types of anchors.

8.4.1.1 Headed Rods. The preferred method of anchorage is the use of headed anchors. These *anchor rods* consist of a rod with a head or nut at the end of the anchor embedded in concrete, where tensile forces (and compressive forces as in the case of base plates supported on leveling nuts) are resisted by a concrete breakout cone or anchor reinforcement provided adjacent to the anchor. This type of anchor requires adequate edge distance to provide the required concrete breakout cone and side-face blowout resistance, unless anchor reinforcement is provided. When nuts are used at the bottom of a threaded rod, it is necessary to prevent rotation of the embedded nut during construction. Methods for preventing rotation are deforming (peening or staking) the threads below the nut (unloaded side of rod), using a second nut as a locknut, or by tightening two nuts to sandwich an anchor/bearing or template plate. Structural nuts are typically heat-treated, and therefore, welding (including tack welding) them is not recommended. In addition, welding of the nut to the anchor rod is not an AWS D1.1/D1.1M prequalified welded joint. However, if welding is required, the material composition and heat treatment process used must be carefully considered, and job-specific specifications according to AISC and AWS D1.1/D1.1M must be developed.

8.4.1.2 Deformed Reinforcing Bar Rods. Deformed reinforcing bars used as anchor rods are typically No. 14 or No. 18 jumbo bars that are larger in diameter and heavier than the corresponding standard reinforcing bar size specified in ASTM A615/A615M (ASTM 2022a) or ASTM A706/A706M (ASTM 2022b). The larger diameter allows for full-depth threads to be cut into the deformed bar. The tension failure mode of these anchors does not follow concrete fracture mechanics as outlined in ACI CODE-318-19, Chapter 17. Rather, a bond stress failure is derived from interaction between the concrete and the deformed rebar, and pullout capacity is achieved by providing adequate development length of the bar. Guidance for this type of design is given in ACI CODE-318-19, Chapter 25, or in ASCE 48-19 (ASTM 2019). The shear failure modes are not addressed in ASCE 48-19 ASTM 2019 and should be addressed as previously described in Sections 8.1.1 and 8.1.2.

Deformed reinforcing bar rods are often used for anchoring multisided (polygon) tubular pole-type structures such as shield wire support masts and H-frame or A-frame dead-end structures to larger-diameter drilled pier foundations. This type of anchorage is typically provided in a preassembled anchor rod cage where the rods are securely held in place by templates.

8.4.1.3 Hooked Rods. Smooth-bar hooked rods are not recommended because of less predictable behavior in tension tests. However, ACI CODE-318-19, Chapter 17, does provide guidance in the design of hooked rods (both deformed and smooth).

8.4.2 Design Considerations for Anchor Steel

The method used to determine the area of anchor steel required to transfer the load from the structure base plate to the foundation depends on the base plate arrangement. One arrangement is the base plate bearing directly on the concrete or bed of nonshrink grout and held in place by nuts that are tightened on top of the base plate. If the base plate is in compression, this arrangement transfers a portion of the shear directly to the foundation through the friction between the base plate and the concrete. However, as discussed in Section 8.3.2, proper grout procedures should be specified, and bolt tightening information should be provided to ensure shear transfer occurs. If the base plate is in tension or subjected to seismic forces, the frictional resistance must be ignored, and anchor rods or shear lugs welded to the bottom of the base plate should be used for transferring the shear to the foundation. Another arrangement is the base plate supported by leveling nuts, which transfers the shear to the concrete by the side-bearing pressure of the anchor rod.

8.4.2.1 Anchor Rods with Base Plate on Concrete or Grout. The following equations are used to determine the maximum strength of the anchor rod on the basis of the provisions of ACI CODE-318-19, Chapter 17. The use of AISC 360-22 (AISC 2022c), Chapter J, to size anchor rods could result in anchor capacities that exceed those of ACI CODE-318-19:

$$\frac{N_{ua}}{\phi N_n} + \frac{V_{ua}}{\phi V_n} \leq 1.2 \tag{8-1}$$

$$\frac{N_{ua}}{\phi N_n} \leq 1.0 \tag{8-2}$$

$$\frac{V_{ua}}{\phi V_n} \leq 1.0 \tag{8-3}$$

where

N_{ua} = Ultimate tensile force (factored design tensile force) per anchor rod (kips or kN);
V_{ua} = Ultimate shear force (factored design shear force) per anchor rod (kips or kN);
ϕ = Strength reduction factor;
N_n = Nominal tensile strength of each anchor rod (kips or kN);
V_n = Nominal shear strength of each anchor rod (kips or kN);
ϕN_n = Design tensile strength (available capacity) of each anchor (kips or kN); and
ϕV_n = Design shear strength (available capacity) of each anchor (kips or kN).

The design tensile strength (ϕN_n) is determined by

$$\phi N_n = \phi f_{uta} A_{se,N} \tag{8-4}$$

where

f_{uta} = Lesser of: f_{ua}, the ASTM-specified tensile strength of the anchor; 1.9 f_{ya}, where f_{ya} = ASTM specified yield strength of the anchor; and 125,000 psi (860 MPa);
$A_{se,N}$ = Effective cross-sectional area of anchor rod in tension (in.^{2} or mm^2); and
ϕ = 0.75 for ductile steel element or 0.65 for brittle steel element.

For threaded rods and headed rods, the effective cross-sectional area $A_{se,N}$ (in.^{2} or mm^2) is given by

$$A_{se,N} = \frac{\pi}{4}\left(d_a - \frac{0.9743}{n_t}\right)^2 \tag{8-5}$$

where d_a is the nominal anchor rod diameter (inch or mm), and n_t is the number of threads per inch or mm of the anchor rod.

For anchors with reduced cross-sectional areas anywhere along the length, the effective cross-sectional area should be provided by the anchor manufacturer.

The ultimate shear force per anchor rod V_{ua} is determined by

$$V_{ua} = \frac{V - \mu N_{cm}}{n_a} \tag{8-6}$$

where

V = Total shear force at the base plate (kips or kN);
N_{cm} = Compressive force on the base plate (kips or kN) concurrent with the shear force V (*Note*: $N_{cm} \geq 0$. If the base plate is subjected to net uplift, $N_{cm} = 0$);
$\mu = 0.4$ the static coefficient of friction between the base plate and the concrete (Figure 8-7); for shear force V that includes seismic load combinations, $\mu = 0$;
n_a = Number of anchor rods at the base plate;
$\mu^* N_{cm} \leq 0.2 f_c' A_c \times 10^{-3}$ or $800 A_c \times 10^{-3}$ (where A_c is in in.^{2}) or $5.5 A_c \times 10^{-3}$ (when A_c is in mm^2);

where f_c' is the specified compressive strength of concrete (psi or MPa), and A_c is the area of concrete section resisting shear transfer (in.^{2} or mm^2).

The design shear strength (ϕV_n) is determined as follows.

For cast-in headed stud anchors with the base plate on concrete without grout,

$$\phi V_n = \phi f_{uta} A_{se,V} \tag{8-7}$$

These types of anchors use an embedded plate in the concrete where equipment (such as transformers or circuit breakers) are welded to the embedded plate and the studs are anchored to the concrete foundation by the heads on the studs. Headed studs welded to an embedded plate should not be used on grout.

For cast-in headed rod anchors with the base plate on concrete without grout,

$$\phi V_n = \phi 0.6 \ f_{uta} \ A_{se,V} \tag{8-8}$$

For cast-in headed rod anchors with the base plate on grout,

$$\phi V_n = \phi 0.48 \ f_{uta} \ A_{se,V} \tag{8-9}$$

where $\phi = 0.65$ for ductile steel elements; $\phi = 0.60$ for brittle steel elements, and $A_{se,V}$ is the effective cross-sectional area of an anchor rod in shear (in.^{2} or mm^2).

For threaded rods and headed rods, the effective cross-sectional area $A_{se,V}$ (in.^{2} or mm^2) is given by

$$A_{se,V} = \frac{\pi}{4}\left(d_a - \frac{0.9743}{n_t}\right)^2 \tag{8-10}$$

8.4.2.2 Base Plate Supported by Anchor Rods with Leveling Nuts. Anchor rods used with leveling nuts should be designed taking into consideration tension, compression, shear, and bending. The design of a base plate supported by anchor rods with leveling nuts differs from the design of a base plate bearing on concrete (or grout) because shear is not resisted by friction between the base plate and the concrete.

The methodology for the design of anchor rods with leveling nuts is described in this section. The moment arm for anchors in bending is determined on the basis of fixity of the rod to the concrete, as shown in Figure 8-9 or 8-10. Each anchor rod is assumed to equally share the shear force from the column base plate. For this assumption to be valid, the leveling nut must be snugged to the bottom of the base plate and the top nut tightened per the turn-of-nut method in Table 8-2.

$$L_{\mathrm{RGD}} = h - t_W - t_n + d_a \quad \text{(Figure 8-9)} \tag{8-11}$$

$$L_{\mathrm{RGD}} = h - t_W - t_n \quad \text{(Figure 8-10)} \tag{8-12}$$

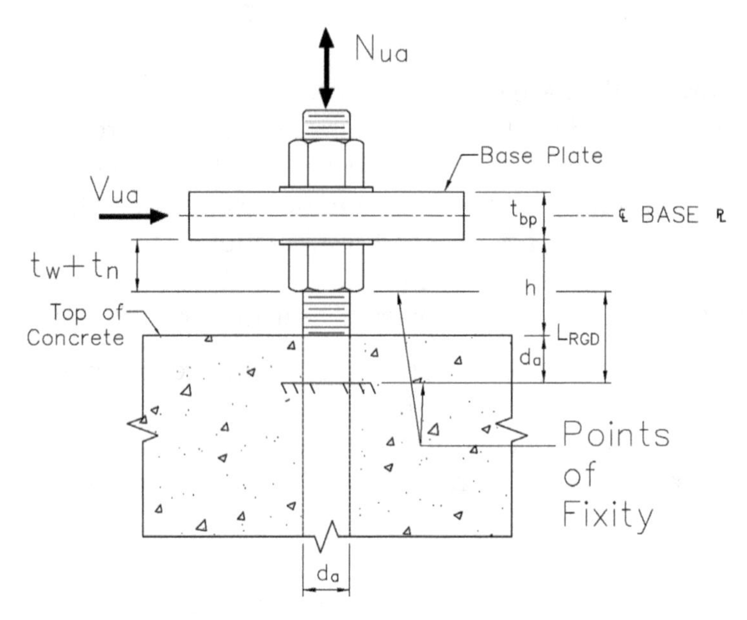

Figure 8-9. Moment arm without a clamping nut.

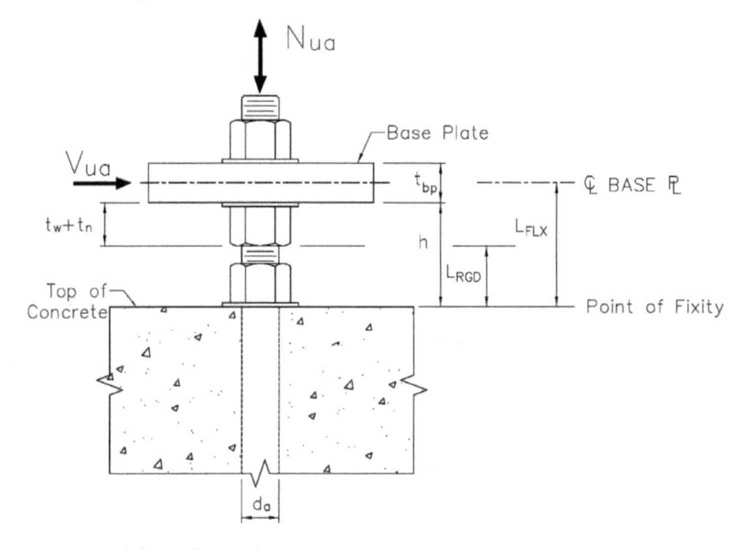

Figure 8-10. Moment arm with a clamping nut.

$$L_{\text{FLX}} = h + \frac{t_{bp}}{2} + d_a \quad \text{(Figure 8-9)} \tag{8-13}$$

$$L_{\text{FLX}} = h + \frac{t_{bp}}{2} \quad \text{(Figure 8-10)} \tag{8-14}$$

where

L_{RGD} = Moment arm for rigid base plate behavior (Figure 8-12);
L_{FLX} = Moment arm for flexible base plate behavior (Figure 8-11);
h = Distance from the bottom of the base plate to the top of concrete;
t_w = Thickness of the washer;
t_n = Thickness of the nut;
t_{bp} = Thickness of the base plate; and
d_a = Diameter of the anchor rod.

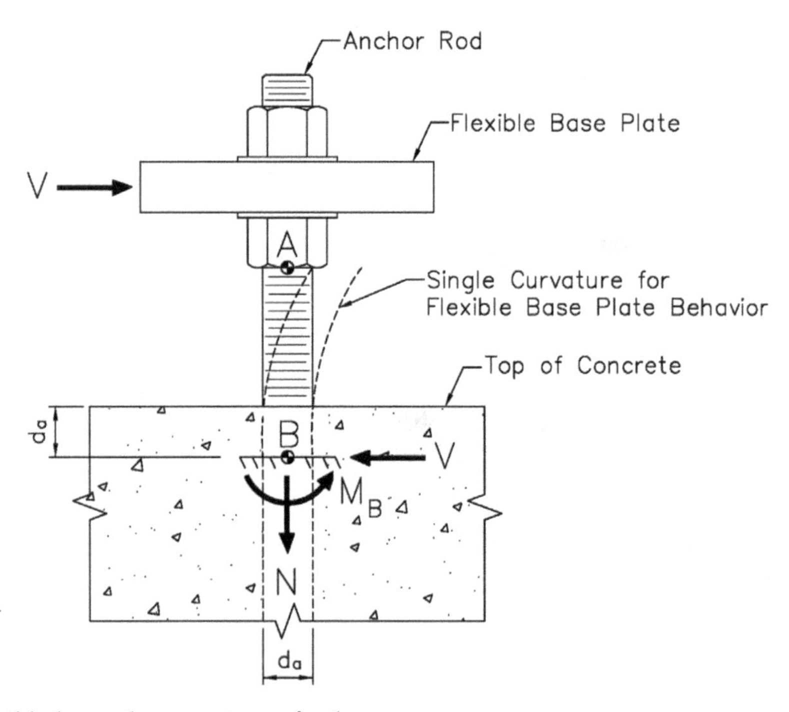

Figure 8-11. Flexible base plate—rotates freely.
Note: The anchor rods are shown elongated and deflection exaggerated to illustrate the deformed shape under a shear load. The gap between the T.O.C. and the base plate should be minimized to limit anchor rod bending.

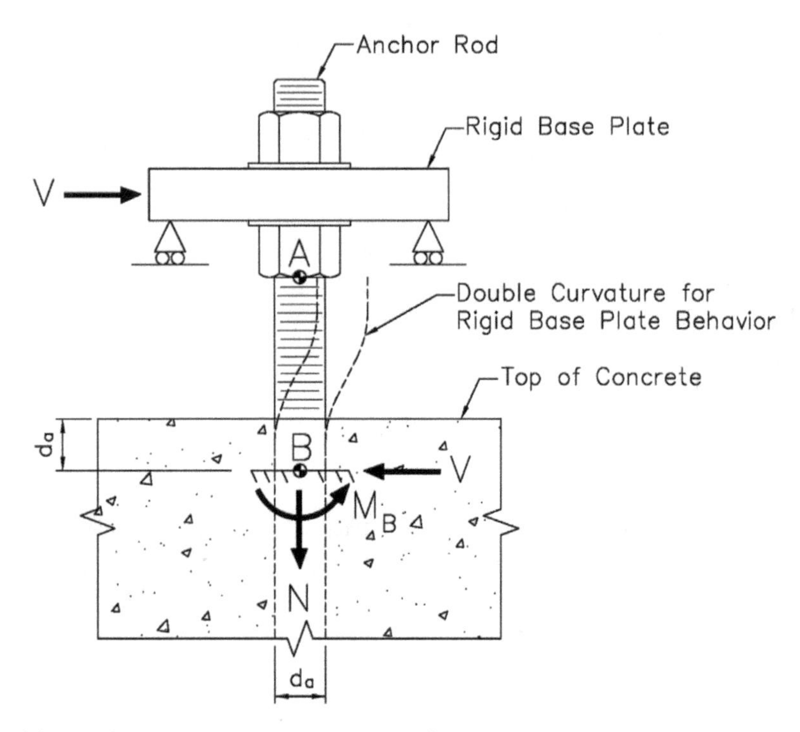

Figure 8-12. Rigid base plate—rotation is restrained.
Note: The anchor rods are shown elongated and deflection exaggerated to illustrate the deformed shape under a shear load. The gap between the T.O.C. and the base plate should be minimized to limit anchor rod bending.

The bending capacity of an individual anchor is given by the following equation:

$$M_0 = (1.2)Sf_u \quad \text{(only when the anchor cross section is constant)} \tag{8-15}$$

where M_0 is the bending capacity of an individual anchor.

$$S = \frac{\pi}{32}\left(d_a - \frac{0.9743}{n_t}\right)^3 \quad \text{(section modulus of the anchor rod)} \tag{8-16}$$

where

d_a = Nominal anchor diameter (in. or mm),
n_t = Number of threads per inch or millimeter, and
f_u = Specified tensile strength of the steel anchor (psi or MPa).

The bending capacity that is reduced owing to an axial force in the anchor is given by the following equation:

$$M_n = M_0\left[1 - \frac{N_{ua}}{\phi_a N_n}\right] \tag{8-17}$$

where

M_n = Reduced bending capacity with an axial force,
N_{ua} = Ultimate axial force in an individual anchor (compression or tension), and
ϕ_a = Strength reduction factor, 0.75 for a ductile steel element and 0.65 for a brittle steel element.

$$N_n = A_{se,N}\, f_{uta} \text{ (the nominal axial capacity of an individual anchor)} \tag{8-18}$$

$$A_{se,N} = \frac{\pi}{4}\left(d_a - \frac{0.9743}{n_t}\right)^2 \tag{8-19}$$

where f_{uta} is the lesser of f_{ua}, the specified tensile strength of the anchor, 1.9 f_{ya}, where f_{ya} is the specified yield strength of the anchor, and 125,000 psi (860 MPa).

The nominal shear force per anchor V_{nm} as limited by the bending moment may be determined by using the following equation:

$$V_{nm} = \frac{M_n}{L} \tag{8-20}$$

where L is 1/2 L_{RGD} for rigid base plate behavior, and L is L_{FLX} for flexible base plate behavior.

The ultimate shear force V_{ua} as limited by the bending moment is determined as

$$V_{ua} \leq \phi_v V_{nm} = \frac{\phi_v M_n}{L} \tag{8-21}$$

where ϕ_v is the strength reduction factor, 0.65, for a ductile steel element and 0.60 for a brittle steel element. Substituting the reduced bending capacity from Equation (8-17) into Equation (8-21) and solving for unity result in Equation (8-23).

$$V_{ua} \leq \frac{\phi_v M_n}{L} = \frac{\phi_v M_0 \left[1 - \frac{N_{ua}}{\phi_a N_n}\right]}{L} \tag{8-22}$$

$$\frac{N_{ua}}{\phi_a N_n} + \frac{V_{ua} L}{\phi_v M_0} \leq 1 \tag{8-23}$$

$$\frac{N_{ua}}{\phi_a N_n} + \frac{V_{ua}}{\phi_v V_n} \leq 1.2 \tag{8-24}$$

Equation (8-24) is the same as Equation (8-1), and it is provided here as an additional check for adequacy of the anchor rod, where

$$V_n = 0.6 f_{uta} A_{se,V} \tag{8-25}$$

$$A_{se,V} = \frac{\pi}{4}\left(d_a - \frac{0.9743}{n_t}\right)^2 \tag{8-26}$$

A rigid base plate should be designed such that the material remains elastic and the yield stress is not exceeded for the ultimate loads (factored or strength level loads). A flexible base plate is one where the plate stresses exceed the yield stress of the material, but the base plate does not rupture. The design of rigid base plates can be found in Section 6.8, "Base Plate Design."

The previous edition of this ASCE MOP 113 (ASCE 2008, Section 7.3.2.2), stated, "If the clearance between the base plate and concrete exceeds 2 times the bolt diameter, then a bending stress and buckling analysis of the bolts is required." The two-times-the-bolt-diameter option is provided in the commentary (C9.3.3) of ASCE 48-19.

A bending stress check of anchor rods is now recommended in this revision of the ASCE MOP 113, regardless of the anchor rod standoff distance, along with a modification to the determination of the location of the inflection point is determined and the ultimate stress is calculated. The stated modifications are offered as a way to provide a more complete design and better align anchorage checks with design standards outside of the utility industry whose structures may share loading similar in nature. It is important to note that, although these changes imply more conservative checks than the previous version of ASCE 113, the net result will, in most cases, show little or no increase in the required anchor rod size. In fact, the lever arm used in anchor bending equations for rigid base plate behavior, as determined in this Section 8.4.2.2, will still produce nearly the same bending value as the previous version of this MOP using a different lever arm geometry (Figure 8-13). The primary difference is that the points of fixity are moved to the bottom of the leveling nut and one diameter into the concrete, as opposed to the previous method at the bottom of the base plate and the top of the concrete.

Consideration should also be given for flexural buckling of anchors on leveling nuts under compressive loads. The procedures outlined in AISC 360-22 (AISC 2022c), Chapter E, "Design of Members for Compression," for flexural buckling of members without slender elements may be used to verify the buckling capacity of anchor rods.

The maximum gap between the bottom of the base plate and the top of the concrete foundation that was used to select the anchor rod diameter should be listed on the construction drawings to ensure that the anchor rods meet the design conditions.

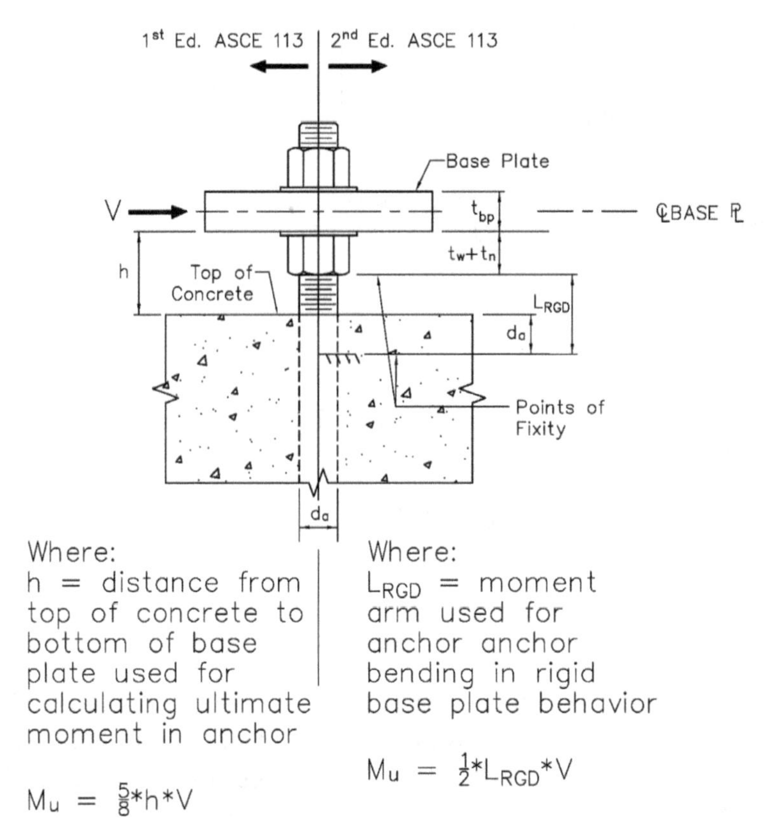

Figure 8-13. Anchor rod lever arm geometry.

8.4.3 Design Considerations for Concrete

The capacity of a cast-in-place anchor rod is limited by either the capacity of the steel anchor rod or the concrete. In the previous section, the strength limitations of the anchor rods were discussed. In this section, the concrete limitations are considered.

8.4.3.1 Capacity of Concrete. The procedure in Chapter 17 of ACI CODE-318-19 is adequate for the type of structures covered by this MOP. In ACI CODE-318-19, the limits on the headed anchor rod diameter mentioned in Chapter 17 are based on the extent of the test data that were used in developing the Chapter 17 provisions. This includes determination of anchor pullout or bond failure, concrete tension breakout (where not resisted by development of adjacent vertical reinforcement as discussed in Section 8.4.3.4), concrete shear breakout, concrete side-face blowout, and splitting.

8.4.3.2 Design of Edge Distance. Chapter 17 of ACI CODE-318-19 should be used to determine the minimum edge distance.

8.4.3.3 Design of Anchor Spacing. Chapter 17 of ACI CODE-318-19 should be used to determine the minimum anchor spacing.

8.4.3.4 Anchor Rod Embedment Length. The embedment for anchor rods should be designed to resist concrete tension breakout or have adequate length to transfer the applied axial force to the foundation reinforcement.

The deformed bar anchor rod transfers the structural loads to the foundation by the mechanical anchorage of the deformations, which bear against the adjacent concrete and prevent the longitudinal movement of the bar with respect to the concrete. Headed anchor rods transfer the structure loads to the foundation through pullout resistance, which is accomplished by bearing of the embedded nut or rod head on the concrete.

Concrete breakout capacity is often inadequate due to the small edge distances often used for substation foundations. Resistance must be provided by transferring the anchor rod axial force to the adjacent rebar. This may be accomplished by treating the vertical reinforcement of the foundation as anchor reinforcement in accordance with ACI CODE-318-19. Sufficient adjacent reinforcement must be provided to transfer the full anchor rod axial force using a strength reduction factor of 0.75. In addition to providing adequate anchor rod embedment, the vertical reinforcement of the foundation should have sufficient length above and below the failure plane extending from the head of the anchor rod to develop the reinforcement in the case of headed rods (Figure 8-14) or to develop the noncontact lap splice with the adjacent rebar in the case of deformed bar anchors rods. The required development length (l_d) of the vertical reinforcement in drilled shafts and spread footings is provided by ACI CODE-318-19, Chapter 25.

8.4.3.5 Concrete Punch Out from Anchor Rods. Anchor rods on leveling nuts without grout are subjected to significant tensile and compressive forces arising primarily from the moment at the base of the column. The downward compressive force is resisted by the headed anchor. Unlike the case where grout provides the means for transferring compressive forces directly to the top of concrete, the ungrouted case requires that the anchor rod itself transfer the downward force into the concrete foundation. If there is insufficient concrete shear capacity beneath the headed anchor that is in compression, a punching shear failure may occur (Figure 8-15). This is analogous to the concrete tension breakout failure presented in ACI CODE-318-19. If the concrete shear capacity is insufficient, anchor reinforcement that intercepts the potential failure surface created by the downward force on the anchor rod should be provided beneath the headed anchor and extending upward into the concrete. Such anchor reinforcement may be designed in a manner similar to an anchor rod in tension. If anchor plates are provided at the embedded end of anchor rods subjected to compression, heavy hex nuts should be installed above and below the anchor plate. Heavy hex nuts (as

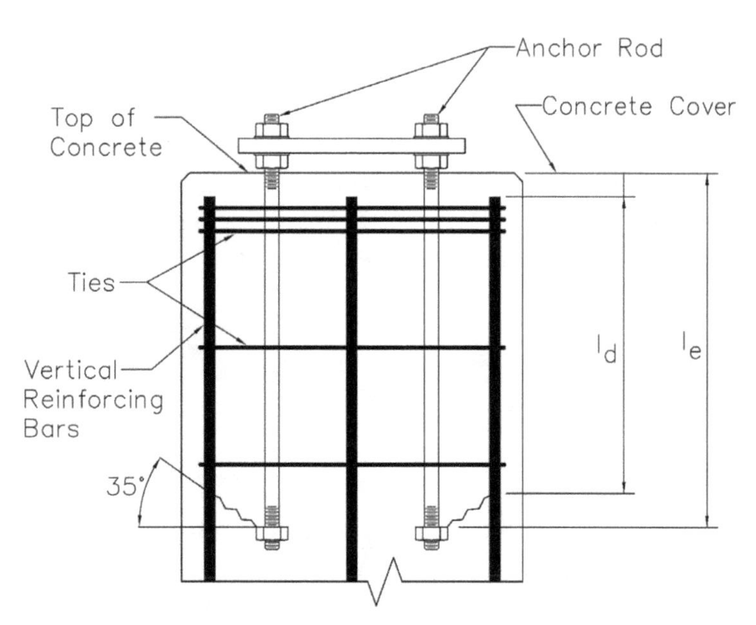

Figure 8-14. Development length of vertical reinforcing steel in drilled shafts and spread footings.

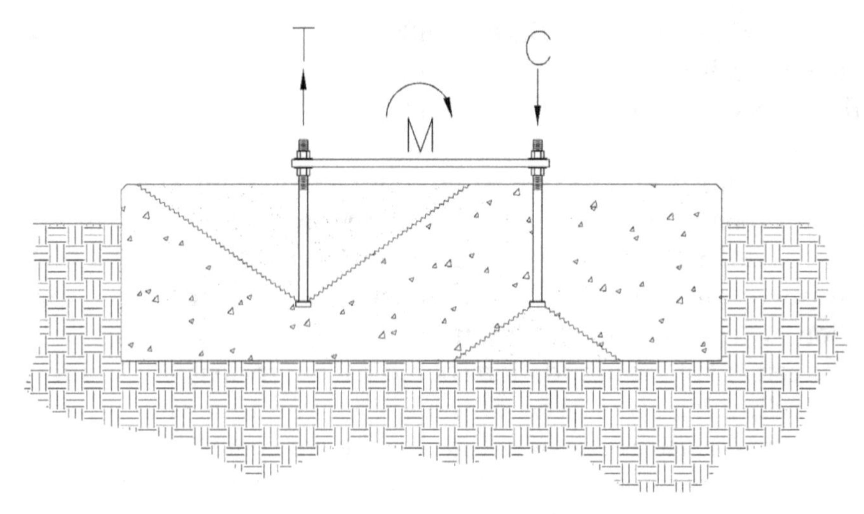

Figure 8-15. Concrete breakout from tension and compression.

opposed to hex nuts) are needed to ensure that there is sufficient thread length to transfer the required axial forces in the anchor rods.

For additional information, consult ACI CODE-318-19.

8.4.3.6 Localized Bearing Failure. Chapter 17 of ACI CODE 318-19 presents procedures for analyzing the failure mode of anchor pullout, which for a deeply embedded headed anchor rod, would present as a localized bearing failure. This can occur for anchors in tension or for anchors in compression as in the case of base plates supported on leveling nuts. Headed anchor rods made from ASTM A36/A36M (ASTM 2019a) or ASTM F1554/F1554M (ASTM 2020b), Grade 36 steel, can normally be used without an anchor-bearing plate, but higher-strength bolt materials may need an anchor-bearing plate to avoid localized bearing failure. An anchor-bearing plate, preferably a round-shaped plate instead of a square one, can be used to increase the headed anchor capacity in accordance with ACI CODE-318-19, Chapter 17. The plate should be sized in accordance with ACI CODE-318-19, Chapter 17. Anchor-bearing plates (or templates used to hold anchor rods in position) with large surface areas can adversely reduce the concrete shear capacity of drilled piers and pedestals by creating an inherently weak plane. A center hole in the template should be used to permit adequate concrete flow around the anchor rods during foundation construction. The projected area of the template should not be included in the effective area of concrete used for computing the nominal concrete shear strength V_c for the cross-section strength of the reinforced concrete pier. Reinforcing steel across the potential plane of weakness may be used to provide sufficient shear resistance. When anchor rods are supported on leveling nuts, anchor plates may require nuts above the bearing plate to transfer downward forces to the concrete by bearing.

8.5 POST-INSTALLED ANCHORS IN CONCRETE

8.5.1 Types and Application

Post-installed anchors include a variety of proprietary anchor rods or anchoring systems that are installed in hardened concrete. Commonly used anchor types include expansion, screw, undercut, and adhesive anchors. These different anchor types derive their load resistance from different mechanisms, for example, expansion anchors depend on expanding shells or wedges to generate frictional resistance, and adhesive anchors depend on the bond between the embedded anchor and the concrete. In addition, there are variations within each type noted previously.

Flexibility in locating the item to be anchored and ease of installation are among the main advantages of post-installed anchors. For example, adhesive anchor systems are often used to bond dowels to existing concrete such that a new concrete element may be connected to it. Post-installed anchors are often used in retrofit situations.

Disadvantages of post-installed anchors include relatively low strength, limits on the available diameters of anchors, difficulty in achieving ductile anchor performance, limitations on the depth of embedment for which testing has been performed, and poorer performance in cracked concrete.

In general, post-installed anchors should be installed in strict adherence to the manufacturer's instructions and in applications for which they are intended. Anchor systems may be sensitive to hole size, depth, cleanliness, presence of moisture, orientation of installation (mainly for adhesive anchors), fire-proofing requirements, type of loading (static, seismic, vibratory, etc.), group/spacing effects, torque requirements, minimum spacing and edge distances to preclude splitting, and other details of installation. Anchors used in outdoor applications should be hot-dip galvanized or made from stainless-steel materials, unless specified otherwise.

8.5.2 Design

In general, the manufacturer of an anchor system would have conducted a test program to develop capacities for various design conditions. Such test programs are also needed to develop the values of parameters required for anchor design in codes and standards such as ACI CODE-318-19 and the International Building Code (IBC). In the United States, the International Code Council (ICC 2018) evaluates manufacturer-furnished data and certifies post-installed anchors for use in various applications, including those governed by building codes.

It is recommended that post-installed anchors be designed in accordance with the requirements of ACI CODE-318-19, Chapter 17, and the applicable portions of Section 8.4.2 of this MOP. For anchors with base plates on leveling nuts, the procedure of Section 8.4.2.2 also applies to post-installed anchors. Anchors should be certified by the International Code Council (ICC) for use in cracked concrete, and if suitable for seismic applications. In many cases, post-installed anchors will be non-ductile.

8.5.3 Installation

To limit the damage to existing reinforcing steel, it is prudent for the engineer to place limitations on the number and orientation of rebar that are allowed to be cut or damaged during installation of the anchors. Pre-installation survey/scans may be used to locate existing rebar. The engineer may also wish to incorporate the means for relocating a post-installed anchor in the field by providing pre-punched alternate rod holes in base plates and anchor brackets. If rebar is encountered, an alternate hole that is offset a few inches away from the original hole may be used, and the original hole abandoned. Instructions for grouting abandoned holes in concrete should also be provided.

Adhesive anchors may be sensitive to the orientation of the hole. Limitations on the orientation of the hole are typically provided in the ICC report or manufacturer's instructions. ACI CODE-318-19 and building codes such as the IBC require that adhesive anchors resisting sustained tension loads and installed in a horizontal or upwardly inclined orientation be installed by personnel certified in accordance with the ACI/CRSI Adhesive Anchor Installation (AAI) program (ACI CPP 681.1-17). ACI CODE-318-19 also requires that the concrete in which adhesive anchors are installed be at least 21 days old.

Anchors should be installed in strict adherence to the manufacturer's instructions. Because of the sensitivity of the performance of anchors to the details of installation, it is recommended that post-installed anchors be subjected to a construction quality assurance program. Inspections should cover installation procedures, use of specified materials, and sample pull-testing for critical applications.

REFERENCES

ACI (American Concrete Institute). 2014. *Guide for tolerance compatibility in concrete construction.* ACI PRC-117.1-14. Detroit: ACI.

ACI. 2019. *Building code requirements for structural concrete (with commentary).* ACI-318-19. Detroit: ACI.

ACI (American Concrete Institute)/CRSI (Concrete Reinforcing Steel Institute). 2017. *Certification policies for adhesive anchor installer.* ACI CPP 681.1-17. Detroit: ACI.

AISC (American Institute of Steel Construction). 2022a. *Code of standard practice for steel buildings and bridges.* AISC 303-22. Chicago: AISC.

AISC. 2022b. *Steel construction manual.* 16th ed. Chicago: AISC.

AISC. 2022c. *Specification for structural steel buildings.* AISC 360-22. Chicago: AISC.

ANSI (American National Standards Institute)/AISC. 2022. *Specification for structural steel buildings.* AISC 360-2022. Chicago: AISC.

ANSI/ASME (American Society of Mechanical Engineers). 2019. *Unified inch screw threads, UN, UNR, and UNJ thread forms.* ANSI/ASME B1.1. New York: ASME.

ANSI/ASME. 2012. *Square, hex, heavy hex, and askew head bolts and hex, heavy hex, hex flange, lobed head, and lag screws (inch series).* ANSI/ASME B18.2.1 (R2021). New York: ASME.

ASCE. 2008. *Substation structure design guide,* MOP 113. Reston, VA: ASCE.

ASCE. 2013. *Anchorage design for petrochemical facilities.* Reston, VA: ASCE.

ASCE. 2015. *Design of latticed steel transmission structures.* ASCE 10-15. Reston, VA: ASCE.

ASCE. 2019. *Design of steel transmission pole structures.* ASCE 48-19. Reston, VA: ASCE.

ASCE. 2022. *Minimum design loads and associated criteria for buildings and other structures.* ASCE 7-22. Reston, VA: ASCE.

ASTM International. 2015. *Standard specification for zinc coating, hot-dip, requirements for application to carbon and alloy steel bolts, screws, washers, nuts, and special threaded fasteners.* ASTM F2329/F2329M. West Conshohocken, PA: ASTM.

ASTM. 2017a. *Standard specification for corrosion protective fastener coatings with zinc rich base coat and aluminum organic/inorganic type.* ASTM F2833. West Conshohocken, PA: ASTM.

ASTM. 2017b. *Standard specification for quenched and tempered alloy steel bolts, studs, and other externally threaded fasteners.* ASTM A354/A354M. West Conshohocken, PA: ASTM.

ASTM. 2018b. *Standard specification for steel bar, carbon and alloy, cold-finished.* ASTM A108. West Conshohocken, PA: ASTM.

ASTM. 2019a. *Standard specification for carbon structural steel.* ASTM A36/A36M. West Conshohocken, PA: ASTM.

ASTM. 2019b. *Standard specification for hardened steel washers inch and metric dimensions.* ASTM F436/F436M. West Conshohocken, PA: ASTM.

ASTM. 2019c. *Standard specification for zinc/aluminum corrosion protective coatings for fasteners.* ASTM F1136/F1136M. West Conshohocken, PA: ASTM.

ASTM. 2019d. *Standard specification for zinc-coated (galvanized) steel bars for concrete reinforcement.* ASTM A767/A767M. West Conshohocken, PA: ASTM.

ASTM. 2020a. *Standard practice for safeguarding against embrittlement of hot-dip galvanized structural steel products and procedure for detecting embrittlement.* ASTM A143. West Conshohocken, PA: ASTM.

ASTM. 2020b. *Standard specification for anchor bolts, steel, 36, 55, and 105-ksi yield strength*. ASTM F1554/F1554M. West Conshohocken, PA: ASTM.

ASTM. 2021a. *Standard specification for carbon and alloy steel nuts*. ASTM A563. West Conshohocken, PA: ASTM.

ASTM. 2021b. *Standard specification for carbon steel bolts and studs, 60,000 psi tensile strength*. ASTM A307. West Conshohocken, PA: ASTM.

ASTM. 2021c. *Standard test methods for determining the mechanical properties of externally and internally threaded fasteners, washers, direct tension indicators, and rivets*. ASTM F606/F606M. West Conshohocken, PA: ASTM.

ASTM. 2022a. *Standard specification for deformed and plain carbon-steel bars for concrete reinforcement*. ASTM A615/A615M. West Conshohocken, PA: ASTM.

ASTM. 2022b. *Standard specification for low-alloy steel deformed and plain bars for concrete reinforcement*. ASTM A706/A706M. West Conshohocken, PA: ASTM.

ASTM. 2023a. *Standard specification for alloy-steel and stainless steel bolting materials for high temperature or high pressure service and other special purpose applications*. ASTM A193/A193M. West Conshohocken, PA: ASTM.

ASTM. 2023b. *Standard specification for carbon steel, alloy steel, and stainless steel nuts for bolts for high pressure or high temperature service, or both*. ASTM A194/A194M. West Conshohocken, PA: ASTM.

ASTM. 2023c. *Standard specification for zinc (hot-dip) on iron and steel hardware*. ASTM A153/A153M. West Conshohocken, PA: ASTM.

AWS (American Welding Society). 2020. *Structural welding code—Steel*. AWS D1.1/D1.1M. Miami: AWS.

FDOT (Florida Department of Transportation). 2007. *Anchor embedment requirements for signal/sign structures*. Rep. No. BD545 RPWO No. 54. Gainesville, FL: University of Florida, FDOT.

ICC (International Code Council). 2018. *International building code*. Washington, DC: ICC.

IEEE (Institute of Electrical and Electronics Engineers). 2018. *Recommended practice for seismic design of substations*. IEEE 693. Piscataway, NJ: IEEE.

USDOT FHWA (Federal Highway Administration). 2015. Vol. 1 of *Steel bridge design handbook: bridge steels and their mechanical properties*. FHWA-HIF-16-002. Washington, DC: USDOT.

USDOT NCHRP (National Cooperative Highway Research Program). 2022. *Fatigue-resistant design of cantilevered signal, sign, and light supports*. NCHRP Rep. No. 469. Washington, DC: USDOT.

CHAPTER 9

QUALITY CONTROL AND QUALITY ASSURANCE

9.1 GENERAL

To ensure quality of the product, good quality control (QC) and quality assurance (QA) programs should be instituted by both the fabricator and the owner. This will assure the owner that the fabricator has the personnel, organization, experience, procedures, knowledge, equipment, capability, and commitment to produce the specified product.

Quality control is the responsibility of the fabricator. The fabricator's QC program should consist of controls and inspections to produce a product within the requirements of the applicable codes, construction standards, specifications, and drawings. This should also include a detailed procedure for contract and project specification review to ensure contract compliance, including a system for requests for information necessary to resolve discrepancies or variations from contract requirements. The fabricator's written QC documents should be clearly defined and available for review and approval by the owner. The owner should also specify any additional requirements to achieve the desired degree of quality.

Quality assurance (QA) is the responsibility of the owner. The QA program is the monitoring and inspection tasks performed by the owner, owner's designated representative, responsible engineer, or third-party inspectors employed by the owner, to provide confidence that the product produced complies with the requirements of the applicable codes, construction standards, specifications, and drawings. It is recommended that shop drawings be reviewed by the owner for compliance with the design requirements and specifications prior to commencing fabrication. Final fabrication drawings may be produced by the fabricator if the fabricator is responsible for the structure design or supplied by the owner on a "fabricate-only" basis if the owner designed the structure. Some utilities design and detail their own structures and supply fabricators with the fabrication drawings. During the manufacturing process, the owner or designated representatives/inspectors should have access to the fabricators' facilities to perform the QA tasks.

The QC/QA programs must be agreed upon between the fabricator and the owner before commencing any fabrication. The extent of QC/QA programs may vary on the basis of initial investigations, audits, the owner's experience, the fabricator's experience, past performance, and the degree of reliability required for the specific job. Depending on the size and scope of the job, the owner should consider periodic witnessing of the product in production.

It is recommended that structures are reviewed during construction to ensure compliance with approved construction drawings, fabrication drawings, and specifications prior to being placed into service. Construction reviews may include special inspections from third-party inspection companies, inspection from in-house staff, and structural observations from the design engineer. Inspections and observations are recommended for foundations, anchorage, and support structures.

9.2 STEEL STRUCTURES

9.2.1 Material

The fabricator should review and inspect all material that is used in the fabrication of the entire structure, all mill test reports for material compliance, all material suppliers for their manufacturing procedures and QC programs, and all welding electrodes. The fabricator should maintain copies of all mill test reports for an agreed upon time period. The owner should review and agree on the fabricator's material specifications, supply sources, material identification procedures, storage, and traceability procedures.

9.2.2 Welding

The fabricators' welding procedures and welder certification programs should be in accordance with the latest revision of the AWS D1.1/D1.1M *Structural Welding Code—Steel* (AWS 2020). All welders performing work should be AWS-certified for the process and position that they are welding.

The owner should establish requirements for the review of, and agreement on, the fabricators' welding procedures for various types of welds and their welding certification programs. All welding procedure specifications (WPS), welder performance qualification test records (WPQR), and supporting documentation should be readily available. Special attention should be paid during the review of the welding procedures to ensure they address all applicable requirements according to AWS D1.1/D1.1M.

9.2.3 Fabrication Inspection

9.2.3.1 Visual Inspection. Visual inspections address the following typical areas:

1. Dimensional correctness,
2. Fabrication straightness,
3. Cleanliness of cuts and welds,
4. Surface integrity at bends,
5. Condition of punched and drilled holes,
6. Hardware fit and length,
7. Weld size and appearance,
8. Overall product workmanship, and
9. Coating thickness and quality.

9.2.3.2 Inspection Methods of Welds. In addition to close visual and dimensional inspection of welds, several methods of nondestructive testing may be used to detect weld discontinuities. They include the following:

1. Magnetic particle testing (MT) is a practical method for detecting tight surface cracks. MT inspections should be in accordance with ASTM E709 (ASTM 2021).

2. Dye penetrant testing (PT) is a reliable method for detecting any cracks or porosity that are open to the test surface. PT inspection should be in accordance with ASTM E165 (ASTM 2023).
3. Ultrasonic testing (UT), using longitudinal angle beams, is the preferred method of determining weld quality in base plates and flange connection welds. It is also reliable in detecting small cracks and internal flaws in other complete penetration welds. UT can also be used to verify that the specified percentage of penetration on partial joint penetration welds has been achieved. UT inspection for laminar defects should also be considered for large T-joints such as base plates and flange joints. AWS D1.1/D1.1M does not provide any specific guidelines for ultrasonic testing of plates less than 5/16 in. (8 mm) thick or for welds using backing bars. The fabricator should follow the procedure established by AWS D1.1/D1.1M in developing a specific inspection procedure. UT inspection should be in accordance with ASTM A435/A435M (ASTM 2017a) and ASTM E164 (ASTM 2019). Complete joint penetration (CJP) welds should be 100% inspected to ensure that the required penetration is achieved. When inspecting CJP welds with backing bars and/or stiffeners, the sequencing of the welding inspection should be considered to avoid interference with the UT inspection of the base weld.
4. Radiographic testing (RT) is used for investigation of failures and provides a permanent record of the test results. However, its use is limited on many of the weld types (e.g., base and flange connections), where it is difficult, if not impossible, to position the film to record the entire weld joint. It is also possible to miss tight cracks that lie normal to the RT source and film.
5. Eddy current testing (ET) techniques have limited application in the determination of weld penetration and the detection of cracks.

9.2.3.3 American Institute of Steel Construction Certification. American Institute of Steel Construction (AISC) offers certification programs to support a quality management system and ensure compliance with standards such as AISC 303 (AISC 2022), AWS D1.1 (AWS 2020), RCSC (2020), and AISC 360 (AISC 2022). The steel fabricator has the option to become AISC certified. These documents provide definitions and information on the roles and responsibilities of steel fabricators and owners of their products. Chapter N of AISC 360 provides requirements for a QA/QC program for structural steel fabricators.

9.2.3.4 Test Assembly. The full or partial assembly of a complicated structure by the fabricator before galvanizing and final shipment may be beneficial in verifying design and detailing correctness. Field construction problems and delays can be minimized if a test assembly of the structure reveals any errors. Missing or mispunched holes can be easily corrected at the fabrication shop instead of causing construction delays if left undone until final assembly.

9.2.3.5 Inspection Reports. Inspection reports should be generated and maintained for all inspection activities, including dimensional correctness, weld quality and size, and cleanliness of cuts. Reports should also be generated for all nondestructive inspection activities. The reports should include the assembly part number, date of inspection, and the inspector's name. The fabricator shall maintain a copy of all inspection reports in accordance with their established record and retention procedures.

9.2.4 Structure Coating

Where painting is required, the system, procedures, and methods of application should be acceptable to both the owner and the fabricator. Also, the system should be suitable for both the product and its intended exposure. The coating supplier will designate the degree of surface preparation required for the materials being used. Available standards for the preparation of metal substrates are a joint effort between the Society for Protective Coatings (SSPC) and the National Association of Corrosion Engineers International (NACE). SSPC and NACE have merged to form the Association for Materials Protection and Performance (AMPP). The coating thickness should be checked to ensure that the minimum dry-film thickness meets both the owner's specification and the paint supplier's specification. In addition, a thorough visual inspection should be made to detect pin holes, cracking, and other undesirable characteristics.

Where galvanizing is required, the procedure and facilities should be agreed upon by the owner and the fabricator. ASTM A123/A123M (ASTM 2017b) for steel products and ASTM A153/A153M (ASTM 2023) for steel hardware are the generally accepted standards for galvanizing the types of steel structures found in substations. After hot-dip galvanizing, nondestructive testing may be specified to ensure that no adverse changes occurred to the finished product, especially on shaft to base plate welds. The magnetic thickness measurement method is usually the test that is used to determine the thickness of the zinc coating.

Where metallizing is required, the procedures and facilities should be in accordance with the coating supplier's recommendations and acceptable to both the fabricator and the owner. Unless there is complete access to the inside of the structure for application of the metalizing coating, metalized structures should be hermetically sealed. The metallized coating should be inspected for thickness by using a magnetic thickness gauge. Also, an adherence test may be made by cutting through the coating with a knife. Bond is considered unsatisfactory if any part of the coating lifts away from the base metal 0.25 in. (6.35 mm) ahead of the cutting blade.

Where bare weathering steel is specified, the need for blast cleaning the steel should be decided and agreed upon by the owner and the fabricator. Blast cleaning is desirable if a clean and uniformly weathered appearance is important in the structure's first years of exposure. In time, even a nonblast-cleaned steel structure develops a uniform oxide coating.

9.3 ALUMINUM STRUCTURES

Many of the requirements, procedures, testing, and handling associated with steel structures are also applicable to aluminum structures, except magnetic particle testing. Refer to Section 9.2.3.2 for methods of nondestructive testing available for welding.

9.3.1 Material

The owner should review and agree on the fabricator's material specifications, supply sources, material identification procedures, storage, and traceability procedures. Mill test reports for all material should be reviewed for compliance with the appropriate ASTM specification by the fabricator. The fabricator should maintain copies of all mill test reports.

9.3.2 Welding

Welding specifications, preparation, procedures, and welder qualifications and inspection should be in accordance with the Aluminum Association's *Aluminum Design Manual* (AA 2020) and AWS D1.2/D1.2M (AWS 2014).

The owner should establish requirements for the review of, and agreement on, the fabricators' welding procedures for various types of welds and their welding certification programs. Welding procedure specifications (WPS) should be prepared for each welding process being used, and the supporting procedure qualification records (PQRs) should be readily available.

9.3.3 Fabrication

The fabricator should have a detailed procedure for contract and project specification review that includes all necessary information for the owner to ensure contract compliance, including a system for requests for information necessary to resolve discrepancies or variations from contract requirements.

The fabricator should follow the guidelines listed as follows:

1. Shearing, sawing, and arc cutting are acceptable methods of cutting. Flame cutting is not acceptable.
2. Holes may be punched if the taper of the hole does not exceed a diametric difference of 0.03125 in. (0.80 mm). Otherwise, subpunching, reaming, or drilling is required.
3. All forming and bending should be carried out cold, unless indicated on the shop drawings.

9.3.4 Inspection

At a minimum, visual inspection of all welded joints is necessary. Additional testing of welded joints, if required, should be agreed upon by the owner and fabricator.

9.3.5 Structure Coating

Aluminum structures are normally supplied with a standard mill finish (i.e., not painted or deglared). If the structures are to be provided deglared, the owner and fabricator should resolve the type of treatment, allowable gloss, and other requirements. Aluminum structures are not ordinarily painted, except in extremely corrosive conditions or if they are in contact with dissimilar materials such as steel. If required, painting should be performed in accordance with the paint fabricator's recommendations.

9.4 CONCRETE STRUCTURES

9.4.1 Reinforced Concrete

Cast-in-place concrete shall be in accordance with the latest applicable and most recent industry standards. These requirements should be based on the following specifications: ACI, ASTM, and the Concrete Reinforcing Steel Institute (CRSI). In addition, reinforced concrete work should conform to the requirements of ACI SPEC-301-20 (ACI 2020) and ACI CODE-318-19 (ACI 2019).

For QA, it is recommended that a concrete mix design submittal be reviewed in accordance with the provisions of ACI SPEC-301-20 prior to authorizing concrete placement.

Under special conditions where corrosion of reinforcement bars is possible, a cathodic protection or protective epoxy coating is recommended to ensure the long-term integrity of the concrete reinforcement elements.

9.4.2 Prestressed Concrete Poles

For prestressed concrete structures, the material should satisfy the requirements as specified by ASCE MOP 123 (ASCE 2012) and the *PCI Manual for Quality Control for Plants and Production of Structural Precast Concrete Products*, PCI MNL-116 (PCI 1999).

9.4.3 Inspection

Chapters 7 and 9 of ASCE MOP 123 provide information on QA and the inspection of concrete structures. These chapters provide recommendations on dimensional tolerances, acceptable pole weight variations, and the sealing of strand ends. Chapter 7 provides additional guidance on the types of inspections and records retention that the concrete fabricator should include as part of its quality program.

9.5 WOOD STRUCTURES

9.5.1 Material and Treatment

The material and treatment of wood members should satisfy the requirements of the ASCE MOP 141 (ASCE 2019), using the applicable standard. Construction of temporary facilities requiring the use of wood should be constructed of Class A Fire Retardant Pressure Impregnated Wood.

9.5.2 Manufacturing and Fabrication

The manufacturing and fabrication of wood members should satisfy the requirements of the ASCE MOP 141 (ASCE 2019). Any special fabrication details or tolerances should be included in the owner's specification.

9.5.3 Inspection

The inspection of wood members should satisfy the requirements of AITC 200-2009 (AITC 2009). Lumber should be inspected for grade stamps for its conformance to the project requirements. Lumber should also be inspected for maximum moisture content.

9.6 SHIPPING

At a minimum, the fabricator should comply with the shipping procedures as listed:

1. Check packaging to minimize shipping damage.
2. Check items to ensure that they have completed specified inspections.
3. Check that specified items are included with the shipment.

Before the start of fabrication, the owner should review the fabricator's methods and procedures for packaging and shipping and agree to the mode of transportation. When receiving materials, all products should be inspected for shipping damage before accepting delivery. If damage is apparent, the owner should immediately notify the fabricator. If the shipments are free on board (FOB) destination, the shipper is responsible for damage repair.

When receiving materials, the owner is also responsible for checking to see that all materials listed on the packing lists are delivered. Where a discrepancy exists, the fabricator should be notified.

9.7 HANDLING AND STORAGE

The fabricator should provide written procedures for handling and storing materials to prevent damage, loss, or deterioration of the structure. The owner should review and approve these procedures before shipping any materials.

If direct-embedded galvanized structures with a below-grade coating are to be stored for an extended period of time, the owner should communicate this to the fabricator. Below-grade coatings exposed to direct sunlight are susceptible to degradation. The fabricator can recommend a coating that has extended life in direct sunlight exposure. The coating of all structures stored for an extended period of time should be inspected prior to use.

At a minimum, stored material should be placed on skids, platforms, or other supports above the ground and away from any vegetation. Decayed or decaying material supports should not be permitted to remain under stored material. Special care should be exercised while storing material. Materials such as steel and aluminum stored within an energized substation, on nonconducting material, can develop an induced electrical potential, and proper precautions should be used for worker safety. Proper initial placement of material sections can increase the efficiency of the final assembly process. Material identification marks should be visible when the material is stacked and should remain legible for the period of time specified by the owner. Equipment should be braced or provided with temporary anchoring in seismic areas.

REFERENCES

AA (Aluminum Association). 2020. *Aluminum design manual, including specification for aluminum structures.* Arlington, VA: AA.

ACI (American Concrete Institute). 2019. *Building code requirements for structural concrete (with commentary).* ACI CODE 318-19. Detroit: ACI.

ACI. 2020. *Specifications for structural concrete.* ACI SPEC-301-20. Detroit: ACI.

AITC (American Institute of Timber Construction). 2009. *Manufacturing quality control systems manual.* AITC 200-2009. Englewood, CO: AITC.

AMPP (Association for Materials Protection and Performance). Accessed October 14, 2023. https://www.ampp.org/home.

ANSI/AISC (American National Standards Institute/American Institute of Steel Construction). 2016a. *Code of standard practice for steel buildings and bridges.* AISC 303. Chicago: AISC.

ANSI/AISC. 2016b. *Specification for structural steel buildings.* AISC 360. Chicago: AISC.

ASCE. 2012. *Prestressed concrete transmission pole structures: Recommended practice for design and installation,* MOP 123. Reston, VA: ASCE.

ASCE. 2019. *Wood pole structures for electrical transmission lines: Recommended practice for design and use,* MOP 141. Reston, VA: ASCE.

ASTM International. 2016. *Standard specification for zinc (hot-dip) on iron and steel hardware.* ASTM A153/A153M. West Conshohocken, PA: ASTM.

ASTM. 2017a. *Standard specification for straight-beam ultrasonic examination of steel plates.* ASTM A435/A435M. West Conshohocken, PA: ASTM.

ASTM. 2017b. *Standard specification for zinc coating (hot-dip galvanized) on iron and steel products.* ASTM A123/A123M. West Conshohocken, PA: ASTM.

ASTM. 2019. *Standard practice for ultrasonic contact examination of weldments.* ASTM E164. West Conshohocken, PA: ASTM.

ASTM. 2021. *Standard guide for magnetic particle examination.* ASTM E709. West Conshohocken, PA: ASTM.

ASTM. 2023. *Standard test method for liquid penetrant examination.* ASTM E165. West Conshohocken, PA: ASTM.

AWS (American Welding Society). 2014. *Structural welding code—Aluminum.* AWS D1.2/D1.2M. Miami: AWS.

AWS. 2020. *Structural welding code—Steel.* AWS D1.1/D1.1M. Miami: AWS.

CRSI (Concrete Reinforcing Steel Institute). Accessed October 14, 2023. https://www.crsi.org.

NACE (National Association of Corrosion Engineers). Accessed October 14, 2023. https://onepetro.org/NACE.

PCI (Precast/Prestressed Concrete Institute). 1999. *Manual for quality control for plants and production of structural precast concrete products.* 4th ed. PCI MNL-116. Chicago: PCI.

RCSC (Research Council on Structural Connections). 2014. *Specification for structural joints using high-strength bolts.* Chicago: Precast/Prestressed Concrete Institute.

SSPC (Society for Protective Coatings). Accessed October 14, 2023. https://sspc.org/.

CHAPTER 10
CONSTRUCTION, MAINTENANCE, AND TESTING

10.1 CONSTRUCTION

The engineer should anticipate construction loads imposed on substation structures and ensure that proper construction methods and quality materials are used to prevent excessive stresses during construction. Examples of construction loads include the load imposed on a concrete foundation while jacking a transformer into place, the load imposed below a trailer's wheels on a cable trench lid or manhole cover during transportation of transformers, or the load imposed on steel structures while lifting them into place.

Outdoor substations are frequently constructed by people with varied levels of experience. Therefore, simplicity of design, detail, and erection with a specific schedule of site inspections should be considered. Inspections should be performed during grading, foundation, and anchor rod placement, structure erection, and initial structure attachment of wires or equipment.

Wind-induced vibrations of unloaded structures also need to be considered during the construction phase. The larger, tapered, tubular structures typically associated with 345 kV substations and higher can be more susceptible to these vibrations because of their longer unbraced member lengths. This can especially be the case for lower-equipment beams found on dead-end structures with long bay widths. These equipment beams are typically lightly loaded and tend to be smaller in size in comparison with the rest of the structure. The combination of smooth constant winds, long bay widths, and lightly loaded members can possibly cause vibrations. Other examples of structures susceptible to vibration include tall static poles (lightning masts), long rigid bus spans, and dead-end structures with large bay widths prior to installation of all lines, equipment, and dampening. During construction, care should be taken to properly weigh down and secure these members until installation is complete.

10.2 MAINTENANCE

The engineer should consider accessibility of equipment for maintenance and operation. Equipment should have provisions for access, that is, crossover platforms or working platforms. These access provisions need to be considered, especially around large transformers with coolant and fire protection piping or in substations where equipment has been raised above grade to protect it from flood damage or for other reasons.

Structures and foundations should be inspected each time the supported equipment is inspected or maintained. These structures should be checked for damage, signs of fatigue, corrosion, loose members, and connection hardware. Also, the foundation should be checked for signs of deterioration such as settlement, cracking, spalling, exposed reinforcing steel, and corrosion of anchor rods. In addition, when equipment is upgraded, structures should be carefully checked for structural adequacy, electrical clearances, damage, and deterioration caused by corrosion. Whenever possible, the interior of hollow tubular members should be periodically inspected for corrosion.

Structures painted before 1975 typically used lead-based paint. All painted structures installed prior to the lead paint ban in 1975 should be checked for lead-based paint. If lead is present, all Occupational Safety and Health Administration (OSHA), US Environmental Protection Agency (EPA 2008), and company requirements should be followed when removing the lead-based paint or doing work that could disturb the paint.

Weathering steel should be inspected periodically for any abnormal oxidation, especially at the member connections.

Animals and insects commonly seek refuge inside substations. Some equipment gives off heat that can provide warmth to various animals. Tall structures, especially lattice structures, become nesting places for birds, whereas hollow structural members are ideal homes for bees and other insects. The engineer should consider detailing structures to omit pockets for animal nests, and adding deterrents such as animal guards, bee screens, or electric fences near equipment and structures. Guidance on animal (and insect) protection/mitigation can be found in IEEE 1264-2022, *IEEE Guide for Animal Mitigation for Electric Power Supply Substations* (IEEE 2022).

10.3 WORKER SAFETY

Applicable OSHA state and local codes must be followed. OSHA 29 CFR 1926, *Subpart M—Fall Protection* (OSHA 2020), and IEEE 1307 (IEEE 2018a) are two sources of information for worker safety on utility structures. Elevated locations on the structure that require workers to move from one location to another (as defined in IEEE 1307-2018) (IEEE 2018a) require fall protection devices such as safety cables or attachment points designed and installed for worker fall restraint.

Other considerations such as grating on stairs or handrail systems should be considered to mitigate common slips, trips, and falls that occur while working on or around substation structures and equipment. Ramps in the substation may require handrails or guardrails.

10.4 FULL-SCALE STRUCTURAL PROOF TESTS

Full-scale structural proof tests are used to verify the structural design capacity and are rarely performed on substation structures. It is not typically cost-effective to perform a full-scale test because substation structures, unlike transmission structures, are not fabricated in large quantities. Full-scale testing should be considered if a particular substation structure is a standard design and will be used multiple times or if that structure uses a unique structural system not typical of current practice. Component testing (e.g., a section of the tower or connections) may be cost-effective for substation structures. If component or full-scale tests are required, they should comply with ASCE 10-15 (ASCE 2015) and ASCE 48-19 (ASCE 2019). For seismic qualification testing, see IEEE 693-2018 (IEEE 2018b).

REFERENCES

ASCE. 2015. *Design of latticed steel transmission structures.* ASCE 10-15. Reston, VA: ASCE.

ASCE. 2019. *Design of steel transmission pole structures.* ASCE 48-19. Reston, VA: ASCE.

EPA (US Environmental Protection Agency). 2008. *Lead-based paint renovation, repair and painting (RRP) rule.* Washington, DC: EPA.

IEEE (Institute of Electrical and Electronics Engineers). 2018a. *Fall protection for electric utility transmission and distribution on poles and structures.* IEEE 1307-2018. Piscataway, NJ: IEEE.

IEEE. 2018b. *Recommended practice for seismic design of substations.* IEEE 693-2018. Piscataway, NJ: IEEE.

IEEE. 2022. *Guide for animal mitigation for electric power supply substations.* IEEE 1264-2022. Piscataway, NJ: IEEE.

OSHA (Occupational Safety and Health Administration). 2020. *OSHA safety and health standards for construction.* 29 CFR 1926, Subpart M. Washington, DC: OSHA.

CHAPTER 11

RETROFIT OF EXISTING SUBSTATION INFRASTRUCTURES

11.1 GENERAL

At times, it becomes necessary to reuse or modify an existing substation structure because of any number of factors. The justification is as varied as the number and types of structures that exist and can often include cost savings, safety concerns, site access, and outage time frames among others. One significant concern when working with existing substation structures relates to changes in the design codes and how to approach the reuse of existing structures that may have been installed decades ago under different codes. This section will act as a guide for the engineer on common drivers and various methods used in the retrofit, analysis, and reuse of an existing substation structure or foundation.

Replacing equipment during upgrades or equipment failures may lead to the reuse of a structure or foundation. Examples include the following:

- Replace the existing transformer and reuse the existing transformer foundation.
- Replace disconnect switches with different disconnect switches.
- Replace the existing oil circuit breaker with a new SF6 gas breaker and reuse the existing foundation.
- Replace the wave trap on the existing stand.
- Move or replace equipment to meet new electrical clearance guidelines.
- Upgrade the size and rating of the rigid bus.
- The addition of equipment to a structure is common and the structure may not have been specifically designed with the added equipment as part of the original design. Examples include the following:
  - Add new disconnect switches to an existing switch rack or a dead-end structure, and
  - Add surge arrestors or CTs to an existing switch rack.
- Adding onto an existing structure during substation expansion is a common occurrence. In some cases, this expansion may have been part of an ultimate or future design, and in others, the expansion may not have been previously considered. Examples include the following:
  - Add a new bay to an existing switch rack or a dead-end structure.
  - Extend the bus run.

Existing structures are sometimes modified to accommodate new equipment. Examples include the following:

- Move or modify bracing on a switch rack to open a bay for new pad-mounted equipment.
- Move or modify bracing to add trusses for new disconnect switches.
- Modify dead-ends to accommodate bay extensions.
- Remove the disconnect switch and use the structure to support the rigid bus.

New conductors are sometimes added, and the existing conductor type and tension are sometimes modified. This results in changes in loads on the existing structures. Examples include the following:

- Add a new transmission or distribution line coming into or leaving from a substation and connecting with an existing structure.
- Change the conductor type or modify the line coming into or leaving from a substation resulting in a different tension load in the existing structures.
- Add a shield wire on an existing mast.

Electrical clearance standards can change, and a structure may need modification and retrofit to meet new clearance requirements.

Over time, access to equipment for maintenance purposes may be impeded or the type of equipment used for maintenance may change, limiting adequate access to switches, current transformers, potential transformers, surge arrestors, and so on. Structures may need relocation or modification to allow for maintenance access.

11.2 ALTERNATIVE METHODS FOR RETROFIT OR REINFORCEMENT OF SUBSTATION INFRASTRUCTURES

11.2.1 Types of Structures That May Require Reinforcement/Repair

Framed structures

- Boxed-type lattice bus structures,
- H-frames,
- A-frames (lattice and modular types), and
- Structures that support switches.

Single shaft/poles structures

- Bus supports,
- Static poles, and
- Equipment supports.

Foundations

- Pier type,
- Spread footing,
- Switchgear foundation,

- Anchor rods, and
- Concrete oil retention pits.

11.2.2 Retrofit Methods

Figure 11-1a to c demonstrate several different methods of retrofitting existing facilities.

Member Replacement or Reinforcement:

Addition of bay(s) to existing framed structures such as A-frames or H-frames: This can impact existing end columns that would become center columns. One potential solution is to replace the existing end column with a new center column and relocate the existing end column, given that the existing foundation is adequate (in usual circumstances, the center column and associated foundation will receive more load).

Addition of stiffeners: Stiffeners can be field-welded to existing columns and beams.

Connection modifications: Some deflection problems can be resolved by changing connection types, given that the structure and its components can handle the resulting loads.

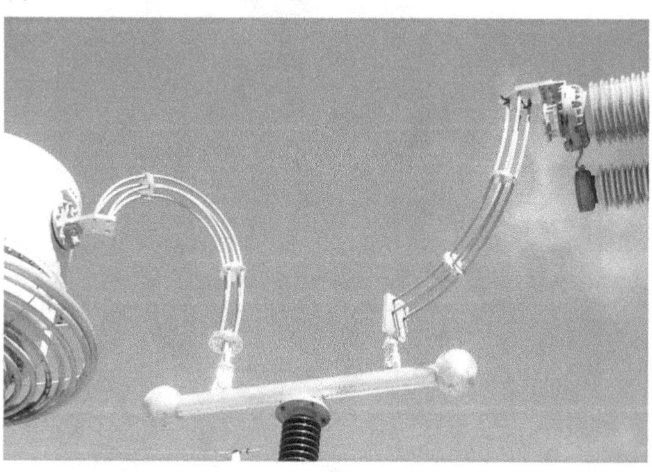

Figure 11-1. (a) 230 kV single-phase transformer anchorage retrofit (silver stubs added), (b) 230 kV disconnect switch support seismic retrofit (bolted-on diagonal bracing), (c) 230 kV equipment connection retrofit to accommodate seismic movement.
Source: Courtesy of Michael Miller.

External member additions: Additional reinforcement can be added via external columns, beams, or a combination thereof.

Carbon fiber reinforcement: This provides tensile as well as compressive strength (Figure 11-2).

Guying and bracing: The height of a single-column support structure may be increased by adding a stub-column connection via a new flange plate, given that the base structure has sufficient strength to handle the additional height.

Change of conductor or addition of static wires: It may be possible to achieve additional line rating by using higher-capacity new conductors such as aluminum conductor steel supported (ACSS) because of its ability to operate at temperatures up to 250 °C (482 °F). For a given ampacity, a smaller ACSS can be used relative to ACSR, typically resulting in lower tension loads being imposed on structures.

Lightning protection: It may be possible to achieve an additional level of lightning protection by using new lightning masts rather than adding additional static wires. It may also be possible to add additional static wires through independent static masts or back-guying the new ones on existing structures. It may be possible to retrofit insulators by replacing them with a stronger design at the same basic impulse level. This can be done frequently without causing any height changes, but bolt circle diameter and shed diameter are sometimes increased.

Foundation improvements (depending on the nature of reinforcement for soil or structural problem): Driven piles, micropiles, or helical piles are added adjacent to the existing foundations. Concrete pad extension can be achieved by connecting the new slab foundation with doweled-in

Figure 11-2. Carbon fiber reinforcement with a drain tube.
Source: Courtesy of Majid R. J. Farahani.

reinforcing bars. Additional bearing capacity may be achieved by undertaking a partial excavation beneath existing foundations and adding a structural rock fill.

11.2.3 Methods of Anchorage Retrofit

Retrofit of anchorage is sometimes desired because the equipment or structure is unanchored, or the anchorage does not have sufficient capacity because of design criteria or the methods applied during the original installation. In some cases, it is desirable to place new or replacement equipment on the same foundation, which then requires the anchorage to be retrofitted or rebuilt. In high seismic hazard areas, utilities may wish to improve the capacity of old anchorage to mitigate the risk of damage from earthquakes. Anchorage retrofits often make use of post-installed (adhesive or mechanical) anchors or welding to existing or new embedded metal or anchor plates/brackets. Anchor plates or brackets to which the equipment or structure can be welded may themselves be anchored to the foundation concrete by post-installed anchors. In some cases, it may be practical to cut out the existing foundation concrete with the anchor plates or brackets grouted in place and replace it with headed studs or cast-in anchor rods. Modifications to equipment bases, supports, and structures may be required to accommodate the new anchorage. This may include brackets or tabs to provide for transfer of forces from the equipment or structure to the anchorages.

11.2.4 Considerations When Retrofitting Steel Structures

With increasing frequency, older-style substation structures are being required to resist higher design loads. These additional loads can be caused by several factors, some of which are given as follows:

1. Changes in code: With regard to loading, some designs may be old enough to have been designed according to a previous version of the *National Electrical Safety Code* (NESC) (IEEE 2023) using lower magnitude loads. With regard to structure capacity, some designs may be old enough to not have been designed according to any standard other than the AISC Manual of Steel Construction. In some cases, the structure may have even been designed before one or both of these were created and could have been designed according to any number of documents in use at the time. For instance, *the American Railway Engineering Association's Definitions, Specifications and Principles of Practice for Railway Engineering* (AREA 1921) contained loading and design information for wire-supporting structures. Also, prior to the standardization of steel structural shapes and creation of AISC, steel mills often had their own product lists and design recommendations.
2. Safety factors may have changed or become more stringent.
3. Changes in usage: Structures that were originally designed for one type or size of equipment may be repurposed for other equipment usage.
4. Structural intermember forces resulting from applied loads determined with current computer analysis programs may be more accurate and often differ from early approximate hand-calculation methods. For instance, analyses had traditionally been performed using first-order static analysis, and structural stability concepts had been captured within allowable stress equations or increased safety factors. Modern designs are more precise with regard to structural analysis, using second-order analyses, direct modeling of

imperfections, more precise load distribution, and so on, often leading to larger design forces. However, this is offset by lower safety factors/higher resistance factors and a more precise calculation of member strength.

When possible, the engineer conducting the assessment of the structure should obtain available calculations, drawings, specifications, records of maintenance or prior alterations, and other relevant information that will aid in evaluating the structure. The dates on these documents may assist in estimating the age and hence the possible grades of materials that may have been used at the time of construction. Prior to performing any structural analysis and design work, a site visit should be undertaken to assess the condition of the structure (condition of coating systems, corrosion, damage, undocumented field modifications, installed equipment, foundation settlement, etc.) and obtain any field measurements that may be required to supplement the information on available drawings.

When considering retrofit of existing structures, several factors should be considered that could impact the overall cost and scope of a project. Some of these factors are

1. *Bolt capacities*: A variety of bolts have been used over the years. Anything from A307 (ASTM 2014a, ASTM 2022), A325 (ASTM 2014b), and A394 (ASTM 2015) can be common. In addition, older structures may have been constructed using rivets. Simply changing fasteners may increase the capacity of the structure.
2. *Steel material*: Prior to 1960, it was common to use A7 (ASTM 1967) steel [$F_y = 33$ ksi (228 MPa)]. In the 1960s, A36 (ASTM 2019) steel [$F_y = 36$ ksi (248 MPa)] became available and is still common today, but A572 (ASTM 2021) Grade 50 [$F_y = 50$ ksi (345 MPa)] steel is widely used. More exotic steels could have been used as well. When checking the capacities of existing structures, using appropriate steel material properties is critical. The structure being modified could have the existing material tested to determine the actual yield strength.
3. *Constructability*: In general, it is easier to add members or strengthen existing members than it is to remove and replace lattice members. Removal and replacement of lattice members, even redundant members, adds a degree of instability to the structure that might be avoidable.

In all cases, the historical performance of the structure should be considered.

Rehabilitation and Retrofit, AISC Design Guide 15, 2nd edition (Brockenbrough and Schuster 2018) is an excellent resource for historical information on commonly available steel items during various time periods, such as steel shapes and materials as well as connector types and grades. In addition, it contains design guidance and examples for many common steel retrofit situations.

11.2.5 Structure Finish and Its Consideration to the Retrofit Process

11.2.5.1 Painted Structures. Tubular-painted structures are usually sealed to protect their interior surfaces. Any alterations that result in developing holes through such structures should be evaluated and protection provided for interior surfaces.

11.2.5.2 Galvanized Structures. Welding on galvanized steel structures produces hazardous heavy metal zinc gases and should therefore be avoided. The zinc content should be removed to bare steel and then welding should be done in a well-ventilated area to protect the welder. A ventilated welding hood should also be used. Once the welding is complete, corrosion protection should be repaired in locations where it was affected or removed. For tubular structures, the impact of exterior welding on the interior surfaces needs to be evaluated and addressed if applicable.

11.2.5.3 Weathering Structures. Weathering structures are typically used for transmission dead-end applications and not often used for equipment support applications. This is in part because of concerns with weathering steel bleeding over electrical connections of such equipment. This can be rectified by painting over the weathering steel structures.

11.2.5.4 Timber Structures. Most substation structures are not constructed of timber material. However, where wood has been utilized, the following guidelines are recommended:

Wood Poles: American National Standard ANSI O5.1-2017 (ANSI 2017) should be followed, including the consideration of fumigants and continuing inspection and maintenance programs. Also, *Wood Pole Structures for Electrical Transmission Lines*, ASCE MOP 141 (ASCE 2019), should be considered.

Heavy Timber Construction (HTC): ASCE has many research documents concerning the repair of large members. Repair and replacement of rotted sections is an important topic that may require careful planning, detailing, and sequencing of construction, which may utilize epoxies along with scarf joints and new laminated pieces.

Guidelines for HTC construction are addressed by the American Institute of Timber Construction *Timber Construction Manual* (AITC 2009).

Wooden structures are susceptible to decay. Embedded wooden poles are particularly prone to decay near the groundline because of accumulating moisture wicked from the soil below, and at the pole top, owing to the tendency of preservative treatments to leach down the pole.

Care must be exercised when reusing a wooden structure to locate any structural defects and determine the remaining structural capacity. Recommendations for wood structure inspection and maintenance can be found in *Rural Utility Service Bulletin*, 1730B-121 (USDA 2013). Inspections can consist of visual evaluation, hammer sounding, borehole testing, or a combination of the three.

If a wooden structure is found to be damaged, it may need to be replaced. Alternatively, several methods exist for repairing or reinforcing a damaged wooden structure. These include carbon fiber reinforcement, application of an internal or external preservative, or addition of a reinforcing truss or a wood splint.

The *National Electrical Safety Code* (IEEE 2023) includes requirements for repair or replacement of deteriorated wood poles based on the remaining capacity of a pole compared to its original capacity.

11.3 ENVIRONMENTAL CONCERNS WHEN RETROFITTING SUBSTATIONS

When retrofitting substations, environmental factors, such as asbestos and soil contamination, should be considered.

11.3.1 Asbestos in Existing Substations

The US Environmental Protection Agency's *Asbestos National Emission Standard for Hazardous Air Pollutants* (EPA 1990) requires utilities to perform certain functions prior to carrying out a demolition or renovation project. Substation structures adjacent to a power plant may have asbestos contamination.

11.3.2 Demolition Activities

Prior to demolishing a structure, utility companies should send out an Asbestos Hazard Emergency Response Act (AHERA) inspector. A certified building inspector should conduct a project inspection survey and collect samples of items that are suspected to contain asbestos. When samples return as positive for asbestos, utility companies should hire an asbestos contractor to perform abatement prior to demolition.

11.3.3 Renovation Activities

If the substation facility being renovated is not a load-supporting structure, then it usually falls under renovation activities. Similar to demolition activities, utilities should dispatch an AHERA-certified building inspector to perform a project inspection survey and collect samples of items that are suspected to contain asbestos. The samples should be analyzed, and when samples return as positive for asbestos, utilities should hire an asbestos contractor to perform abatement prior to renovation.

11.3.4 Soil Contamination in Existing Substations

Depending on each substation's design (structures and equipment installed) and maintenance activities, materials that may contaminate soils include, but are not limited to, asbestos, battery electrolyte (acidic or alkaline), diesel fuel, gasoline, lead paint, mercury from broken lamps, mineral oil, polychlorinated biphenyls, propane, SF_6 decomposition by-products or residue (e.g., HF, SF_4, SiF_4, SO_2F_2, SO_2, SOF_2, $S2F_{10}$, metals), and other chemicals or materials brought into a substation.

Federal, state, and local environmental regulations address the issues of proper handling, management, transport, and disposal of each of these contaminants if found in soils or other surrounding areas. For all projects at existing substations, whether it involves renovations, demolitions, or full decommissioning, any discoloration, odors, or staining observed in soils or other areas should be evaluated by the utility environmental department as appropriate. Historical spill records, if available, can also provide insights into possible contamination.

11.4 ENHANCING SECURITY AND RESILIENCE OF ELECTRICAL SUBSTATIONS

Attacks on electrical substations have raised questions and concerns about the safety and protection of these facilities.

Any number of security improvements in an existing substation may be necessary for increasing security. Briefings by the Department of Homeland Security Office of Intelligence and Analysis have recommended taking the following steps as a means of improving security for electrical substation facilities:

1. Improve security coordination with local and federal law enforcements.
2. Utilities can provide law enforcements with GPS coordinates of critical facilities.
3. Each utility will identify which of their substations are considered to be critical to their operations.

4. Limit view access into the substation facilities by using slats through chain link perimeters or other similar means (impact of air flow through the substation should be considered).
5. Install infrared cameras.
6. Install mobile cameras and supporting lighting fixtures.
7. Install bullet-resistant barriers.

Guidance on enhancing the security of substations can be found in IEEE 1402-2021, *IEEE Guide for Physical Security of Electric Power Substations* (IEEE 2021).

11.5 RETROFIT DESIGN CONSIDERATIONS

A rational approach should be taken when evaluating the impacts of load variations or structural modifications to existing structures. A few of the many changes to design criteria that have occurred over the years include

1. Prevalent use of Ultimate Strength Design (LRFD) instead of the Allowable Strength (Stress) ASD approach;
2. Changes to load factors in load combinations;
3. Use of 3-second gust wind speed maps that are based on risk categories and associated return periods [with Ultimate Strength Design (USD or LRFD)] instead of 50-year fastest-mile wind speed maps and importance factors [with Allowable Strength Design (ASD)]. This MOP recommends the 300-year wind speed maps for strength-level loads;
4. Significant changes to the seismic provisions (seismic zone–based design parameters replaced with risk-targeted ground motion design parameters that vary with latitude and longitude), increase in seismic demand in many locations, changes to design and detailing provisions;
5. Changes in material specifications (such as AISC and ACI) pertaining to structural element strength (capacity) determination; and
6. ACI CODE 318-19 (ACI 2019) provisions for anchorage design have changed significantly.

With the changes to element forces and stresses or structural demand (*D*), and element strengths or capacity (*C*), a logical parameter for evaluation would be the demand to capacity ratio (*D*/*C*). It is important to note that not all changes result in increased loading or reduced structural element capacities. In fact, the use of LRFD may, in some cases, result in lower *D*/*C* ratios, eliminating the need for a retrofit. Changes to available strength equations in material specification are often introduced as refinements to original equations, which may result in increased or reduced structural element capacities. With the many design changes, one possible rational approach for evaluating structural modifications would be to implement a step-by-step process that starts with the design provisions in this MOP as follows:

1. Determine structure loads and load combination for the structure and equipment in its new configuration in accordance with Chapter 3, "Loading Criteria for Substation Structures."
2. Perform a structural analysis to determine the structural element forces and stresses (*D*).
3. Determine structural element capacities (*C*) in accordance with the material specification listed in this MOP.

4. Determine the D/C ratios; if $D/C \leq 1.0$, then no retrofit is required.
5. If $D/C > 1.0$, the utility should decide on an acceptable limit for D/C that exceeds 1.0. It is not uncommon to set a D/C of 1.05 (more than 5%) for gravity loads and a 1.10 (more than 10%) for loads in combination with lateral loads (wind, seismic) as acceptable limits, but this decision has to be made by the utility with an understanding of the required reliability and associated risks. Alternatively, the utility may establish lower return periods for wind and seismic loading, with a D/C limit of 1.0 as acceptance criteria for the structural evaluation of existing structures requiring modifications. It is also important to ensure that all serviceability limits such as deflection limits and minimum electrical clearances are maintained.

Using design codes and material specifications in effect at the time of original structure design to evaluate the adequacy of a structure with load or structural modifications should be carefully considered for the following reasons:

1. Current load requirements may have increased from previous codes and standards, especially seismic load requirements. It is expected that the current codes and standards reflect a better understanding of load and element strength requirements, and using older versions may be unconservative in some situations [e.g., anchor design approach in ACI CODE 318-19 (ACI 2019)].
2. Some older structures may have been designed for lateral loads because of wind and wire/conductor tensions only without any consideration to seismic loading.
3. Using an ASD approach may not be beneficial for achieving a possible reduction of D/C ratios that may be realized using the LRFD approach.
4. Design calculations and construction specifications may not always be available for older structures (calculations if available should be checked for accuracy and completeness). However, structural drawings may be available, with some indicating the design codes and standards. If new calculations have to be generated, then it is advisable to develop these in accordance with current codes and standards or provisions described in this MOP. These calculations will provide the utility with updated D/C ratios for the structural elements to make an informed decision on the requirements for any retrofit work.

Evaluating for increase in load without element strength (capacity) considerations should be carefully considered for the following reasons:

1. Accepting an increase in equipment load on the basis of a predetermined percentage increase from an existing load without reviewing prior designs may not be conservative. The structure supporting the existing load may already be at its design limit or may already be subjected to undocumented load increases from the past.
2. Structures may not have been designed to all applicable loads such as seismic loads.
3. If the original design was based on an ASD approach, then any advantage of the reduced D/C ratios using the LRFD approach may not be realized.

It is up to the utility to choose a level of reliability and resilience that is appropriate for the infrastructure to be modified as a function of its role in power system operations and supporting economics. Modifications to existing structures or foundations that result in structural load variation or structural response behavior alteration could be analyzed in accordance with one of the following as appropriate:

1. Current recommendations of this MOP, when possible.

2. Most recent utility standard with which the structure has been previously brought into compliance when combined with a rational approach to review demand-to-capacity ratios of the load-carrying structural elements supporting the modifications as discussed in this chapter.
3. Utility standard that was in effect at the time of the original installation, if no previous structural upgrades have been made to more current standards, when combined with a rational approach to review the demand to capacity ratios of the load-carrying structural elements supporting the modifications as discussed in this chapter.

In earthquake-sensitive regions, it may be required to meet additional IEEE or vendor requirements for corresponding earthquake qualification of certain substation equipment or forfeit-associated warranties. For earthquake requirements, refer to IEEE 693-2018 (IEEE 2018) and Section 6.7 of this MOP.

Any analysis using today's contemporary modeling tools may result in members of existing structures not meeting the intent of the code or standard for which it was originally designed. Such discoveries should be corrected in a safe and timely manner.

As discussed in the anchorage section of this MOP, a set of design criteria should be established for the design of the new anchor system. These criteria should also define the demand loads and anchor forces. Depending on the desired performance objectives and design criteria, ductile or nonductile anchor performance may be required or used for anchorage design. The constraints of the existing conditions such as quality or thickness of concrete, or available edge distance for anchor rods, may dictate that a nonductile design be used for anchor rods.

The engineer should ensure the existence of an adequate load path from the equipment or structure into the anchor system. For framed structures or supports, verification of the load path is fairly straightforward. However, this may not be the case for some types of equipment such as transformers. Uplift forces resulting from overturning may pose particular challenges. Jacking pads or lifting eyes on transformer tanks provide a useful location for attachment of anchorage elements. The engineer should exercise caution when welding on a transformer or oil circuit breaker tank, as the heat of welding may adversely affect the insulating oil. If the weldability of old steel is in question, it is prudent to perform a chemical composition analysis to develop an appropriate Weld Procedure Specification.

The reuse of existing anchors is frequently an attractive alternative to retrofitting anchorages, because costly foundation modifications are avoided or reduced in scope. In such cases, the engineer should consider the following:

1. Material condition of anchor rods and reinforcing steel;
2. Type of anchor installed, embedment depth, type of head;
3. Details of reinforcing; and
4. Concrete strength and condition.

Nondestructive examination techniques may be employed to assess the material condition of the anchor rods, reinforcing steel, and concrete beneath the concrete surface. Sample pull-testing of anchor rods may be considered to establish allowable loads for use in design. When retrofitting concrete foundations, welding to existing reinforcing steel should typically not be performed, unless weldability has been established. New reinforcing steel to be welded should conform to ASTM A706/A706M (ASTM 2016) and be welded in accordance with AWS D1.4 (AWS 2018). In addition, potential damage to existing concrete caused by heat generated from welding operations should be considered.

11.6 INSTALLATION

Retrofit of existing structures should include a prior inspection of structurally relevant areas. Any significantly corroded or significantly damaged members should be repaired or replaced as part of the retrofit. Care should be taken to appropriately phase the project, if necessary, to maintain structural adequacy during all phases of construction.

Where new bays are to be added to existing dead-end structures, it may be required to unload all or part of existing structures to align new member connections to existing ones.

REFERENCES

ACI (American Concrete Institute). 2019. *Building code requirements for structural concrete (with commentary).* ACI CODE-318. Detroit: ACI.

AITC (American Institute of Timber Construction). 2012. *Timber construction manual.* 6th edition. AITC 200. Englewood, CO: AITC.

ANSI (American National Standards Institute). 2017. *Wood poles—Specifications and dimensions.* ANSI O5.1-2017. New York: ANSI.

AREA (American Railway Engineering Association). 1921. *Definitions, specifications, and principles of practice for railway engineering.* Chicago: AREA.

ASCE. 2019. *Wood pole structures for electrical transmission lines,* MOP 141. Reston, VA: ASCE.

ASTM International. 1967. *Specification for steel for bridges and buildings.* ASTM A7-67. West Conshohocken, PA: ASTM.

ASTM. 2014a. *Standard specification for carbon steel bolts and studs, 60,000 psi tensile strength.* ASTM A307-14. West Conshohocken, PA: ASTM.

ASTM. 2014b. *Standard specification for structural bolts, steel, heat treated, 120/105 ksi minimum tensile strength (withdrawn 2016).* ASTM A325-14. West Conshohocken, PA: ASTM.

ASTM. 2015. *Standard specification for steel transmission tower bolts, zinc-coated and bare.* ASTM A394-08. West Conshohocken, PA: ASTM.

ASTM. 2016. *Standard specification for low-alloy steel deformed and plain bars for concrete reinforcement.* ASTM A706/A706M. West Conshohocken, PA: ASTM.

ASTM. 2019. *Standard specification for carbon structural steel.* ASTM A36/A36M. West Conshohocken, PA: ASTM.

ASTM. 2021. *Standard specification for high-strength low-alloy columbium-vanadium structural steel.* ASTM A572/A572M. West Conshohocken, PA: ASTM.

ASTM. 2022. *Standard specification for carbon steel bolts and studs, 60,000 psi tensile strength.* ASTM A307-22. West Conshohocken, PA: ASTM.

AWS (American Welding Society). 2018. *Structural welding code—Steel reinforcing bars.* AWS D1.4/D1.4 M. Miami: AWS.

Brockenbrough, R. L., and J. Schuster. 2018. *AISC design guide 15, rehabilitation and retrofit guide: A reference for historic shapes and specifications.* 2nd ed. Chicago: AISC.

EPA (US Environmental Protection Agency). 1990. *National emission standards for hazardous air pollutants: Asbestos NESHAP revision: Final rule.* Washington, DC: EPA.

IEEE. 2018. *Recommended practice for seismic design of substations.* IEEE 693-2018. Piscataway, NJ: IEEE.

IEEE. 2021. *Guide for physical security of electric power substations.* IEEE 1402-2021. Piscataway, NJ: IEEE.

IEEE. 2023. *National electrical safety code.* ANSI C2. Piscataway, NJ: IEEE.

USDA (US Department of Agriculture). 2013. *Wood pole inspection and maintenance. Rural Utility Service Bulletin* 1730B-121. Washington, DC: USDA.

CHAPTER 12
OIL CONTAINMENT AND BARRIER WALLS

12.1 GENERAL

This chapter is intended to provide utility engineers with information to develop their own substation design criteria/standards for

- Secondary and tertiary oil containments: Secondary containment is intended to catch most if not all oil, and tertiary containment is meant to catch any oil prior to leaving the property;
- Firewalls;
- Sound walls;
- Ballistic walls; and
- Blast walls.

In addition to other information discussed in this chapter, substation oil containment and barrier walls should allow access to equipment during any incidents, regular inspection, maintenance, and future replacement of substation equipment and structures. Aesthetic consideration should be given where walls are highly visible by the public.

12.2 OIL CONTAINMENT

12.2.1 General

Substations that have oil-filled operational equipment with an aggregate capacity of 1,320 gal. (5,000 L) or more, and that could reasonably be expected to discharge harmful quantities of oil into or upon navigable waters or adjoining shorelines, must have a spill prevention control and countermeasure (SPCC) plan prepared in accordance with the EPA's Code of Federal Regulations, Title 40 Section 112 (API 1989). Navigable waters, as applied in this context, are defined in the Code of Federal Regulations, 40 CFR 112. One of the fundamental requirements of the SPCC rule is the provision of secondary containment to prevent a discharge from reaching navigable waters or adjoining shorelines. However, the rule allows for a plan that relies on emergency response including a written commitment of manpower, equipment, and materials, as a means of secondary containment for oil-filled operational equipment. Although such a plan is suitable for certain facilities, under such an approach, there is still a risk of a catastrophic release from substation equipment reaching navigable waters or adjoining shorelines. A discharge to a navigable waterway or adjoining shorelines is a violation of the Clean Water Act, which could result in significant fines in addition to remediation costs.

Therefore, where feasible, utilities, have deployed the addition of physical secondary and tertiary containment at substations as an effective risk management strategy.

12.2.2 Containment Systems

Oil containment systems can be designed for individual equipment, as common containment for multiple pieces of equipment, or for the entire substation. This section describes common types of containment systems; however, there may also be many other systems or a hybrid of systems. In addition, fire-quenching systems may be used in conjunction with the containment system. For further information regarding containment and fire quenching [IEEE 979-2012 (IEEE 2012), IEEE 980-2013 (IEEE 2013a)].

12.2.2.1 Substation Mat. The substation mat consists of the base material used throughout the substation and can be designed to act as an oil containment system. Typically, substation mats are designed to prevent the release of oil from outside the substation. Factors that affect an oil containment system include equipment distance to the fence, void ratio of the base material, and spill response time for cleanup. An impervious liner can also be installed underneath the mat to prevent the vertical spread of any released oil.

12.2.2.2 Oil Retention Pit. The oil retention pit can be located around the equipment. If space is constrained within the substation, the utility can also install a collection pit around the equipment and pipe the oil to a retention pit located in a less-congested location. If the surrounding soils are porous, the oil retention pit can be lined with a layer of clay, concrete, plastic, or a rubber liner. See Figure 12-1 for an example.

Figure 12-1. Reinforced concrete open pit containment.
Source: Courtesy of CenterPoint Energy.

12.2.2.3 Oil Absorbents and Oil Solidifiers. Oil absorbents and oil solidifiers are typically oleophilic hydrophobic polymers that absorb and react with oil and can be useful in retrofit situations where space and construction considerations are a concern. As the polymer comes into contact with oil, the medium swells or reacts with the oil to increase its viscosity and to stop the release of the oil. Under normal operating conditions, the polymer will still allow water to drain through the system. Oil absorbents and oil solidifiers can be used as a standalone containment system or can be used in conjunction with other systems such as berms and dikes and oil retention pits. These materials can also be contained in storm-water control valves, pillows, pads, socks, or various other systems.

12.2.2.4 Berms and Dikes. Berms and dikes can be constructed from many different materials such as concrete, soil, or prefabricated plastic or fiberglass panels. Liners, oil filtration materials, oil absorbents, and oil solidifiers can be used in conjunction with primary material to prevent the migration of oil. When designing berms and dikes, the engineer should consider equipment type and size, available space, site-specific soil conditions, level of protection desired, and access for equipment maintenance. Depending on the oil containment design and utility needs, berms and dikes can be installed around equipment and/or around the substation perimeter. When installed around the equipment perimeter, they should encompass all of the oil-containing appurtenances (e.g., transformer radiators).

12.2.2.5 Self-Extinguishing Oil Containment. Although the containment of oil spillage is necessary from an environmental standpoint, it may cause an inadvertent fire risk if the event causing the spillage has a source of ignition. This is often the case with live substation equipment.

One strategy to both collect the oil and extinguish a fire in the oil is to have a remote oil retention basin that is connected to the equipment pad via a reinforced concrete pipe. The site is graded such that liquids on the pad flow toward the basin via the pipe by means of gravity.

The pipe is sized (length and diameter) such that it will extinguish the burning oil by limiting the oxygen.

12.2.3 Oil Retention Drainage

The engineer should consider solutions to remove collected rainwater. Some examples, as described further in IEEE 980-2013 (IEEE 2013a), include

- Oil storage tanks,
- Oil–water separators,
- Oil-detection-triggered sump pumps,
- Gravity oil–water separators,
- Oil–water stop valves, and
- Oil-absorbing polymer beads.

12.2.4 Design Considerations

Oil containment should be designed to comply with the requirements of the SPCC plan. The design may also consider the possibility of unacceptable damage to nearby control enclosures, equipment, or structures.

The need for oil containment should be determined on a case-by-case basis. IEEE 979-2012 (IEEE 2012) and IEEE 980-2013 (IEEE 2013a) are references that may be used in evaluating the need, type, and size of oil containment.

Some additional considerations include the transformer installation method and physical site constraints. Typical "Jack-and-skid" installation or crane installation methods often impose different space constraints.

Environmental considerations can also have an impact on the choice of the oil containment system. Existing yards can have soils that have already become contaminated through the life cycle of the substation. It is recommended that the utility engineer coordinate with their internal environmental resources, as needed, to determine what the impact of large excavation can have on the overall project cost and familiarize themselves with their organization's policies and methods for such remediation.

12.3 TYPES OF BARRIER WALLS

12.3.1 General

This section will introduce engineers to various barrier walls used in electric substations and their relevant design considerations. These barrier walls and their functions are described as follows:

- Fire walls are used to shield adjacent structures, equipment, or the public from potential fire hazards arising from substation equipment or facilities.
- Sound walls are used to reduce noise impact caused by operational sounds generated as a result of substation equipment installation and operation.
- Ballistic or blast walls are used to protect substation equipment from external criminal intrusion/sabotage or equipment malfunction.

Utilities often do site-specific studies to examine the vulnerabilities of critical substations. These studies can be used to develop design criteria specific to site-security requirements and can be used to determine the appropriate type of barrier wall. The barrier walls can be a combination of one or more of the wall types described in this section and should be designed with ease of maintenance in mind.

Barrier walls should be designed to meet the provisions in Chapter 3, "Loading Criteria for Substation Structures," for wind, earthquake, and dead loads in addition to the recommendations provided in the following sections.

12.3.2 Firewalls

Firewalls are commonly constructed of conventional cast-in-place reinforced concrete, masonry block, or composite materials (Figure 12.2).

A properly designed equipment firewall should be capable of withstanding extreme temperatures resulting from an oil fire that can be in the range of 1,700 to 2,200 °F (927 to 1,204 °C) and that can burn for a considerable amount of time as local fire departments may not be equipped to handle such a fire. Design recommendations for firewall fire rating, separation distances, and minimum heights are provided in IEEE 979-2012 (IEEE 2012) and NFPA 850-2020 (NFPA 2020). In addition, fire ratings for various wall materials and associated thicknesses are listed in Table 721.1(2) of the *International Building Code*, Volume 1 (ICC 2012). Further information is available from various commercial code testing organizations for other fire-rated wall assemblies.

Figure 12-2. Transformer protection fire wall.
Source: Courtesy of the US Department of Energy.

Firewalls are often installed when acceptable equipment separation distances cannot be met. Minimum separation distance recommendations are provided in IEEE 979-2012 (IEEE 2012). The firewall design should minimize interference with proper ventilation of nearby equipment and include consideration for maintenance access. Furthermore, control cables can be affected if a fire is allowed to propagate. Care should be taken in the placement and/or protection of these cables.

12.3.3 Sound Walls

Equipment within a substation can generate noise, which can be a nuisance to the surrounding community. Transformers, in particular, can generate a humming noise. Jurisdictions (city, county, state, and federal) may have sound-level requirements that are applicable to the particular substation site under consideration.

The following is a list of references for evaluating sound level and associated project impacts.

- ANSI/ASA S12.9-2012, Part 5 (ANSI/ASA 2012);
- ANSI/IEEE C57.12.90 (IEEE 2015); and
- IEEE 1127-2013 (IEEE 2013b).

12.3.4 Ballistic Walls and Blast Walls

To develop an effective and economical ballistic wall, the following parameters should be defined in consultation and coordination with the utility's security specialist:

1. Identify a range of threats and a distance from which these can be launched against the substation equipment. These may range from various firearms to explosives.
2. Choose an acceptable caliber of bullet, a range, and a position from which it is fired. This is a function of the security plans determined by individual utilities.

3. Identify a permissible level of damage to the facilities that are being protected.
4. Define the line of sight from which the barrier wall is to protect the equipment. This is typically the anticipated positions from which the equipment could be fired upon.
5. Try to not prevent both air flow and drainage.

Ballistic walls should be designed with nonflammable material if possible. Examples of materials used for ballistic walls include

- Masonry block,
- Precast firewalls,
- Carbon fiber sheets,
- Kevlar,
- Layered fiberglass,
- Steel plate, and
- Combination of any of the above.

Several standards are available for ballistic testing, some of them include

- Underwriters Laboratories 752 (UL/ANSI 2005),
- National Institute of Justice 0108.01 (NIJ 1985), and
- US Department of State SD-STD-02.01 (DOS 2003).

For the design of blast walls, information is available in specialized building design standards; these include ASCE 59-11, *Blast Protection of Buildings* (ASCE 2011), and UFC 4-010-01, *Minimum Antiterrorism Standards for Buildings* (DOD 2018).

In some cases, the utility may decide to install screen walls to block the line of site to critical infrastructure. This can be done with ballistic-or nonballistic-rated screen walls.

Guidance on ballistic walls can be found in IEEE 1402-2021, *Guide for Physical Security of Electric Power Substations* (IEEE 2021).

REFERENCES

ANSI/ASA (American National Standards Institute/Acoustical Society of America). 2012. *Quantities and procedures for description and measurement of environmental sound—Part 5: Sound level description for determination of compatible land use.* ANSI/ASA S12.9-2012. Sewickley, PA: ASA.

API (American Petroleum Institute). 1989. *Suggested procedure for development of spill prevention control and countermeasure plans.* Title 40 Code of Federal Regulations, Part 112, Oil Pollution Prevention. Washington, DC: API.

ASCE. 2011. *Blast protection of buildings.* ASCE 59-11. Reston, VA: ASCE.

DOD (US Department of Defense). 2018. *Minimum antiterrorism standards for buildings.* UFC 4-010-01. Washington, DC: DOD.

DOS (US Department of State). 2003. *Test method for crash testing of perimeter barrier and gates.* SD-STD-02.01. Washington, DC: DOS.

ICC (International Code Council). 2012. *International building code: Code and commentary.* Table 721.1(2). Washington, DC: ICC.

IEEE (Institute of Electrical and Electronics Engineers). 2012. *Guide for substation fire protection.* IEEE 979-2012. Piscataway, NJ: IEEE.

IEEE. 2013a. *Guide for containment and control of oil spills in substations*. IEEE 980-2013. Piscataway, NJ: IEEE.

IEEE. 2013b. *Guide for the design, construction and operation of electric power substations for community acceptance and environmental compatibility*. IEEE 1127-2013. Piscataway, NJ: IEEE.

IEEE. 2015. *Standard test code for liquid-immersed distribution, power, and regulating transformers*. IEEE C57.12.90. Piscataway, NJ: IEEE.

IEEE. 2021. *Guide for physical security of electric power substations*. IEEE 1402-2021. Piscataway, NJ: IEEE.

NFPA (National Fire Protection Association). 2020. *Recommended practice for fire protection for electric generating plants and high voltage direct current converter stations*. NFPA 850-2020. Quincy, MA: NFPA.

NIJ (National Institute of Justice). 1985. *Ballistic resistant protective materials: Standard*. NIJ 0108.01. Washington, DC: US Department of Justice.

UL (Underwriters Laboratories)/ANSI (American National Standards Institute). 2005. *Standard for bullet-resisting equipment*. UL 752. Northbrook, IL: UL.

US Code of Federal Regulations (CFR), Title 40 Protection of Environment, Washington, DC: EPA, Office of Emergency Management.

APPENDIX A
EXAMPLES

The examples given in this appendix are generic and are meant to show the use of the equations from the various chapters. The locations of the structures are not given or implied. The materials chosen, wind speeds, and ice thicknesses with concurrent wind were used to illustrate the equations from the chapters. These examples are not meant to limit the scope of possible load combinations on the structures chosen. The utility owner may direct the designer to consider additional loads or a combination of loads, such as construction and maintenance loads. When computing the structure height factor K_z, the highest point of the structure or equipment was used for simplicity. All units used in these examples are US customary units. The conversion to metric units is left to the reader.

A.1 LOAD DEVELOPMENT FOR A THREE-PHASE BUS SUPPORT STRUCTURE (FIGURE A-1a)

Determine the basic design inputs and develop loading for the three-phase bus support structure shown in the following figures, for the following load cases:

- Extreme wind loads (Section 3.1.5),
- Combined ice and wind loads (Section 3.1.6),
- Seismic loads (Section 3.1.7),
- Short-circuit loads (Section 3.1.8), and
- Deflection loads (Section 3.1.11).

Structure Data and Geometry:

The structure will be composed of steel using 8 in. HSS steel sections for the columns and an 8 in. W-section for the beam, with the following basic geometric information:

$H_{str} = 18.50$ ft   Height of structure to the rigid bus

$\text{Beam}_{depth} = 8.0$ in.   Main beam depth

$\text{Column}_{depth} = 8.0$ in.   Column depth

$DL_{beam} = 30.00$ lb/ft   Main beam weight

$DL_{column} = 30.00$ lb/ft   Column weight

Phase Spacing $= 10.00$ ft   Phase spacing of the rigid bus

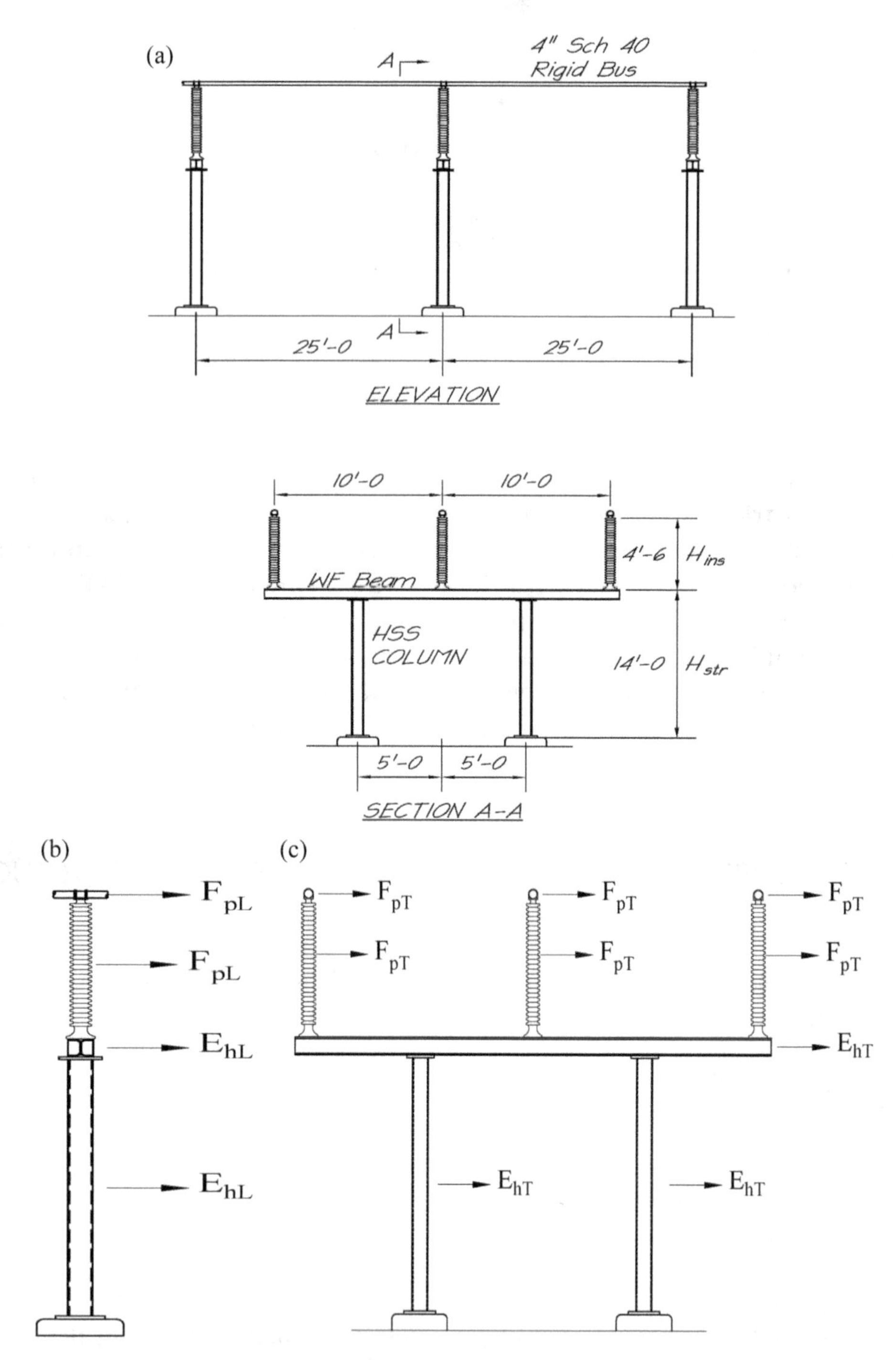

Figure A-1. (a) Three-phase bus support structure, (b) seismic forces in the longitudinal direction of rigid bus, (c) seismic forces in the transverse frame direction of the bus support.

Rigid Bus Data:

The rigid bus will be composed of seamless pipe. A typical choice for substations is Aluminum 6063-T6 alloy, ANSI Schedule 40 pipe because of its excellent mechanical and electrical properties. This example will use a nominal 4 in. rigid bus size with the following basic geometric information:

$d_{RB} = 4.50$ in. Outside diameter of rigid bus

$DL_{RB} = 3.73$ lb/ft Weight of 4 in. rigid bus and 795 MCM ACSR conductor

$\text{Trib}_{\text{span}} = 25.00\,\text{ft}\left(\frac{5}{4}\right) = 31.25\,\text{ft}$ Tributary rigid bus span length, repetitive span, center support, according to IEEE 605 (IEEE 2008) recommendations

Insulator and Bus Fitting Data:

The insulator will be a T.R. 288 and will use a fixed bus fitting for connection between the insulator and rigid bus.

$H_{\text{ins}} = 4.50$ ft Height of insulator, T.R. 288

$d_{\text{ins}} = 9.69$ in. Outside shed diameter, T.R. 288

$DL_{\text{ins}} = 200.00$ lb Weight of (1) T.R. 288

$\text{AREA}_{\text{ins}} = H_{\text{ins}} d_{\text{ins}} = 3.63\ \text{ft}^2$ Projected wind area, T.R. 288

$H_{\text{fit}} = 5.00$ in. Height of rigid bus fitting

$DL_{\text{fit}} = 6.25$ lb Weight of rigid bus fitting

A.1.1 Extreme Wind Loads (Section 3.1.5)

Calculate the extreme wind pressure for use in design with USD and ASD load combinations: The values for C_f are taken from Table 3-8.

Coefficients and Definitions:

$Q = 0.00256$ Air density factor, sea level, 59 °F (Section 3.1.5.1)

$K_Z = 2.41\left(\frac{18.5}{2,460}\right)^{2/9.8} = 0.89$ [(Equation(3-4)] Terrain exposure coefficient, Exposure C (Section 3.1.5.2.1) at 18.5 ft

$K_d = 1.00$ Wind directionality factor (Section 3.1.5)

$K_{zt} = 1.00$ Topographic factor (Section 3.1.5)

$V_{\text{MRI\_300}} = 95.0$ mph Basic wind speed, 300-year MRI (Section 3.1.5.3)

$V_{\text{MRI\_100}} = 85.0$ mph Basic wind speed, 100-year MRI (Section 3.1.5.3)

$G_{\text{SRF}} = 1.00$ Gust response factor (structure) (Section 3.1.5.5, Table 3-6)

$C_{f\_rnd} = 0.90$ Force coefficient, circular cross section (Table 3-8)

$C_{f\_\text{bus}} = 1.00$ Force coefficient, rigid bus (Table 3-8)

$C_{f\_\text{avg}} = 1.60$ Force coefficient, structural shapes (Table 3-8)

$C_{f\_\text{sq}} = 2.00$ Force coefficient, square/rectangular section (Table 3-8)

The following wind pressures are used with USD load combinations.
USD wind force on insulators:

$$F_{\text{USD\_}rnd} = QK_Z K_d K_{zt}(V_{\text{MRI\_300}})^2 G_{\text{SRF}} C_{f\_rnd} AREA_{\text{ins}} = 67.25\ \text{lb}$$

USD wind force on the rigid bus:

$$F_{\text{USD\_bus}} = QK_Z K_d K_{zt}(V_{\text{MRI\_300}})^2 G_{\text{SRF}} C_{f\_\text{bus}} d_{\text{RB}} \text{Trib}_{\text{span}} = 240.97\ \text{lb}$$

USD wind pressure on the WF beam structure surface:

$$P_{\text{USD\_str\_sq}} = QK_Z K_d K_{zt}(V_{\text{MRI\_300}})^2 G_{\text{SRF}} C_{f\_\text{avg}} = 32.90\ \text{psf}$$

USD wind pressure on the HSS (flat) column structure surface:

$$P_{USD\_str\_sq} = QK_Z K_d K_{zt} (V_{MRI\_300})^2 G_{SRF} C_{f\_sq} = 41.13 \text{ psf}$$

The following wind pressures are used with ASD load combinations.
ASD wind force on the insulators:

$$F_{ASD\_rnd} = QK_Z K_d K_{zt} (V_{MRI\_100})^2 G_{SRF} C_{f\_rnd} \text{AREA}_{ins} = 53.84 \text{ lb}$$

ASD wind force on the rigid bus:

$$F_{ASD\_bus} = QK_Z K_d K_{zt} (V_{MRI\_100})^2 G_{SRF} C_{f\_bus} d_{RB} \text{Trib}_{span} = 192.91 \text{ lb}$$

ASD wind pressure on the WF beam structure surface:

$$P_{ASD\_sw} = QK_Z K_d K_{zt} (V_{MRI\_100})^2 G_{SRF} C_{f\_avg} = 26.34 \text{ psf}$$

ASD wind pressure on the HSS column (flat) structure surfaces:

$$P_{ASD\_str\_sq} = QK_Z K_d K_{zt} (V_{MRI\_100})^2 G_{SRF} C_{f\_sq} = 32.92 \text{ psf}$$

A.1.2 Combined Ice and Wind Loads (Section 3.1.6)

Calculate the combined ice and wind pressure for use in design with USD and ASD load combinations:

Coefficients and Definitions:

All factors previously defined remain applicable here. Only new parameters are defined subsequently.

$V_{MRI\_WI\_100} = 40.0$ mph Concurrent wind speed with ice, 100-year MRI (Figure 3-6b)

$t_{ICE\_100} = 0.50$ in. Radial ice thickness, 100-year MRI (Figure 3-6a)

$V_{MRI\_WI\_50} = 40.0$ mph Concurrent wind speed with ice, 50-year MRI (Figure 3-6b)

$t_{ICE\_50} = 0.25$ in. Radial ice thickness, 50-year MRI (Figure 3-6d)

$\gamma_{ICE} = 56.0 \text{ lb/ft}^3$ Density of ice (Section 3.1.6.1)

The following wind pressures and ice loads are used with Ultimate Strength Design (USD) load combinations.

USD combined wind force and ice load on the insulators:

$$F_{USD\_WI\_ins} = QK_Z K_d K_{zt} (V_{MRI\_WI\_100})^2 G_{SRF} C_{f\_rnd} H_{ins} (d_{ins} + 2t_{ICE\_100})$$

$$F_{USD\_WI\_ins} = 13.15 \text{ lb}$$

$$F_{USD\_ICE\_ins} = \gamma_{ICE} \frac{\pi[(d_{ins} + 2t_{ICE-100})^2 - d_{ins}{}^2]}{4} H_{ins} = 28.01 \text{ lb}$$

USD combined wind force and ice load on the rigid bus:

$$F_{USD\_WI\_bus} = QK_Z K_d K_{zt} (V_{MRI\_WI\_100})^2 G_{SRF} C_{f_{bus}} \text{Trib}_{span} (d_{RB} + 2t_{ICE-100})$$

$$F_{USD\_WI\_bus} = 52.21 \text{ lb}$$

$$F_{USD\_ICE\_bus} = \gamma_{ICE} \frac{\pi[(d_{RB} + 2t_{ICE-100})^2 - d_{RB}{}^2]}{4} \text{Trib}_{span} = 95.45 \text{ lb}$$

The following wind pressures and ice loads are used with ASD load combinations.
ASD combined wind force and ice load on the insulators:

$$F_{\text{ASD\_WI\_ins}} = QK_Z K_d K_{zt} (V_{\text{MRI\_WI\_50}})^2 G_{\text{SRF}} C_{f\_rnd} H_{\text{ins}} (d_{\text{ins}} + 2t_{\text{ICE}-50})$$

$$F_{\text{ASD\_WI\_ins}} = 12.54 \text{ lb}$$

$$F_{\text{ASD\_ICE\_ins}} = \gamma_{\text{ICE}} \frac{\pi[(d_{\text{ins}} + 2t_{\text{ICE}-50})^2 - d_{\text{ins}}{}^2]}{4} H_{\text{ins}} = 13.66 \text{ lb}$$

ASD combined wind force and ice load on the rigid bus:

$$F_{\text{ASD\_WI\_bus}} = QK_Z K_d K_{zt} (V_{\text{MRI\_WI\_50}})^2 G_{\text{SRF}} C_{f_{\text{bus}}} \text{Trib}_{\text{span}} (d_{\text{RB}} + 2t_{\text{ICE}-50})$$

$$F_{\text{ASD\_WI\_bus}} = 47.47 \text{ lb}$$

$$F_{\text{ASD\_ICE\_bus}} = \gamma_{\text{ICE}} \frac{\pi[(d_{\text{RB}} + 2t_{\text{ICE}-50})^2 - d_{\text{RB}}{}^2]}{4} \text{Trib}_{\text{span}} = 45.34 \text{ lb}$$

A.1.3 Seismic Loads (Section 3.1.7)

Calculate the seismic load effect for use in design with both USD and ASD load combinations. The seismic load effect will be calculated for the structure and the insulator/rigid bus according to the recommendations of this guide.

The basic seismic load effect consists of both a horizontal and a vertical application of loading, as shown in the following:

$$E = E_h + E_v$$

and

$$E = E_h - E_v$$

Site-Specific Seismic Parameters and Given Quantities—For Site Class D:

Coefficients and Definitions:

$S_S = 0.59$ Mapped ground motion spectral response, short period

$S_1 = 0.18$ Mapped ground motion spectral response (1 s)

$I_e = 1.25$ Seismic importance factor (Table 3-14) (this bus support structure was assumed to be essential to operation)

$I_{mv} = 1.50$ Modal contribution factor (Section 3.1.7.6.2) (multimode participation was assumed)

$S_{MS} = 0.78$ Maximum considered earthquake spectral response acceleration parameter, at short period (Section 3.1.7.2)

$S_{M1} = 0.37$ Maximum considered earthquake spectral response acceleration parameter, at 1 s, (Section 3.1.7.2)

$S_{DS} = 0.52$ Design spectral response acceleration parameter, at short period [Equation (3-9)]

$S_{D1} = 0.25$ Design spectral response acceleration parameter, at 1-second period [Equation (3-10)]

This three-phase bus support structure with a single beam and two columns behaves as a simple cantilever in the longitudinal bus direction (Figure A-1b) and as a moment frame transverse to the bus (Figure A-1c). Parameters for R, Ω_0, and C_d were chosen from Table 3-13 for the two directions. It was assumed that the bus did not provide longitudinal stiffness for this example. The seismic load development will be derived in both directions.

Simple Cantilever Direction—Weak Axis of Frame (Table 3-13)

$R_L = 1.00$ Response modification coefficient

$\Omega_{0L} = 1.00$ Overstrength factor

$C_{dL} = 1.00$ Deflection amplification factor

Moment Frame Direction—Strong Axis of Frame

$R_T = 2.00$ Response modification coefficient

$\Omega_{0T} = 2.00$ Overstrength factor

$C_{dT} = 2.00$ Deflection amplification factor

$I_e = 1.25$ Seismic importance (Table 3-14)

Lateral Seismic Forces—Equivalent Lateral Force Procedure—Structure (Section 3.1.7.6.2)

$V = E_h = C_s W(I_{mv})$ [Equation (3-11)]

Note: For these examples, E_h will be used for base shear V in Equation (3-11)

$C_s = \dfrac{S_{DS}}{(R/I_e)} = 0.65$ Seismic response coefficient [Equation (3-12)]

Note: The C_s value from Equation (3-12) should be checked for the outer-bound maximum and minimum values of Equations (3-13), (3-14), or (3-15). This check will not be performed here.

$V = E_h = C_s W(I_{mv}) = 0.98W$ Basic seismic lateral force [Equation (3-11)]

Lateral Seismic Forces—Equivalent Lateral Force Procedure—Rigid Bus/Insulator (Section 3.1.7.11)

$F_P = 1.6 S_{DS} I_e W_p = 1.05 W_p$ Basic seismic lateral force, for the rigid bus and insulators [Equation (3-36)]

Vertical Seismic Forces (Section 3.1.7.7)

$E_V = 0.80 S_{DS} D = 0.42D$ Basic seismic vertical force [Equation (3-19a)]

Basic Seismic Load Effects (Section 3.1.7.8.1)

Terminology:

- "W" = used for weight of the element.
- "hL" = used for orthogonal longitudinal direction—horizontal plane
- "hT" = used for orthogonal transverse direction—horizontal plane
- "v" = used for orthogonal direction—vertical plane
- D = dead load

Structure:

Basic seismic load effects for application to the structure itself, used for structural design of main members. The structure will be analyzed for the seismic load vectors in the three orthogonal directions in each of the following equations:

$E = E_{hL} + 0.4E_{hT} \pm 0.4E_v = (0.98W) + (0.39W) \pm (0.17W)$ [Equation (3-20)]

$E = 0.4E_{hL} + E_{hT} \pm 0.4E_v = (0.39W) + (0.98W) \pm (0.17W)$ [Equation (3-21)]

$E = 0.4E_{hL} + 0.4E_{hT} \pm E_v = (0.39W) + (0.39W) \pm (0.42W)$ [Equation (3-22)]

Seismic Force Magnitudes to be applied in Equations (3-20), (3-21), and (3-22):

$0.98W = 0.98DL_{\text{beam}} = 29.43$ lb/ft Magnitude of longitudinal load (3-20)

$0.39W = 0.39DL_{\text{beam}} = 11.77$ lb/ft Magnitude of transverse load (3-20)

$0.17W = 0.17DL_{\text{beam}} = \pm 5.02$ lb/ft Magnitude of vertical load (3-20)

$0.39W = 0.39DL_{\text{beam}} = 11.77$ lb/ft Magnitude of longitudinal load (3-21)

$0.98W = 0.98DL_{\text{beam}} = 29.43$ lb/ft Magnitude of transverse load (3-21)

$0.17W = 0.17DL_{\text{beam}} = \pm 5.02$ lb/ft Magnitude of vertical load (3-21)

$0.39W = 0.39DL_{\text{beam}} = 11.77$ lb/ft Magnitude of longitudinal load (3-22)

$0.39W = 0.39DL_{\text{beam}} = 11.77$ lb/ft Magnitude of transverse load (3-22)

$0.42W = 0.42DL_{\text{beam}} = \pm 12.56$ lb/ft Magnitude of vertical load (3-22)

Seismic Load Effect with Overstrength Factor, Ω_0—(Section 3.1.7.8.2)

Seismic load effects for application to the structure itself, and analysis results used for all structural designs of connections and anchorage.

$E_m = \Omega_{0L}E_{hL} + 0.4\Omega_{0T}E_{hT} \pm 0.4E_v = (0.98W) + (0.78W) \pm (0.17W)$ (3-27)

$E_m = 0.4\Omega_{0L}E_{hL} + \Omega_{0T}E_{hT} \pm 0.4E_v = (0.39W) + (1.96W) \pm (0.17W)$ (3-28)

$E_m = 0.4\Omega_{0L}E_{hL} + 0.4\Omega_{0T}E_{hT} \pm E_v = (0.39W) + (0.78W) \pm (0.42W)$ (3-29)

Seismic Force Magnitudes (columns and beam) to be applied to Equations (3-27), (3-28), and (3-29):

$0.98W = 0.98DL_{\text{member}} = 29.43$ lb/ft Magnitude of longitudinal load (3-27)

$0.78W = 0.78DL_{\text{member}} = 23.54$ lb/ft Magnitude of transverse load (3-27)

$0.17W = 0.17DL_{\text{member}} = \pm 5.02$ lb/ft Magnitude of vertical load (3-27)

$0.39W = 0.39DL_{\text{member}} = 11.77$ lb/ft Magnitude of longitudinal load (3-28)

$1.96W = 1.96DL_{\text{member}} = 58.85$ lb/ft Magnitude of transverse load (3-28)

$0.17W = 0.17DL_{\text{member}} = \pm 5.02$ lb/ft Magnitude of vertical load (3-28)

$0.39W = 0.39DL_{\text{member}} = 11.77$ lb/ft Magnitude of longitudinal load (3-29)

$0.78W = 0.78DL_{\text{member}} = 23.54$ lb/ft Magnitude of transverse load (3-29)

$0.42W = 0.42DL_{\text{member}} = \pm 12.56$ lb/ft Magnitude of vertical load (3-29)

Rigid Bus and Insulator:

Basic seismic load effects for application to the rigid bus or insulator used for structural design of main members. *Note:* In this case, the horizontal seismic force "E_h" is replaced with "F_p" for the Horizontal Seismic Component Force (3.1.7.11) applied to the insulators and rigid bus in the longitudinal and transverse directions.

$E = F_{PL} + 0.4F_{PT} \pm 0.4E_v = (1.05W) + (0.42W) \pm (0.17W)$ (3-20)

$$E = 0.4F_{PL} + F_{PT} \pm 0.4E_v = (0.42W) + (1.05W) \pm (0.17W) \quad (3\text{-}21)$$

$$E = 0.4F_{PL} + 0.4F_{PT} \pm E_v = +(0.42W) + (0.42W) \pm (0.42W) \quad (3\text{-}22)$$

Seismic Force Magnitudes—Insulator to be applied in Equations (3-20), (3-21), and (3-22):

$1.05W = 1.05DL_{\text{ins}} = 209.25$ lb Magnitude of longitudinal load (3-20)

$0.42W = 0.42DL_{\text{ins}} = 83.70$ lb Magnitude of transverse load (3-20)

$0.17W = 0.17DL_{\text{ins}} = \pm 33.48$ lb Magnitude of vertical load (3-20)

$0.42W = 0.42DL_{\text{ins}} = 83.70$ lb Magnitude of longitudinal load (3-21)

$1.05W = 1.05DL_{\text{ins}} = 209.25$ lb Magnitude of transverse load (3-21)

$0.17W = 0.17DL_{\text{ins}} = \pm 33.48$ lb Magnitude of vertical load (3-21)

$0.42W = 0.42DL_{\text{ins}} = 83.70$ lb Magnitude of longitudinal load (3-22)

$0.42W = 0.42DL_{\text{ins}} = 83.70$ lb Magnitude of transverse load (3-22)

$0.42W = 0.42DL_{\text{ins}} = \pm 83.70$ lb Magnitude of vertical load (3-22)

Seismic Force Magnitudes—Rigid Bus to be applied in Equations (3-20), (3-21), and (3-22):

$1.05W = 1.05DL_{RB}\text{Trib}_{\text{span}} = 121.96$ lb Magnitude of longitudinal load (3-20)

$0.42W = 0.42DL_{RB}\text{Trib}_{\text{span}} = 48.78$ lb Magnitude of transverse load (3-20)

$0.17W = 0.17DL_{RB}\text{Trib}_{\text{span}} = \pm 19.51$ lb Magnitude of vertical load (3-20)

$0.42W = 0.42DL_{RB}\text{Trib}_{\text{span}} = 48.78$ lb Magnitude of longitudinal load (3-21)

$1.05W = 1.05DL_{RB}\text{Trib}_{\text{span}} = 121.96$ lb Magnitude of transverse load (3-21)

$0.17W = 0.17DL_{RB}\text{Trib}_{\text{span}} = \pm 19.51$ lb Magnitude of vertical load (3-21)

$0.42W = 0.42DL_{RB}\text{Trib}_{\text{span}} = 48.78$ lb Magnitude of longitudinal load (3-22)

$0.42W = 0.42DL_{RB}\text{Trib}_{\text{span}} = 48.78$ lb Magnitude of transverse load (3-22)

$0.42W = 0.42DL_{RB}\text{Trib}_{\text{span}} = \pm 48.78$ lb Magnitude of vertical load (3-22)

Seismic Load Effect with Overstrength Factor (Section 3.1.7.8.2)
Seismic load effects for application to the rigid bus or insulator are shown below. The analysis results are used for all structural designs of connections and anchorage.

$$E_m = \Omega_{0L}F_{PL} + 0.4\Omega_{0T}F_{PT} \pm 0.4E_v = (1.05W) + (0.84W) \pm (0.17W) \quad (3\text{-}27)$$

$$E_m = 0.4\Omega_{0L}F_{PL} + \Omega_{0T}F_{PT} \pm 0.4E_v = (0.42W) + (2.09W) \pm (0.17W) \quad (3\text{-}28)$$

$$E_m = 0.4\Omega_{0L}F_{PL} + 0.4\Omega_{0T}F_{PT} \pm E_v = (0.42W) + (0.84W) \pm (0.42W) \quad (3\text{-}29)$$

Seismic Force Magnitudes—Insulator to be applied in Equations (3-27), (3-28), and (3-29):

$1.05W = 1.05DL_{\text{ins}} = 209.25$ lb Magnitude of longitudinal load (3-27)

$0.84W = 0.84DL_{\text{ins}} = 167.40$ lb Magnitude of transverse load (3-27)

$0.17W = 0.17DL_{\text{ins}} = \pm 33.48$ lb Magnitude of vertical load (3-27)

$0.42W = 0.42DL_{\text{ins}} = 83.70$ lb Magnitude of longitudinal load (3-28)

$2.09W = 2.09DL_{\text{ins}} = 418.51$ lb Magnitude of transverse load (3-28)

$0.17W = 0.17DL_{\text{ins}} = \pm 33.48$ lb Magnitude of vertical load (3-28)

$0.42W = 0.42DL_{\text{ins}} = 83.70$ lb Magnitude of longitudinal load (3-29)

$0.84W = 0.84DL_{\text{ins}} = 167.40$ lb Magnitude of transverse load (3-29)

$0.42W = 0.42DL_{\text{ins}} = \pm 83.70$ lb Magnitude of vertical load (3-29)

Seismic Force Magnitudes—Rigid Bus to be applied in Equations (3-27), (3-28), and (3-29):

$1.05W = 1.05DL_{RB}\text{Trib}_{\text{span}} = 121.96$ lb Magnitude of longitudinal load (3-27)

$0.84W = 0.84DL_{RB}\text{Trib}_{\text{span}} = 97.56$ lb Magnitude of transverse load (3-27)

$0.17W = 0.17DL_{RB}\text{Trib}_{\text{span}} = \pm 19.51$ lb Magnitude of vertical load (3-27)

$0.42W = 0.42DL_{RB}\text{Trib}_{\text{span}} = 48.78$ lb Magnitude of longitudinal load (3-28)

$2.09W = 2.09DL_{RB}\text{Trib}_{\text{span}} = 243.91$ lb Magnitude of transverse load (3-28)

$0.17W = 0.17DL_{RB}\text{Trib}_{\text{span}} = \pm 19.51$ lb Magnitude of vertical load (3-28)

$0.42W = 0.42DL_{RB}\text{Trib}_{\text{span}} = 48.78$ lb Magnitude of longitudinal load (3-29)

$0.84W = 0.84DL_{RB}\text{Trib}_{\text{span}} = 97.56$ lb Magnitude of transverse load (3-29)

$0.42W = 0.42DL_{RB}\text{Trib}_{\text{span}} = \pm 48.78$ lb Magnitude of vertical load (3-29)

Lateral Seismic Forces—Equivalent Lateral Force Procedure Transverse Direction—Structure (Section 3.1.7.6.2)

$V = E_h = C_s W(I_{mv})$ Equation (3-11)

$C_s = \dfrac{S_{DS}}{(R/I_e)} = 0.33$ Seismic response coefficient Equation (3-12)

Note: The C_s value from Equation (3-12) should be checked for the outer-bound maximum and minimum values of Equations (3-13), (3-14), or (3-15). However, this check will not be performed here.

$V = E_H = C_s W(I_{mv}) = 0.49W$ Basic seismic lateral force Equation (3-11)

Lateral Seismic Forces—Equivalent Lateral Force Procedure—Rigid Bus/Insulator (Section 3.1.7.11)

$F_P = 1.6S_{DS}I_e W_p = 1.05W_p$ Basic seismic lateral force for rigid bus and insulators Equation (3-36)

Vertical Seismic Forces (Section 3.1.7.7)

$E_V = (0.80)S_{DS}D = 0.42D$ Basic seismic vertical force Equation (3-19a)

Structure:

Basic seismic load effects applied to the structure itself, and analysis results used for structural design of main members.

$E = E_{hL} + 0.4E_{hT} \pm 0.4E_v = (0.49W) + (0.20W) \pm (0.17W)$ (3-20)

$E = 0.4E_{hL} + E_{hT} \pm 0.4E_v = (0.20W) + (0.49W) \pm (0.17W)$ (3-21)

$E = 0.4E_{hL} + 0.4E_{hT} \pm E_v = (0.20W) + (0.20W) \pm (0.42W)$ (3-22)

Seismic Force Magnitudes to be applied in Equations (3-20), (3-21), and (3-22):

$0.49W = 0.49DL_{\text{beam}} = 14.71$ lb/ft Magnitude of longitudinal load (3-20)

$0.20W = 0.20DL_{\text{beam}} = 5.89$ lb/ft Magnitude of transverse load (3-20)

$0.17W = 0.17DL_{\text{beam}} = \pm 5.02$ lb/ft Magnitude of vertical load (3-20)

$0.20W = 0.20DL_{\text{beam}} = 5.89$ lb/ft Magnitude of longitudinal load (3-21)

$0.49W = 0.49DL_{\text{beam}} = 14.71$ lb/ft Magnitude of transverse load (3-21)

$0.17W = 0.17DL_{\text{beam}} = \pm 5.02$ lb/ft Magnitude of vertical load (3-21)

$0.20W = 0.20DL_{\text{beam}} = 5.89$ lb/ft Magnitude of longitudinal load (3-22)

$0.20W = 0.20DL_{\text{beam}} = 5.89$ lb/ft Magnitude of transverse load (3-22)

$0.42W = 0.42DL_{\text{beam}} = \pm 12.56$ lb/ft Magnitude of vertical load (3-22)

Seismic Load Effect with Overstrength Factor (Section 3.1.7.8.2)
Seismic load effects for application to the structure itself, analysis results used for all structural designs of connections and anchorage.

$E_m = \Omega_{0L}E_{hL} + 0.4\Omega_{0T}E_{hT} \pm 0.4E_v = (0.49W) + (0.39W) \pm (0.17W)$ (3-27)

$E_m = 0.4\Omega_{0L}E_{hL} + \Omega_{0T}E_{hT} \pm 0.4E_v = (0.20W) + (0.98W) \pm (0.17\ W)$ (3-28)

$E_m = 0.4\Omega_{0L}E_{hL} + 0.4\Omega_{0T}E_{hT} \pm E_v = (0.20W) + (0.20W) \pm (0.42\ W)$ (3-29)

Seismic Force Magnitudes (columns and beam) to be applied in Equations (3-27), (3-28), and (3-29):

$049W = 0.49DL_{\text{member}} = 14.71$ lb/ft Magnitude of longitudinal load (3-27)

$0.39W = 0.39DL_{\text{member}} = 11.77$ lb/ft Magnitude of transverse load (3-27)

$0.17W = 0.17DL_{\text{member}} = \pm 5.02$ lb/ft Magnitude of vertical load (3-27)

$0.20W = 0.20DL_{\text{member}} = 5.89$ lb/ft Magnitude of longitudinal load (3-28)

$0.98W = 0.98DL_{\text{member}} = 29.43$ lb/ft Magnitude of transverse load (3-28)

$0.17W = 0.17DL_{\text{member}} = \pm 5.02$ lb/ft Magnitude of vertical load (3-28)

$0.20W = 0.20DL_{\text{member}} = 5.89$ lb/ft Magnitude of longitudinal load (3-29)

$0.20W = 0.20DL_{\text{member}} = 5.89$ lb/ft Magnitude of transverse load (3-29)

$0.42W = 0.42DL_{\text{member}} = \pm 12.56$ lb/ft Magnitude of vertical load (3-29)

Rigid Bus and Insulator:

Basic seismic load effects for application to the rigid bus or insulator used for structural design of main members. *Note:* In this case, the Horizontal Seismic Force "E_h" is replaced with "F_p" for the Horizontal Seismic Component Forces (3.1.7.11) applied to the insulators and rigid bus.

$E = F_{PL} + (0.4F_{PT}) \pm (0.4E_v) = (1.05W) + (0.42W) \pm (0.17W)$ (3-20)

$E = (0.4F_{PL}) + F_{PT} \pm (0.4E_v) = (0.42W) + (1.05W) \pm (0.17W)$ (3-21)

$E = (0.4F_{PL}) + (0.4F_{PT}) \pm E_v = (0.42W) + (0.42W) \pm (0.42W)$ (3-22)

Seismic Force Magnitudes—Insulator, to be applied in Equations (3-20), (3-21), and (3-22):

$1.05W = 1.05DL_{\text{ins}} = 209.25$ lb Magnitude of longitudinal load (3-20)

$0.42W = 0.42DL_{\text{ins}} = 83.70$ lb Magnitude of transverse load (3-20)

$0.17W = 0.17DL_{\text{ins}} = \pm 33.48$ lb Magnitude of vertical load (3-20)

$0.42W = 0.42DL_{\text{ins}} = 83.70$ lb Magnitude of longitudinal load (3-21)

$1.05W = 1.05DL_{\text{ins}} = 209.25$ lb Magnitude of transverse load (3-21)

$0.42W = 0.42DL_{\text{ins}} = \pm 33.48$ lb Magnitude of vertical load (3-21)

$0.42W = 0.42DL_{\text{ins}} = 83.70$ lb Magnitude of longitudinal load (3-22)

$0.42W = 0.42DL_{\text{ins}} = 83.70$ lb Magnitude of transverse load (3-22)

$0.42W = 0.42DL_{\text{ins}} = \pm 83.70$ lb Magnitude of vertical load (3-22)

Seismic Force Magnitudes—Rigid Bus, to be applied in Equations (3-20), (3-21), and (3-22):

$1.05W = 1.05DL_{RB}\text{Trib}_{\text{span}} = 121.96$ lb Magnitude of longitudinal load (3-20)

$0.42W = 0.42DL_{RB}\text{Trib}_{\text{span}} = 48.78$ lb Magnitude of transverse load (3-20)

$0.17W = 0.17DL_{RB}\text{Trib}_{\text{span}} = \pm 19.51$ lb Magnitude of vertical load (3-20)

$0.42W = 0.42DL_{RB}\text{Trib}_{\text{span}} = 48.78$ lb Magnitude of longitudinal load (3-21)

$1.05W = 1.05DL_{RB}\text{Trib}_{\text{span}} = 121.96$ lb Magnitude of transverse load (3-21)

$0.17W = 0.17DL_{RB}\text{Trib}_{\text{span}} = \pm 19.51$ lb Magnitude of vertical load (3-21)

$0.42W = 0.42DL_{RB}\text{Trib}_{\text{span}} = 48.78$ lb Magnitude of longitudinal load (3-22)

$0.42W = 0.42DL_{RB}\text{Trib}_{\text{span}} = 48.78$ lb Magnitude of transverse load (3-22)

$0.42W = 0.42DL_{RB}\text{Trib}_{\text{span}} = \pm 48.78$ lb Magnitude of vertical load (3-22)

Seismic Load Effect with Overstrength Factor (Section 3.1.7.8.2)
Seismic load effects for application to all structural designs of connections and anchorage of the rigid bus or insulator.

$E_m = \Omega_{0L}F_{PL} + 0.4\Omega_{0T}F_{PT} \pm 0.4E_v = (1.05W) + (0.84W) \pm (0.17W)$ (3-27)

$E_m = 0.4\Omega_{0L}F_{PL} + \Omega_{0T}F_{PT} \pm 0.4E_v = (042W) + (2.09W) \pm (0.17W)$ (3-28)

$E_m = 0.4\Omega_{0L}F_{PL} + 0.4\Omega_{0T}F_{PT} \pm E_v = (0.42W) + (0.84W) \pm (0.42W)$ (3-29)

Seismic Force Magnitudes—Insulator, to be applied in Equations (3-27), (3-28), and (3-29):

$1.05W = 1.05DL_{\text{ins}} = 209.25$ lb Magnitude of longitudinal load (3-27)

$0.84W = 0.84DL_{\text{ins}} = 167.40$ lb Magnitude of transverse load (3-27)

$0.17W = 0.17DL_{\text{ins}} = \pm 33.48$ lb Magnitude of vertical load (3-27)

$0.42W = 0.42DL_{\text{ins}} = 83.70$ lb Magnitude of longitudinal load (3-28)

$2.09W = 2.09DL_{\text{ins}} = 418.51$ lb Magnitude of transverse load (3-28)

$0.17W = 0.17DL_{\text{ins}} = \pm 33.48$ lb Magnitude of vertical load (3-28)

$0.42W = 0.42DL_{\text{ins}} = 83.70$ lb Magnitude of longitudinal load (3-29)

$0.84W = 0.84DL_{\text{ins}} = 167.40$ lb Magnitude of transverse load (3-29)

$0.42W = 0.42DL_{\text{ins}} = \pm 83.70$ lb Magnitude of vertical load (3-29)

Seismic Force Magnitudes—Rigid Bus, to be applied in Equations (3-27), (3-28), and (3-29):

$1.05W = 1.05DL_{RB}\text{Trib}_{\text{span}} = 121.96$ lb Magnitude of longitudinal load (3-27)

$0.84W = 0.84DL_{RB}\text{Trib}_{\text{span}} = 97.56$ lb Magnitude of transverse load (3-27)

$0.17W = 0.17DL_{RB}\text{Trib}_{\text{span}} = \pm 19.51$ lb Magnitude of vertical load (3-27)

$0.42W = 0.42DL_{RB}\text{Trib}_{\text{span}} = 48.78$ lb Magnitude of longitudinal load (3-28)

$2.09W = 2.09DL_{RB}\text{Trib}_{\text{span}} = 243.91$ lb Magnitude of transverse load (3-28)

$0.17W = 0.17DL_{RB}\text{Trib}_{\text{span}} = \pm 19.51$ lb Magnitude of vertical load (3-28)

$0.42W = 0.42DL_{RB}\text{Trib}_{\text{span}} = 48.78$ lb Magnitude of longitudinal load (3-29)

$0.84W = 0.84DL_{RB}\text{Trib}_{\text{span}} = 97.56$ lb Magnitude of transverse load (3-29)

$0.42W = 0.42DL_{RB}\text{Trib}_{\text{span}} = \pm 48.78$ lb Magnitude of vertical load (3-29)

A.1.4 Short-Circuit Loads (Section 3.1.8)

Simplified static short-circuit force on rigid conductors: (IEEE 605, IEEE 2008, and Section 3.1.8.1):

$I_{SC} = 20,000$ RMS Fault Current (amps)

$D = 10.00$ Conductor center-to-center spacing (ft)

$K_f = 1.00$ Mounting structure flexibility factor, 3-PH bus support. IEEE 605 (IEEE 2008) Figure 20

$\frac{X}{R} = 10.77$ Specified X/R ratio, given

$f = 60$ 60 Hz, given

$c = \frac{X}{R}\left(\frac{1}{2\pi f}\right) = 0.03$ Time constant of the circuit [Equation (3-43)]

$D_f = \frac{1 + e^{-1/2fc}}{2} = 0.87$ Calculated decrement factor [Equation (3-42)]

$D_{fsq} = D_f^{\,2} = 0.76$ Decrement factor squared

$\Gamma = 0.866$ Constant for fault type, 3-Phase, "B," IEEE 605 (IEEE 2008), Table 13

$F_{SC} = \frac{3.6\Gamma(I_{SC}^{\,2})}{10^7 D} = 12.47\ \text{lb/ft}$ Basic short-circuit force equation, uncorrected, Equation (3-40)

$F_{SC\_COR} = K_f D_{fsq} F_{SC} = 9.51\ \text{lb/ft}$ Corrected short-circuit force, Equation (3-41)

$SC_{bus} = F_{SC\_COR}\text{Trib}_{span} = 297.34\ \text{lb}$ Short-circuit force on rigid bus

A.1.5 Deflection Case Wind Loads (Section 3.1.11)

Calculate the deflection wind case pressure for use in design with 3.1.11.1 and 3.1.11.2 load combinations:

Note: The values for C_f are taken from Table 3-8.

Coefficients and Definitions:

$Q = 0.00256$ Air density factor, sea level, 59 °F (Section 3.1.5.1)

$K_Z = 0.85$ Terrain exposure coefficient, Exposure C (Section 3.1.5.2.1)

$K_d = 1.00$ Wind directionality factor (Section 3.1.5)

$K_{zt} = 1.00$ Topographic factor (Section 3.1.5)

$V_{DEFL} = 70.0\ \text{mph}$ Operational wind speed for deflection (Section 3.1.11.1)

$V_{MRI\_WI\_100} = 40.0\ \text{mph}$ Operational ice and concurrent wind speed (Section 3.1.11.2)

$t_{ICE\_5} = 0.10\ \text{in.}$ Operational radial ice thickness, 5-year MRI (Section 3.1.11.2)

$G_{SRF} = 1.00$ Gust response factor (Structure) (Section 3.1.5.5, Table 3-6)

$C_{f\_rnd} = 0.90$ Force coefficient, circular cross section (Table 3-8)

$C_{f\_bus} = 1.00$ Force coefficient, rigid bus and insulators (Table 3-8)

$C_{f\_avg} = 1.60$ Force coefficient, average structural shapes (Table 3-8)

$C_{f\_sq} = 2.00$ Force coefficient, square/rectangular section (Table 3-8)

The operational wind pressures are used with Section 3.1.11.1, "Wind Load for Deflection Calculations."

Wind Force on Insulators:

$$F_{DEFL\_rnd} = QK_Z K_d K_{zt}(V_{DEFL})^2 G_{SRF} C_{f\_rnd}\text{AREA}_{ins} = 34.87\ \text{lb}$$

Wind Force on the Rigid Bus:

$$F_{\mathrm{DEFL\_bus}} = QK_Z K_d K_{zt} (V_{\mathrm{DEFL}})^2 G_{\mathrm{SRF}} C_{f\_\mathrm{bus}} d_{\mathrm{RB}} \mathrm{Trib}_{\mathrm{span}} = 124.95 \text{ lb}$$

Wind Pressure on the WF Beam Structure Surface:

$$P_{\mathrm{DEFL\_str\_sq}} = QK_Z K_d K_{zt} (V_{\mathrm{DEFL}})^2 G_{\mathrm{SRF}} C_{f\_\mathrm{avg}} = 17.06 \text{ psf}$$

Wind Pressure on HSS Column (Flat) Structure Surfaces:

$$P_{\mathrm{DEFL\_str\_sq}} = QK_Z K_d K_{zt} (V_{\mathrm{DEFL}})^2 G_{\mathrm{SRF}} C_{f\_\mathrm{sq}} = 21.32 \text{ psf}$$

The following operational wind pressures are used with Section 3.1.11.2, "Load Cases."
Wind Force on Insulators:

$$F_{\mathrm{ICE\_}rnd} = QK_Z K_d K_{zt} \left(V_{\mathrm{MRIWI100}}\right)^2 G_{\mathrm{SRF}} C_{frnd} H_{\mathrm{ins}} \left(d_{\mathrm{ins}} + 2t_{\mathrm{ICE}_5}\right) = 11.62 \text{ lb}$$

Wind Force on the Rigid Bus:

$$F_{\mathrm{ICE\_bus}} = QK_Z K_d K_{zt} \left(V_{\mathrm{MRIWI100}}\right)^2 G_{\mathrm{SRF}} C_{f\mathrm{bus}} \left(d_{RB} + 2t_{\mathrm{ICE}_5}\right) \mathrm{Trib}_{\mathrm{span}} = 42.61 \text{ lb}$$

Wind Pressure on Average Structure Surfaces:

$$F_{\mathrm{ICE\_avg}} = QK_Z K_d K_{zt} (V_{\mathrm{MRI\_WI\_100}})^2 G_{\mathrm{SRF}} C_{f\_\mathrm{avg}} (\mathrm{Beam}_{\mathrm{depth}} + 2t_{\mathrm{ICE}_5}) = 3.81 \text{ lb/ft}$$

Wind Pressure on HSS Column (Flat) Structure Surfaces:

$$P_{\mathrm{ICE\_str\_sq}} = QK_Z K_d K_{zt} (V_{\mathrm{MRI\_WI\_100}})^2 G_{\mathrm{SRF}} C_{f\_\mathrm{sq}} (\mathrm{Column}_{\mathrm{depth}} + 2t_{\mathrm{ICE}_5}) = 4.76 \text{ lb/ft}$$

Result: All basic loading has been calculated for use in design.

A.2 LOAD DEVELOPMENT FOR A DEAD-END STRUCTURE EXAMPLE

Determine the basic design inputs and develop loading for the dead-end structure shown in Figure A-2, for the following load cases:

1. Extreme Wind Loads (Section 3.1.5),
2. Combined Ice and Wind Loads (Section 3.1.6), and
3. NESC District Loading—Heavy Loading (IEEE 2023).

Note: Wire tension loads should be derived using sag-tension calculations or computer software.

Geometry and Span Data:

$H = 30.0$ ft Height of the structure

$\mathrm{Span}_{\mathrm{wire}} = 200$ ft Wind span length

Conductor and Shield Wire:

$d_{\mathrm{ACSR}} = 1.108$ in. Diameter of the 795 MCM ACSR Conductor

$DL_{\mathrm{ACSR}} = 1.904$ lb/ft Weight of the 795 MCM ACSR Conductor

$d_{\mathrm{SW}} = 0.375$ in. Diameter of the 3/8 in. EHS shield wire

$DL_{\mathrm{SW}} = 0.273$ lb/ft Weight of the 3/8 in. EHS shield wire

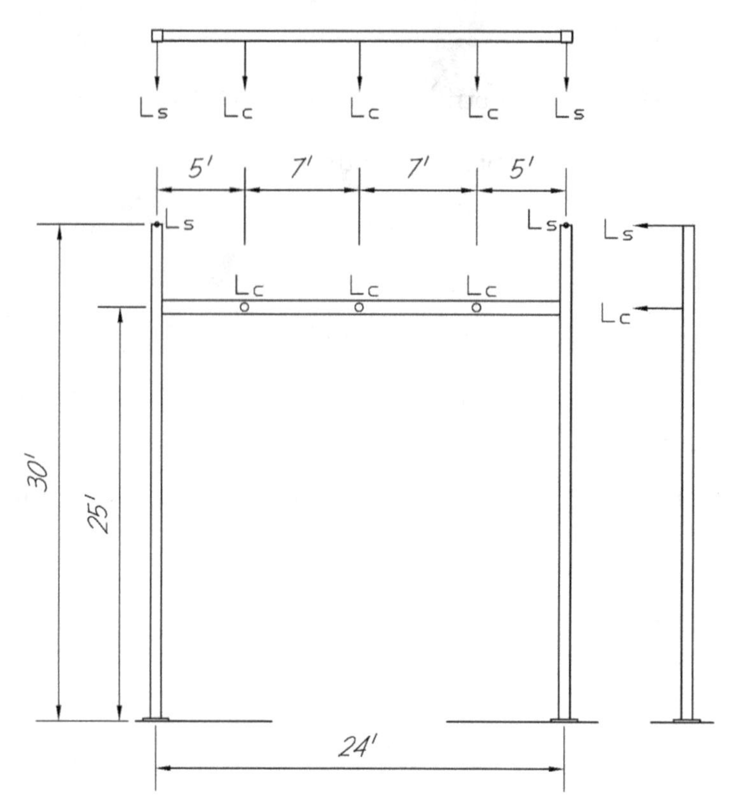

Figure A-2. Dead-end structure.

Insulator Information:

Note: Assume that seven insulator bells will be used for the conductor, and the shield wire will be directly attached to the mast.

$L_{INS} = 7.00$ in. Length of (1) insulator bell

$d_{INS} = 7.00$ in. Outside shed diameter of (1) insulator bell

$DL_{INS} = 9.00$ lb Weight of (1) insulator bell

$n_{INS} = 7.0$ Number of insulators

$\text{AREA}_{INS} = L_{INS} d_{INS} n_{INS} = 2.38 \text{ ft}^2$ Projected wind area

$DL_{CHAIN} = DL_{INS} n_{INS} = 63$ lb Weight of the insulator chain

A.2.1 Extreme Wind Loads (Section 3.1.5)

Calculate the extreme wind pressure for use in design with USD and ASD load combinations: The values for C_f are taken from Table 3-8.

Coefficients and Definitions:

$Q = 0.00256$ Air density factor, sea level, 59 °F (3.1.5.1)

$K_Z = 2.41\left(\dfrac{30}{2{,}460}\right)^{2/9.8} = 0.98$ Terrain exposure coefficient, Exposure C (3.1.5.2.1)

Note: The structure height (30 ft) was used to determine K_z for simplicity

$K_d = 1.00$ Wind directionality factor (3.1.5)

$K_{zt} = 1.00$ Topographic factor (3.1.5)

$V_{MRI\_300} = 95.0$ mph Basic wind speed, 300-year MRI (3.1.5.3)

$V_{\mathrm{MRI\_100}} = 85.0$ mph Basic wind speed, 100-year MRI (3.1.5.3)

$G_{\mathrm{SRF}} = 0.85$ Gust response factor (structure) (3.1.5.5.1, Equation 3-5)

$G_{\mathrm{WRF}} = 0.81$ Gust response factor (wire) (3.1.5.5.2, Equation 3-6)

$C_{f\_rnd} = 0.90$ Force coefficient, structural circular pipes (Table 3-8)

$C_{f\_\mathrm{bus}} = 1.00$ Force coefficient, rigid bus, conductor, and insulators (Table 3-8)

$C_{f\_\mathrm{sq}} = 2.00$ Force coefficient, square/rectangular section (Table 3-8)

The following wind pressures are used with the USD load combinations.
USD Wind Force on Insulators:

$$F_{\mathrm{USD\_bus}} = QK_Z K_d K_{zt} (V_{\mathrm{MRI\_300}})^2 G_{\mathrm{WRF}} C_{f\_\mathrm{bus}} \mathrm{AREA}_{\mathrm{INS}} = 43.70 \text{ lb}$$

USD Wind Force on the Conductor:

$$F_{\mathrm{USD\_bus}} = QK_Z K_d K_{zt} (V_{\mathrm{MRI\_300}})^2 G_{\mathrm{WRF}} C_{f\_\mathrm{bus}} \frac{\mathrm{Span}_{\mathrm{wire}}}{2} d_{\mathrm{ACSR}} = 169.38 \text{ lb}$$

USD Wind Force on the Shield Wire:

$$F_{\mathrm{USD\_sw}} = QK_Z K_d K_{zt} (V_{\mathrm{MRI\_300}})^2 G_{\mathrm{WRF}} C_{f\_\mathrm{bus}} \frac{\mathrm{Span}_{\mathrm{wire}}}{2} d_{\mathrm{SW}} = 57.33 \text{ lb}$$

USD Wind Pressure on Square/Rectangular Structure Surfaces:

$$F_{\mathrm{USD\_str\_sq}} = QK_Z K_d K_{zt} (V_{\mathrm{MRI\_300}})^2 G_{\mathrm{SRF}} C_{f\_\mathrm{sq}} = 38.64 \text{ psf}$$

USD Wind Pressure on Round Structure Surfaces:

$$F_{\mathrm{USD\_str\_}rnd} = QK_Z K_d K_{zt} (V_{\mathrm{MRI\_300}})^2 G_{\mathrm{SRF}} C_{f\_rnd} = 15.65 \text{ psf}$$

The following wind pressures are used with the ASD load combinations.
ASD Wind Force on Insulators:

$$F_{\mathrm{ASD\_bus}} = QK_Z K_d K_{zt} (V_{\mathrm{MRI\_100}})^2 G_{\mathrm{WRF}} C_{f\_\mathrm{bus}} \mathrm{AREA}_{\mathrm{INS}} = 34.98 \text{ lb}$$

ASD Wind Force on the Conductor:

$$F_{\mathrm{ASD\_bus}} = QK_Z K_d K_{zt} (V_{\mathrm{MRI\_100}})^2 G_{\mathrm{WRF}} C_{f\_\mathrm{bus}} \frac{\mathrm{Span}_{\mathrm{wire}}}{2} d_{\mathrm{ACSR}} = 135.60 \text{ lb}$$

ASD Wind Force on the Shield Wire:

$$F_{\mathrm{ASD\_sw}} = QK_Z K_d K_{zt} (V_{\mathrm{MRI\_100}})^2 G_{\mathrm{WRF}} C_{f\_\mathrm{bus}} \frac{\mathrm{Span}_{\mathrm{wire}}}{2} d_{\mathrm{SW}} = 45.89 \text{ lb}$$

ASD Wind Pressure on Square/Rectangular Structure Surfaces:

$$F_{\mathrm{ASD\_str\_sq}} = QK_Z K_d K_{zt} (V_{\mathrm{MRI\_100}})^2 G_{\mathrm{SRF}} C_{f\_\mathrm{sq}} = 30.93 \text{ psf}$$

ASD Wind Pressure on Round Structure Surfaces:

$$F_{\mathrm{ASD\_str\_}rnd} = QK_Z K_d K_{zt} (V_{\mathrm{MRI\_100}})^2 G_{\mathrm{SRF}} C_{f\_rnd} = 13.92 \text{ psf}$$

A.2.2 Combined Ice and Wind Loads (Section 3.1.6)

Calculate the combined ice and wind pressure for use in design with USD and ASD load combinations:

Coefficients and Definitions:

All factors previously defined remain applicable here. Only new parameters are defined here. Refer to Section 3.1.6 for all values shown subsequently.

Temperature = 5 °F Figure 3-6c

$V_{MRI\_WI\_100} = 40.0$ mph Concurrent wind speed with ice, 100-year MRI, Figure 3-6b

$t_{ICE\_100} = 0.50$ in. Radial ice thickness, 100-year MRI, Figure 3-6a

$V_{MRI\_WI\_50} = 40.0$ mph Concurrent wind speed with ice, 50-year MRI, Figure 3-6b

$t_{ICE\_50} = 0.25$ in. Radial ice thickness, 50-year MRI, Figure 3-6d

$\gamma_{ICE} = 56.0$ lb/ft³ Density of ice (Section 3.1.6.1)

The following wind pressures and ice loads are used with the USD load cases.

USD Combined Wind Force and Ice Load on Insulators:

For this example, ice diameter and weight were applied to the insulator to illustrate the process.

$$F_{USD\_WI\_ins} = QK_Z K_d K_{zt}(V_{MRI\_WI\_100})^2 G_{WRF} C_{f\_bus}(L_{INS})(n_{INS})(d_{INS} + 2t_{ICE-100})$$

$$F_{USD\_WI\_ins} = 8.85 \text{ lb}$$

$$F_{USD\_ICE\_ins} = \gamma_{ICE} \frac{\pi[(d_{INS} + 2t_{ICE-100})^2 - d_{INS}{}^2]}{4}(L_{INS})(n_{INS}) = 18.71 \text{ lb}$$

USD Combined Wind Force and Ice Load on Conductors:

$$F_{USD\_WI\_cond} = QK_Z K_d K_{zt}(V_{MRI\_WI\_100})^2 G_{WRF} C_{f\_bus} \frac{\text{Span}_{wire}}{2}(d_{ACSR} + 2t_{ICE-100})$$

$$F_{USD\_WI\_cond} = 57.13 \text{ lb}$$

$$F_{USD\_ICE\_cond} = \gamma_{ICE} \frac{\pi[(d_{ACSR} + 2t_{ICE-100})^2 - d_{ACSR}{}^2]}{4} \frac{\text{Span}_{wire}}{2} = 98.23 \text{ lb}$$

USD Combined Wind Force and Ice Load on the Shield Wire:

$$F_{USD\_WI\_sw} = QK_Z K_d K_{zt}(V_{MRI\_WI\_100})^2 G_{WRF} C_{f\_bus} \frac{\text{Span}_{wire}}{2}(d_{SW} + 2t_{ICE-100})$$

$$F_{USD\_WI\_sw} = 37.27 \text{ lb}$$

$$F_{USD\_ICE\_sw} = \gamma_{ICE} \frac{\pi[(d_{SW} + 2t_{ICE-100})^2 - d_{SW}{}^2]}{4} \frac{\text{Span}_{wire}}{2} = 53.45 \text{ lb}$$

The following wind pressures are used with the ASD load combinations:

ASD Combined Wind Force and Ice Load on Insulators:

$$F_{ASD\_WI\_ins} = QK_Z K_d K_{zt}(V_{MRI\_WI\_50})^2 G_{WRF} C_{f\_rnd}(L_{INS})(n_{INS})(d_{INS} + 2t_{ICE\text{-}50})$$

$$F_{ASD\_WI\_ins} = 7.47 \text{ lb}$$

$$F_{ASD\_ICE\_ins} = \gamma_{ICE} \frac{\pi[(d_{INS} + 2t_{ICE\text{-}50})^2 - d_{INS}{}^2]}{4}(L_{INS})(n_{INS}) = 9.04 \text{ lb}$$

ASD Combined Wind Force and Ice Load on Conductors:

$$F_{\text{ASD\_WI\_cond}} = QK_Z K_d K_{zt}(V_{\text{MRI\_WI\_50}})^2 G_{\text{WRF}} C_{f_{\text{bus}}} \frac{\text{Span}_{\text{wire}}}{2}(d_{\text{ACSR}} + 2t_{\text{ICE}-50})$$

$$F_{\text{ASD\_WI\_cond}} = 43.58 \text{ lb}$$

$$F_{\text{ASD\_ICE\_cond}} = \gamma_{\text{ICE}} \frac{\pi[(d_{\text{ACSR}} + 2t_{\text{ICE-50}})^2 - d_{\text{ACSR}}{}^2]}{4} \frac{\text{Span}_{\text{wire}}}{2} = 41.48 \text{ lb}$$

ASD Combined Wind Force and Ice Load on the Shield Wire:

$$F_{\text{ASD\_WI\_sw}} = QK_Z K_d K_{zt}(V_{\text{MRI\_WI\_50}})^2 G_{\text{WRF}} C_{f\_\text{bus}} \frac{\text{Span}_{\text{wire}}}{2}(d_{\text{SW}} + 2t_{\text{ICE-50}})$$

$$F_{\text{ASD\_WI\_sw}} = 23.71 \text{ lb}$$

$$F_{\text{ASD\_ICE\_sw}} = \gamma_{\text{ICE}} \frac{\pi[(d_{\text{SW}} + 2t_{\text{ICE-50}})^2 - d_{\text{SW}}{}^2]}{4} \frac{\text{Span}_{\text{wire}}}{2} = 19.09 \text{ lb}$$

A.2.3 NESC District Loading—Heavy Loading (IEEE 2023)

Calculate the wind pressure and ice load using NESC Rule 250B:

Definitions:

$W_{\text{NESC\_250}B} = 4.00$ psf Wind Pressure, Reference NESC Table 250-1

$t_{\text{ICE\_NESC\_250}B} = 0.50$ in. Ice Thickness, Reference NESC Table 250-1

Temperature $= 0°\text{F}$ Temperature, Reference NESC Table 250-1

NESC Combined Wind Force and Ice Load on Insulators:

$$F_{\text{NESC\_ins}} = W_{\text{NESC}_{250B}}(L_{\text{INS}})(n_{\text{INS}})[d_{\text{INS}} + 2(t_{\text{ICE\_NESC\_250}B})]$$

$$F_{\text{NESC\_ins}} = 10.89 \text{ lb}$$

$$F_{\text{NESC\_ICE\_ins}} = \gamma_{\text{ICE}} \frac{\pi\left[\left(d_{\text{INS}} + 2\left[t_{\text{ICE}_{\text{NESC250}B}}\right]\right)^2 - d_{\text{INS}}{}^2\right]}{4}(L_{\text{INS}})(n_{\text{INS}}) = 18.71 \text{ lb}$$

NESC Combined Wind Force and Ice Load on Conductors:

$$F_{\text{NESC\_cond}} = W_{\text{NESC\_250}B} \frac{\text{Span}_{\text{wire}}}{2}[d_{\text{ACSR}} + 2(t_{\text{ICE}_{\text{NESC250}B}})]$$

$$F_{\text{NESC\_cond}} = 70.27 \text{ lb}$$

$$F_{\text{NESC\_ICE\_cond}} = \gamma_{\text{ICE}} \frac{\pi[(d_{\text{ACSR}} + 2(t_{\text{ICE\_NESC\_250}B}))^2 - d_{\text{ACSR}}{}^2]}{4} \frac{\text{Span}_{\text{wire}}}{2} = 98.23 \text{ lb}$$

NESC Combined Wind Force and Ice Load on the Shield Wire:

$$F_{\text{NESC\_sw}} = W_{\text{NESC\_250}B} \frac{\text{Span}_{\text{wire}}}{2}[d_{\text{SW}} + 2(t_{\text{ICE\_NESC\_250}B})]$$

$$F_{\text{NESC\_sw}} = 45.83 \text{ lb}$$

$$F_{\text{NESC\_ICE\_sw}} = \gamma_{\text{ICE}} \frac{\pi[(d_{\text{SW}} + 2(t_{\text{ICE\_NESC\_250}B}))^2 - d_{\text{SW}}{}^2]}{4} \frac{\text{Span}_{\text{wire}}}{2} = 53.45 \text{ lb}$$

The following load factors from NESC Table 253-1 are applicable to the loads derived here:

$NESC_{LF\_WIND\_TRANS} = 2.50$ Transverse wind load factor.

$NESC_{LF\_WIND\_TENSION} = 1.65$ Transverse wind—wire tension load factor.

$NESC_{LF\_WIND\_DL} = 1.50$ Transverse wind—vertical dead load factor.

Result: All basic loading has been calculated for use in design.

A.3 ANCHOR ROD ON LEVELING NUTS DESIGN EXAMPLE

An analysis using the calculated loads was done where the shear loads and axial loads were determined and used for this example. Determine whether a 1.0 in. diameter F1554 Grade 36 anchor is adequate to carry the applied loading. Given the ultimate factored load per anchor, a rigid base plate defined as given in Chapter 6, and the bottom of the base plate is 2.0 in. from the top of the concrete (Figure A-3).

Note: Additional concrete checks would be required in accordance with ACI CODE-318-19 requirements.

Geometry from Figure A-3:

$h = 2.0$ in. Gap between the bottom of a rigid base plate and the top of the concrete

$t_w = 0.177$ in. (washer)

$t_n = \frac{63}{64}$ in. (nut)

Anchor Inputs and Properties:

$f_{ya} = 36$ ksi Yield strength of anchor

$f_u = 58$ ksi Tensile strength of anchor

$E = 29,000$ ksi Elastic Modulus of steel

$d_a = 1.0$ in. Nominal anchor diameter

$n_t = 8$ Number of threads per inch

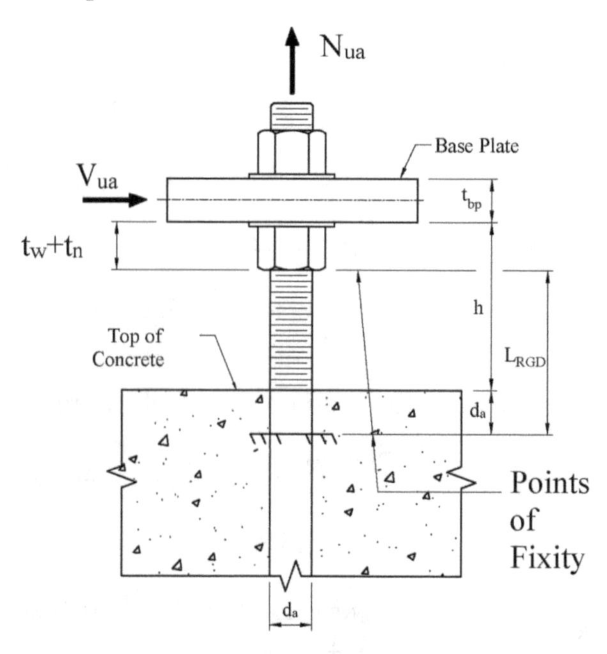

Figure A-3. Anchor rod on a leveling nut.

Controlling Factored Design Reaction Per Anchor Rod:

$N_{ua} = 4.90$ kip Ultimate tensile force (factored) per anchor rod.

$V_{ua} = 2.10$ kip Ultimate shear force (factored) per anchor rod.

Calculate Anchor Properties:

$L_{\mathrm{RGD}} = h - t_{\mathrm{w}} - t_{\mathrm{n}} + \mathrm{d}_{\mathrm{a}} = 1.84$ in. Moment arm without clamping nut, Equation (8-11).

$A_{se,N} = \frac{\pi}{4}\left(d_a - \frac{0.9743}{n_t}\right)^2 = 0.61 \text{ in.}^2$ Effective cross-sectional area, Equation (8-19)

$A_{se,V} = \frac{\pi}{4}\left(d_a - \frac{0.9743}{n_t}\right)^2 = 0.61 \text{ in.}^2$ Effective cross-sectional area, Equation (8-26)

$S = \frac{\pi}{32}\left(d_a - \frac{0.9743}{n_t}\right)^3 = 0.07 \text{ in.}^3$ Section modulus, Equation (8-16)

Calculate the Capacity of a Given Anchor:

$$f_{uta} = \text{Minimum of}\,(f_u, 1.9 f_{ya}, 125 \text{ ksi}) = 58 \text{ ksi}$$

$N_n = (A_{se,N})\,(f_{uta}) = 35.13$ kip Nominal axial anchor capacity, Equation (8-18)

$V_n = 0.6\,(A_{se,V})(f_{uta}) = 21.08$ kip Nominal shear anchor capacity, Equation (8-25)

$\Phi a = 0.75$ Strength reduction factor, axial, ductile steel (ACI CODE 318-19)

$\Phi v = 0.65$ Strength reduction factor, shear, ductile steel (ACI CODE 318-19)

$M_o = 1.2\,(S)\,f_u = 4.63$ kip-in. Bending capacity, Equation (8-15)

$M_n = M_o\left(1 - \frac{N_{ua}}{\Phi a(N_n)}\right) = 3.77$ kip-in. Reduced bending capacity, Equation (8-17).

$L = \frac{1}{2} L_{\mathrm{RGD}} = 0.919$ in. Moment arm for anchorage

$V_{nm} = \frac{M_n}{L} = 4.098$ kip Nominal shear force, Equation (8-20)

Verify the Adequacy of the Given Anchor:

Combined stress interaction with bending considered. The value must be below unity (1.0).

$\frac{N_{ua}}{\Phi a N_n} + \frac{V_{ua} L}{\Phi v M_o} = 0.828$ Equation (8-23)

Combined stress interaction with bending ignored. The value must be below 1.2.

$\frac{N_{ua}}{\Phi a N_n} + \frac{V_{ua}}{\Phi v V_n} = 0.339$ Equation (8-24)

Result: The proposed 1.0 in. diameter F1554 Grade 36 anchor is adequate to carry the applied loading.

A.4 BASE PLATE DESIGN EXAMPLE

Determine the required thickness of a Grade 36 steel base plate to resolve the anchor rod loading from the previous example, using the Figure A-4:

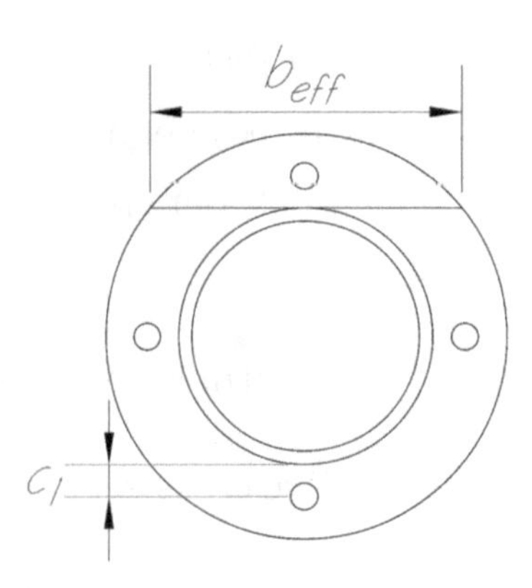

Figure A-4. Base plate effective bending plane.

Geometry from Figure:

$c_1 = 2.0$ in. $b_{\text{eff}} = 8.0$ in.

Anchor Inputs and Properties:

$f_y = 36$ ksi Yield strength of anchor

$BL_1 = N_{UA} = 4.90$ kip Ultimate anchor load for the rod shown in Figure A-3

Calculate the Minimum Thickness of the Base Plate:

$$t_{\min} = \sqrt{\frac{6}{b_{\text{eff}} f_y}(BL_1 c_1)} = 0.452 \text{ in.}$$ Minimum required base plate thickness, Equation (6-4)

Result: A minimum 0.5 in. thick Grade 36 steel base plate is adequate to resist the applied loading.

A.5 DEFLECTION EXAMPLES

The following three examples of deflection analysis look at both Class "B" bus supports and Class "A" switch stands, as discussed in Chapter 4. For these examples, a fixed connection to the foundation is assumed for simplicity. The deflections are tabulated for three deflection load cases for demonstration purposes. The rotations given are about the axis indicated and are oriented using the right-hand rule. All units are US customary inches, kips, and degrees. The conversion to metric units is left to the designer.

A.5.1 Deflection Example, Three-Phase Bus Support Stand—Class "B" Structure

A typical three-phase bus support structure and select joints are shown in Figure A-5. The structure is 18 ft high and the beam extends 6.5 ft from the center column (13 ft long beam).

Load 1—Combined ice and concurrent wind in the X-direction

Load 2—Deflection level wind in the Y-direction

Load 3—Deflection level wind in the X-direction

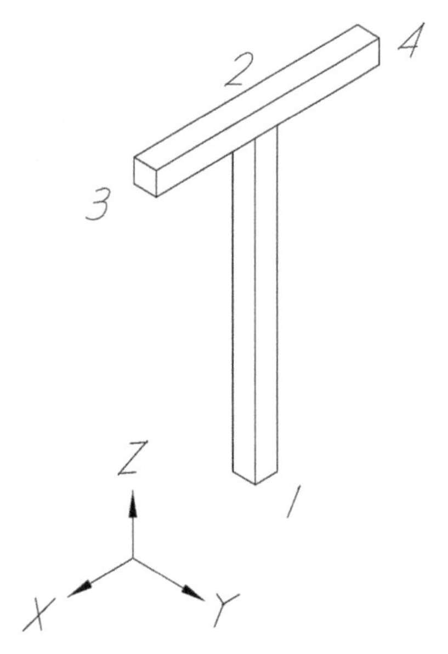

Figure A-5. Three-phase bus support stand with selected joint displacements.

Table A-1 shows the displacement and rotation results from a finite-element analysis of the example structure for the three load cases.

Horizontal Deflection of a Vertical Member

Joint 2 (the top of the column) has a maximum displacement of 1.640 in. in this case. Class "B" structure vertical members are limited to 1/100 of the height for maximum horizontal deflection. The structure height is 18 ft (216 in.) in this case.

$$\frac{216 \text{ in.}}{100} = 2.160 \text{ in. allowable} > 1.640 \text{ in. actual therefore, OK}$$

Vertical Deflection of a Horizontal Member

Gross versus net deflection is critical for horizontal members. For Load Case 3, Joint 4 (the tip of the beam) has a maximum gross vertical deflection of −0.927 in.; Joint 3, on the opposite end of the beam, has a maximum gross vertical deflection of 0.789 in. To "straighten out" the

Table A-1. Operational Displacements and Rotations for Example A.5.1.

Joint	Load	X-displ. (in.)	Y-displ. (in.)	Z-displ. (in.)	X-rotation (degrees)	Y-rotation (degrees)	Z-rotation (degrees)
2	1	0.511	—	0.001	—	−0.197	—
2	2	—	1.230	−0.001	0.471	—	—
2	3	1.640	—	−0.001	—	−0.630	—
3	1	0.511	—	0.181	—	−0.109	—
3	2	—	1.290	−0.069	0.471	0.070	0.055
3	3	1.640	—	0.789	—	−0.560	—
4	1	0.511	—	−0.355	—	−0.285	—
4	2	—	1.290	−0.069	0.471	−0.070	−0.055
4	3	1.640	—	−0.927	—	−0.700	—

Note: 1 in. = 25.4 mm.

column to calculate the net deflection (deflection of the beam relative to the top of the column), the top of the column rotation must be backed out. The Y-rotation of joint 2 (top of the column) for Load Case 3 is 0.630 degrees. To find the deflection caused by the top of the column rotation,

$$\tan(0.630°) = \frac{\text{vertical deflection of joint 4}}{78 \text{ in.}}$$

Vertical deflection of Joint 4 (caused by column rotation) = 0.858 in.

To obtain the net deflection, the Joint 4 gross deflection of negative 0.927 in. is reduced by 0.858 in. (for column rotation), resulting in a net deflection of 0.069 in. As a check, the Joint 3 deflection of positive 0.789 in. subtracting 0.858 in. is a net deflection of 0.069 in. This method works only with a symmetrical and evenly loaded structure. For nonsymmetrical structures, the net deflection has to be calculated on the left and right beams independently.

The net vertical deflection of a horizontal member is limited to 1/200 for a Class "B" structure:

$$6.5\text{ft}\left(12\frac{\text{in.}}{\text{ft}}\right) = 78 \text{ in.}$$

$$\frac{78 \text{in.}}{200} = 0.390 \text{ in. allowable} > 0.069 \text{ in. actual therefore, OK}$$

In addition, for loading 1, Joint 4 has a maximum gross vertical deflection of −0.355 in. Joint 3 has a maximum gross vertical deflection of 0.181 in. To "straighten out" the column to calculate the net deflection, the top of the column rotation must be backed out. The top of the column rotation for Load Case 1 is 0.197 degrees.

$$\tan(0.197°) = \frac{\text{vertical deflection of joint 4}}{78 \text{ in.}}$$

Vertical deflection of joint 4 caused by rotation = 0.268 in.

The net vertical deflection (or the deflection caused by the flex of the beam) is approximately

$$0.355 \text{ in.} - 0.268 \text{ in.} = 0.087 \text{ in.}$$

$$\frac{78 \text{ in.}}{200} = 0.39 \text{ in. allowable} > 0.087 \text{ in., therefore, OK}$$

Horizontal Deflection of a Horizontal Member

For Loading 2, the gross displacement in the Y-direction for Joints 3 and 4 is 1.290 in. For Joint 2, the top of the column, the gross displacement is 1.230 in. The net displacement is

$$1.290 \text{ in.} - 1.230 \text{ in.} = 0.060 \text{ in.}$$

$$\frac{78 \text{ in.}}{100} = 0.780 \text{ in. allowable} > 0.060 \text{ in., therefore, OK}$$

A.5.2 Deflection Example, Three-Phase Switch Support Stand, Single Column—Class "A" Structure

A typical three-phase switch support stand with select joints is shown in Figure A-6 for various deflection load cases. The structure is 17.5 ft high and has a platform that is 3 ft wide by 12 ft long that is centered on the column.

Load 1—Deflection level wind in the Y-direction

Load 2—Deflection level wind in the X-direction

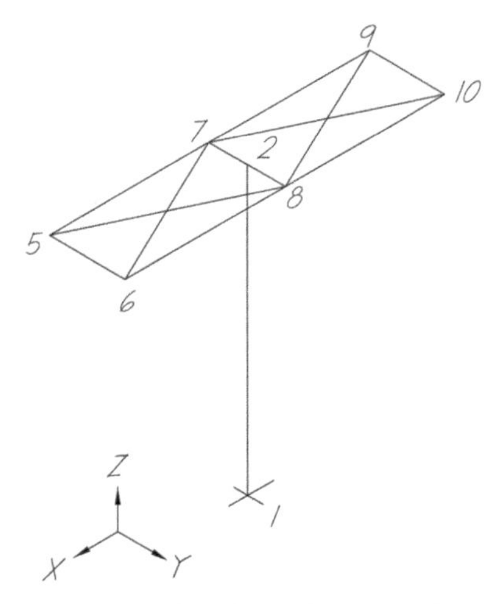

Figure A-6. Single-column switch stand with selected joints shown.

Table A-2. Operational Displacements and Rotations for Example A.5.2.

Joint	Load	X-displ. (in.)	Y-displ. (in.)	Z-displ. (in.)	X-rotation (degrees)	Y-rotation (degrees)	Z-rotation (degrees)
2	1	—	1.660	—	0.648	—	—
6	2	1.380	—	0.631	0.006	−0.492	—
7	1	—	1.660	0.200	0.642	—	—
8	1	—	1.660	−0.207	0.654	—	—
8	2	1.380	—	−0.003	0.006	−0.537	—
10	1	—	1.660	−0.248	0.654	−0.045	0.002
10	2	1.380	—	−0.718	0.006	−0.045	—

Note: 1 in. = 25.4 mm.

Table A-2 provides the displacement and rotation results from a finite-element analysis of the example structure for the two load cases.

Horizontal Deflection of a Vertical Member

Joint 2 (the top of the column) has a maximum displacement of 1.660 in. Class "A" structure vertical members are limited to 1/100 of the height for maximum horizontal deflection. The height of this structure is 17.5 ft.

$$17.5\text{ ft}\left(12\frac{\text{in.}}{\text{ft}}\right) = 210\text{ in.}$$

$$\frac{210\text{ in.}}{100} = 2.100\text{ in. allowable} > 1.660\text{ in., therefore, OK}$$

Vertical Deflection of a Horizontal Member

Joint 2 to Joint 8 support member

For Loading 1, Joint 8 has a maximum gross vertical deflection of −0.207 in. Joint 7, on the opposite side of the column, has a maximum gross vertical deflection of 0.200 in. To "straighten out" the column to calculate the net deflection (deflection of the support member relative to the top of the column), the top of the column rotation must be backed out. The X-rotation angle of Joint 2 for loading 1 is 0.648 degrees. To find the deflection caused by the top of the column rotation,

$$\tan(0.648°) = \frac{\text{vertical deflection of Joint 8 } \textit{caused by} \text{ column rotation}}{18 \text{ in.}}$$

$$\text{vertical deflection of Joint 8 } \textit{caused by} \text{ column rotation} = 0.204 \text{ in.}$$

To obtain the net deflection, we take Joint 8 gross deflection of negative 0.207 in. and add back 0.204 in. (column rotation) to obtain a net deflection of negative 0.003 in.

The net vertical deflection of a horizontal member is limited to L/200 for a Class "A" structure:

$$1.5 \text{ ft}\left(12\frac{\text{in.}}{\text{ft}}\right) = 18 \text{ in.}$$

$$\frac{18 \text{ in.}}{200} = 0.090 \text{ in. allowable} > 0.003 \text{ in.}, \text{therefore, OK}$$

Vertical Deflection of a Horizontal Member

Joint 8 to Joint 10 support member (72 in. long)

For Loading 2, Joint 10 has a maximum gross vertical deflection of −0.718 in. Joint 8 has a maximum gross vertical deflection of −0.003 in. Joint 6 has a maximum gross vertical deflection of 0.631 in. To find the net deflection of Joint 10, we need to take out the rotation of Joint 8 and the deflection of Joint 8. Y-rotation for Joint 8 for Load Case 2 is 0.537 degrees.

$$\tan(0.537°) = \frac{\text{vertical deflection of Joint 10}}{72 \text{ in.}}$$

$$\text{vertical deflection of Joint 10 caused by rotation} = 0.675 \text{ in.}$$

The net vertical deflection, or deflection caused by the flex of the channel, is 0.718 in. −0.675 in. −0.003 in. = 0.040 in.

$$\frac{72 \text{ in.}}{200} = 0.360 \text{ in. allowable} > 0.040 \text{ in., therefore, OK}$$

Horizontal Deflection of a Horizontal Member

Joint 8 to Joint 10 Support Member (72 in. Long)

For Loading 1, Joint 10 has a gross horizontal deflection of 1.660 in. However, Joint 8 has a gross horizontal deflection of 1.660 in. as well. Therefore, the net horizontal deflection for Load Case 1 of Joint 10 is zero.

To check horizontal clearances to adjacent equipment or structures, the gross deflection at Joint 10 would be used. The gross deflection at Joint 10 would also be used to check for sufficient slack in the jumpers.

A.5.3 Deflection Example, Three-Phase Switch Support Stand, Double Column—Class "A" Structure

A typical three-phase switch stand with select joints is shown in Figure A-7. The structure is 12 ft 3 in. high and has a platform that is 8 ft wide by 25 ft long on two columns, spaced 12 ft apart.

Load 1—Combined Ice and Concurrent Wind in the Y-direction

Load 2—Deflection Level Wind in the Y-direction

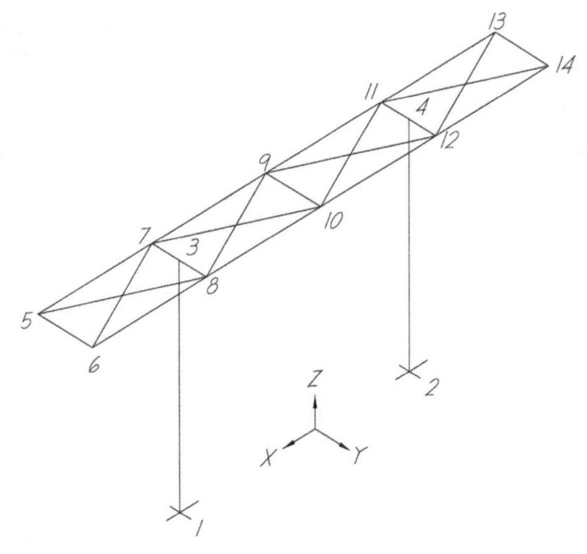

Figure A-7. Two-column switch stand with selected joints shown.

Table A-3. Operational Displacements and Rotations for Example A.5.3.

Joint	Load	X-displ.	Y-displ. (in.)	Z-displ. (in.)	X-rotation (degrees)	Y-rotation (degrees)	Z-rotation (degrees)
4	1	—	0.403	0.004	0.300	−0.001	—
4	2	—	0.639	−0.002	0.406	−0.001	—
11	2	—	0.639	0.327	0.007	−0.001	—
12	1	—	0.403	−0.290	0.357	−0.223	—
12	2	—	0.639	−0.363	0.439	−0.127	−0.001
14	1	—	0.403	−0.742	0.357	−0.374	−0.004
14	2	—	0.640	−0.621	0.439	−0.214	0.026

Note: 1 in. = 25.4 mm.

Table A-3 is the displacement and rotation results from a finite-element analysis of the example structure for the two load cases.

Horizontal Deflection of a Vertical Member

Joint 4 (the top of the column) has a maximum displacement of 0.639 in. Class "A" structure vertical members are limited to 1/100 of the height for maximum horizontal deflection. The height of this structure is 12.25 ft.

$$12.25\text{ ft}\left(12\frac{\text{in.}}{\text{ft}}\right) = 147\text{ in.}$$

$$\frac{147\text{ in.}}{100} = 1.470\text{ in. allowable} > 0.639\text{ in., therefore, OK}$$

Vertical Deflection of a Horizontal Member

Joint 4 to Joint 12 Support Member

For Load Case 2, Joint 12 has a maximum gross vertical deflection of −0.363 in. Joint 11, on the opposite side of the column, has a maximum gross vertical deflection of 0.327 in. for Load Case 2. To "straighten out" the column to calculate the net deflection (deflection of the support member relative to the top of the column), the top of the column rotation must be backed out. The X-rotation angle of Joint 4 for Load Case 2 is 0.406 degrees. To find the deflection caused by the top of the column rotation,

$$\tan(0.406°) = \frac{\text{vertical deflection of Joint 12 } \textit{caused by} \text{ column rotation}}{48 \text{ in.}}$$

$$\text{vertical deflection of joint 12 } \textit{caused by} \text{ to column rotation} = 0.340 \text{ in.}$$

To obtain the net deflection, we take the Joint 12 gross deflection of negative 0.363 in. and add back 0.340 in. (column rotation) to obtain a net deflection of negative 0.023 in.

The net vertical deflection of a horizontal member is limited to $L/200$ for a class "A" structure:

$$4 \text{ ft}\left(12 \frac{\text{in.}}{\text{ft}}\right) = 48 \text{ in.}$$

$$\frac{48 \text{ in.}}{200} = 0.240 \text{ in. allowable} > 0.023 \text{ in. therefore, OK}$$

Vertical Deflection of a Horizontal Member

Joint 12 to Joint 14 Support Member

For Load Case 1, Joint 14 has a maximum gross vertical deflection of −0.742 in. Joint 12 has a maximum gross vertical deflection of −0.290 in. To find the net deflection of Joint 14, we need to take out the rotation of Joint 12 and the vertical deflection of Joint 12. Y-rotation for Joint 12 for load case 1 is 0.223 degrees.

$$\tan(0.223°) = \frac{\text{vertical deflection of joint 14}}{78 \text{ in.}}$$

$$\text{vertical deflection of Joint 14 } \textit{caused by} \text{ rotation} = 0.304 \text{ in.}$$

The net vertical deflection, or deflection *caused by* the flex of the channel, is 0.742 in. −0.304 in. −0.290 in. = 0.148 in.

$$\frac{78 \text{ in.}}{200} = 0.39 \text{ in. allowable} > 0.148 \text{ in., therefore, OK}$$

Horizontal Deflection of a Horizontal Member

Joint 12 to Joint 14 Support Member

For Load Case 2, Joint 14 has a gross horizontal deflection of 0.639 in. Joint 12 has a gross horizontal deflection of 0.640 in. Therefore, the net horizontal deflection for Load Case 2 is 0.001 in., essentially zero.

A.6 DYNAMIC ANALYSIS OF A STEEL LATTICE RACK-TYPE STRUCTURE (FIGURE A-8)

Overview

This example presents the calculation for the seismic demand for this steel structure. Both equivalent lateral force and dynamic analysis methods are shown for comparison only.

Figure A-8. Steel rack–type structure.
Source: Courtesy of Brian Low.

The purpose of this structure is to interconnect the circuit breaker to the substation. To serve its purpose, this structure supports conductor loads, switches, insulators, and electrical connections. All units are in US customary units. The conversion to metric units is left to the engineer.

This analysis model (Figure A-9) is simplified to consider the structure only and not the conductor attached to adjacent structures. The loads from the conductor, insulators, switches, and electrical connections are incorporated in the analysis through load or mass assignments at attachment points, and therefore, details of these components are not modeled. As a result, the dynamic effects of these components are not represented in this analysis for simplicity. Using engineering judgment, the following assumptions were selected:

- Structure geometry and mass distribution is the primary influence on structure behavior,
- Short-circuit loads are typically assessed in rigid bus configurations and not in dead-end structures with flexible conductors,
- Conductor modeled by applying static tension and dead load for the conductor and insulators. Section 3.1.7.12 provides more information regarding flexible wire seismic force considerations,
- Assumed multimode contributions for this structure, and
- 5% Damped response spectrum (Figure A-10).

Weight and tension load of equipment and structure

Total Weight of Steel Structure, Overhead Wire, Switches, and Insulators = 9.9 kips

Tension Load:

1 kip maximum tension (Section 3.1.7.12)

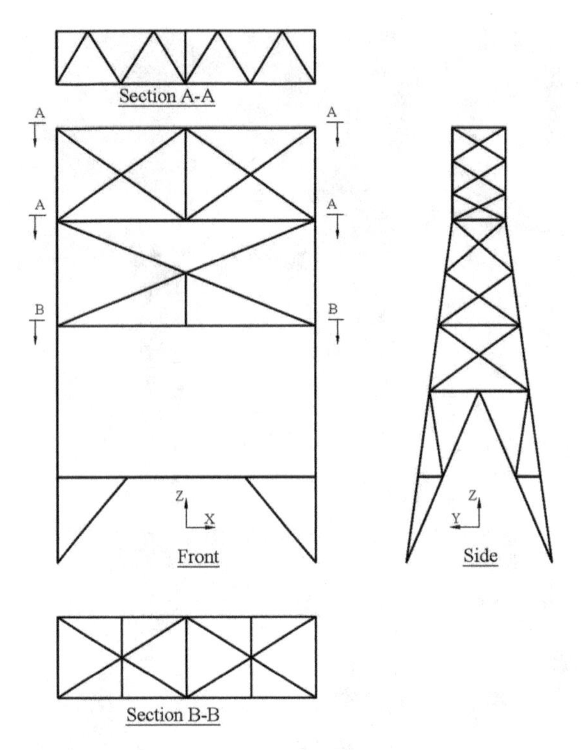

Figure A-9. Structural model of a frame.

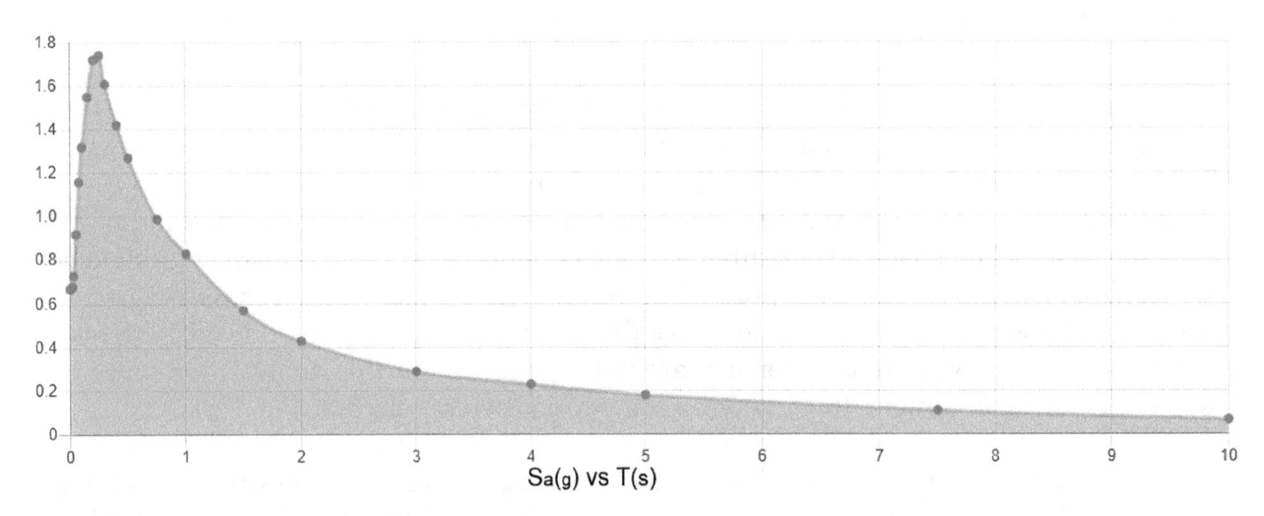

Figure A-10. 5% Damped response spectrum.

Seismic Ground Motion Parameters:

5% damped response spectrum parameters (Figure A-10) from ASCE 7-22 Hazard Tool

Site Class: C—Very Dense Soil and Soft Rock

$S_{DS} = 1.671$ [Numeric seismic design value at 0.2 s spectral acceleration]

$S_{D1} = 0.99$ [Numeric seismic design value at 1.0 s]

$T_S = S_{D1}/S_{DS} = 0.59$ s

$T_L = 12$ s

Table A-4. Seismic Design Coefficients and Factors (Section 3.1.7.4).

Structure type	Steel lattice portal frame (X-direction)	Steel braced frame (Y-direction)
R	3	1.5
Ω_0	1.5	1
C_d	3	1.5
I_e (Section 3.1.7.5)	1.25	
I_{mv} (Section 3.1.7.6.2)	1.5	

Seismic design coefficients and factors (Table A-4):
Equivalent Lateral Force Procedure Base Shear:
Conservatively assume $T = T_s = 0.59$ s

	Steel lattice portal frame (X-direction)	Steel braced frame (Y-direction)
$C_s = \dfrac{S_{DS}}{\left(\dfrac{R}{I_e}\right)}$ [Equation (3-12)]	$\dfrac{1.671\text{g}}{\left(\dfrac{3}{1.25}\right)} = 0.70\text{g}$	$\dfrac{1.671\text{g}}{\left(\dfrac{1.5}{1.25}\right)} = 1.39\text{g}$
Equation (3-12) C_s Controls		
C_s need not exceed: $C_s = \dfrac{S_{D1}}{T\left(\dfrac{R}{I_e}\right)}$ [Equation (3-13)]	$\dfrac{0.99\text{g}}{0.59\text{s}\left(\dfrac{3}{1.25}\right)} = 0.70\text{g}$	$\dfrac{0.99\text{g}}{0.59\text{s}\left(\dfrac{1.5}{1.25}\right)} = 1.40\text{g}$
C_s should not be less than $C_s = 0.044 S_{DS} I_e \geq 0.03$ [Equation (3-14)]	$C_s = 0.044 \times 1.671 \times 1.25 = 0.09$ g	$C_s = 0.044 \times 1.671 \times 1.25 = 0.09$ g
For $S_1 \geq 0.6$g, C_s should not be less than $C_s = \dfrac{0.8 S_1}{R/I_e}$ [Equation (3-15)]	$C_s = \dfrac{0.8 \times 1.06\text{g}}{3/1.25} = 0.35\text{g}$	$C_s = \dfrac{0.8 \times 1.06\text{g}}{1.5/1.25} = 0.71\text{g}$
$V = C_s W I_{mv}$ [Equation (3-11)]	$V = 0.70$ g $\times$ 9.9 k $\times$ 1.5 = 10.4 kips	$V = 1.39$ g $\times$ 9.9 k $\times$ 1.5 = 20.6 kips

USD load combinations (Table 3-18)

Case 4: $1.1D + 1.0E + 0.75SC + 1.1T_{EA}$

Case 8: $0.9D + 1.0E + 0.75SC + 1.1T_{EA}$

System Modeling (Figure A-9)

The system is modeled in three dimensions with nodes and members. The legs are modeled as continuous frame members, whereas all other members are modeled as trusses (shown as shortened members in Figure A-11). The switches and electrical connections are applied as masses, shown in Figure A-12. The conductor tension is applied as a static load. The weight of the dead-end insulators and weight span of the conductor, along with the electrical connections down to the breaker, are applied as dead loads, shown in Figure A-12. The seismic design coefficients were obtained from Table A-4).

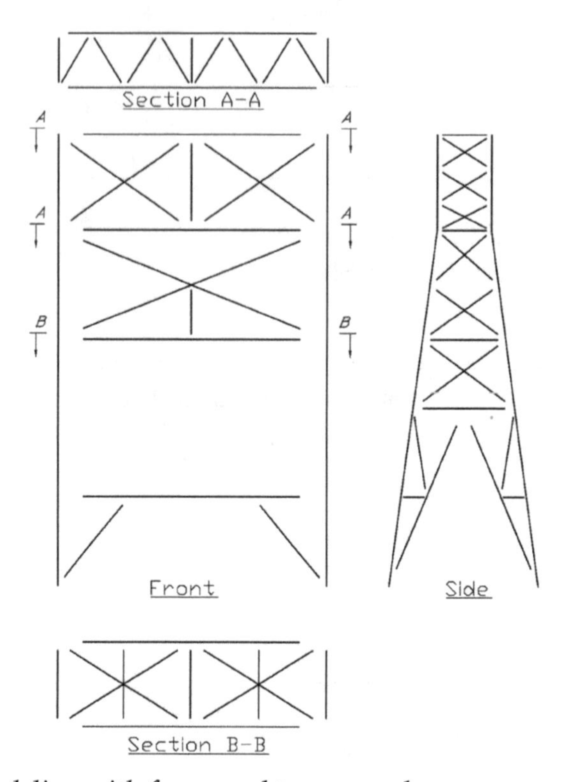

Figure A-11. Structure modeling with frame and truss members.

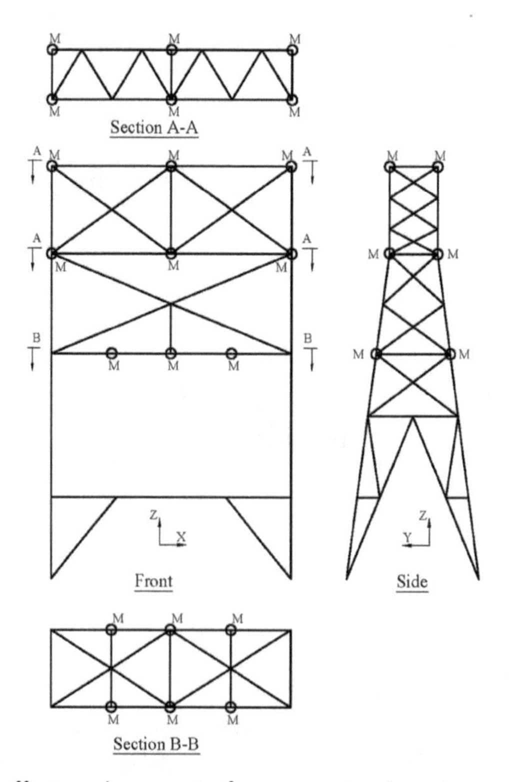

Figure A-12. External mass effect assignments for computer input.

Modal Properties:

The modal properties of the structure are given in Table A-5. Thirty modes were included to obtain a combined modal mass participation of at least 90% of the actual mass in each orthogonal horizontal direction of response.

Results:

The value of each force-related design parameter of interest for each mode of response should be computed using the properties of each mode and the design response spectrum defined in Section 3.1.7.3 divided by the quantity R/I_e, as described in Section 3.1.7.6.3.

The results of the element forces can now be used to analyze the structure in accordance with appropriate material capacities.

Table A-5. Modal Participating Mass Ratios.

	Frequency	Modal participating mass ratios					
Mode	Hz	*UX* (%)	*UY* (%)	*UZ* (%)	Sum *UX* (%)	Sum *UY* (%)	Sum *UZ* (%)
1	1.76	95	0	0	95	0	0
2	2.80	0	1	0	95	1	0
3	2.80	0	0	0	95	1	0
4	4.43	0	1	0	95	2	0
5	5.35	0	1	0	95	2	0
6	6.30	0	0	0	96	2	0
7	6.30	0	0	0	96	2	0
8	7.25	0	6	0	96	8	0
9	7.26	0	0	0	96	8	0
10	7.38	0	69	0	96	77	0
11	9.77	0	0	0	96	77	0
12	10.26	0	0	0	96	77	0
13	10.52	0	0	0	96	77	0
14	12.31	0	0	0	96	77	0
15	12.57	0	0	0	96	77	0
16	12.61	0	0	0	96	77	0
17	12.64	0	11	0	96	88	0
18	14.57	0	0	0	96	88	0
19	14.60	0	1	0	96	89	0
20	14.89	0	0	0	96	89	0
21	17.00	0	0	0	96	89	0
22	17.01	0	0	0	96	89	0
23	19.05	0	0	0	96	89	0
24	19.08	0	0	0	96	89	0
25	21.30	0	1	0	96	90	0
26	21.42	3	0	0	99	90	0
27	21.95	0	0	0	99	90	0
28	23.82	0	0	0	99	90	0
29	24.01	0	7	0	99	97	1
30	25.14	0	1	6	99	98	6

The displacements obtained from an equivalent lateral force elastic analysis should be multiplied by the quantity C_d/I_e, as described in Section 3.1.7.9. The displacement can then be utilized for verification of sufficient slack between structures, as described in Section 3.1.7.12.

Discussion of Results:

The assumed period for the equivalent lateral force analysis was 0.59 s (or 1.69 Hz). The modeled primary frequencies in the X-direction (Figure A-13) is $f = 1.76$ Hz (period = 0.57 s) and in the Y-direction (Figure A-14) is $f = 7.38$ Hz (period = 0.14 s). The difference in the structure period will impact how the seismic demand is calculated.

Comparison of Base Shear:

To illustrate how equivalent lateral force and dynamic analysis can impact results, a comparison of base shears is presented in Table A-6. Note that the dynamic analysis base shear is less than that as calculated by the equivalent lateral force method. The largest contribution to this is the assumed multimode contributions (I_{mv}) considered in the equivalent lateral force analysis compared to the calculated effects in the dynamic analysis.

Other cases may differ, and the structure designer must use engineering judgement when considering which type of analysis to use. Contributing factors to this decision include, but not limited to, structural irregularities, fundamental natural period, structure height, and economics.

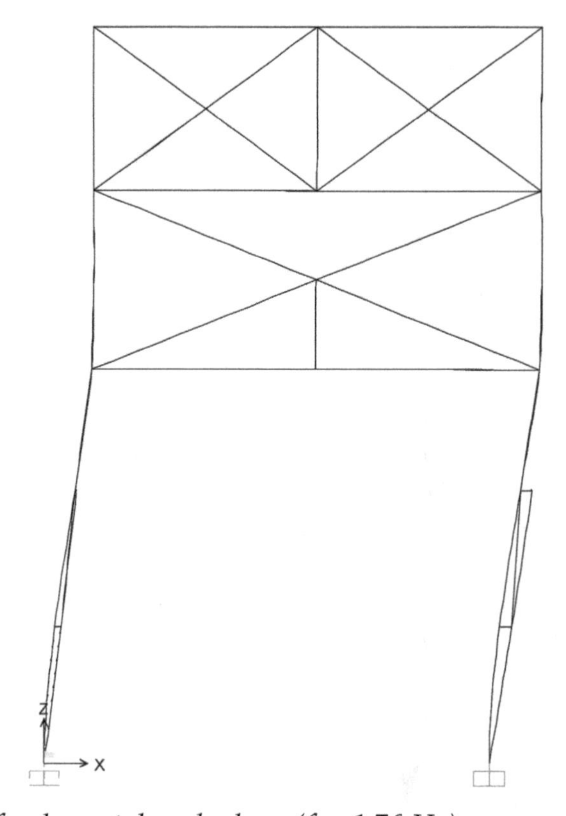

Figure A-13. X-direction fundamental mode shape ($f = 1.76$ Hz).

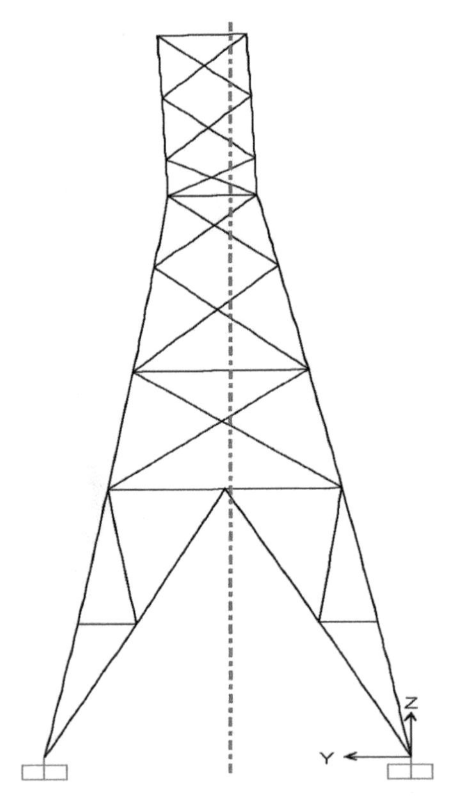

Figure A-14. Y-direction fundamental mode shape ($f = 7.38$ Hz).

Table A-6. Base Shear Using ELF and Dynamic Analysis.

	Calculated base shear (kips)	
Loading direction	Steel lattice portal frame (X-direction)	Steel braced frame (Y-direction)
Equivalent lateral force	10.4	20.6
Dynamic	6.4	12.7

REFERENCES

ACI (American Concrete Institute). 2019. *Building code requirements for structural concrete (with commentary).* ACI CODE-318-19. Detroit: ACI.

ASCE. 2022. *Minimum design loads and associated criteria for buildings and other structures.* ASCE 7-22. Reston, VA: ASCE.

IEEE (Institute of Electrical and Electronics Engineers). 2023. *National electrical safety code.* ANSI C2-2023. Piscataway, NJ: IEEE.

APPENDIX B
SHORT-CIRCUIT FORCES

B.1 SHORT-CIRCUIT CURRENT

According to Ohm's law, the current (I) of a circuit is equal to the voltage (V) divided by the resistance (R)

$$I = V/R$$

Ohm's law can be generalized for applying it to AC circuits and expanded to include capacitance, inductance, and resistance. These components make up the impedance (Z) (Wang 2018)

$$I = V/Z$$

System voltage is assumed at a maximum to yield worst-case fault current. The resulting fault current is then impacted by the equivalent impedance of the system as measured at the fault location. This is referred to as the Thevenin equivalent impedance. The more impedance between the source and the fault, the lower the resulting fault current. Conversely, lower impedance results in higher fault current. This explains why a substation close to a power source with fewer impeding components separating it from the source has a larger fault current.

The maximum available fault current on the lower voltage side of the transformer is limited by the Thevenin equivalent impedance of the system, as well as the impedance of the transformer itself and the conductive components to the fault (e.g., buswork) and their ability to transmit the high fault current.

Using the nominal current and fault variables described in this Appendix, we can examine the fault current waveform with respect to time, the resulting magnetic field in the conductor, and the interaction between magnetic fields.

Figure B-1 depicts a typical fault current waveform as assumed by IEEE 605-2008 (IEEE 2008) and other documents (CIGRE 1996, Reichenstein and Gomez 1985, Hartman 1985). This waveform represents the fault current in a single conductor electrically located at a far distance from a generation source. The total fault current wave form consists of two components, an AC, steady–state function, and a decaying DC function (CIGRE 1996). The decaying DC component shifting the fault current from a steady state is referred to as the "DC offset" or "fault asymmetry" (Reichenstein and Gomez 1985). The DC component decays with time on the basis of the ratio of the reactance to the resistance (X/R) of the power system. Because the fault current also has an AC oscillating component, the total fault current oscillates with the frequency of the system (60 Hz in the United States).

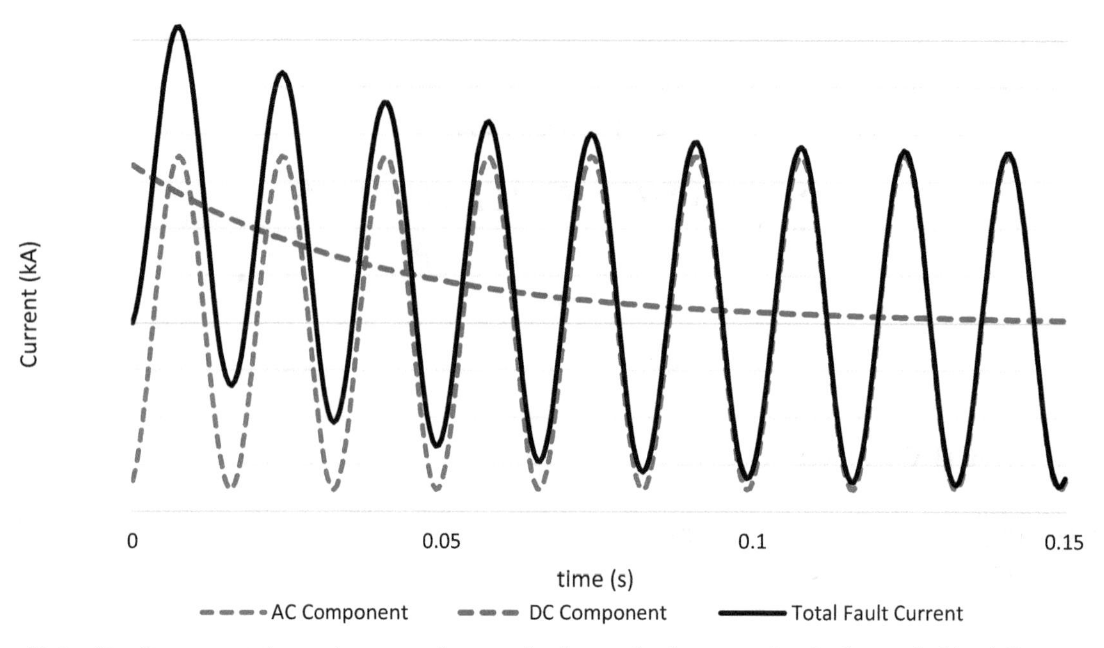

Figure B-1. Fault current in a given conductor during a fault comprised of a periodic AC component and an aperiodic decaying DC component.

The electrical design variables that are used to determine the fault current should be provided by the electrical engineer.

The general equation for fault current in a given conductor is

$$i(t) = \underbrace{\sqrt{2}I_{\text{RMS}} * [\sin(2\pi ft + \phi - \theta)}_{\text{AC Component}} - \underbrace{e^{(-t/c)} * \sin(\phi - \theta)}_{\text{DC Component}}] \tag{B-1}$$

where I_{RMS} is the fault current (root mean squared).

This value is typically determined through a fault study performed by electrical engineers. In general, a maximum fault current value is selected, which would reflect the ultimate demand on the substation. The root mean squared value is equal to the peak value of the AC component divided by $\sqrt{2}$. Thus, multiplying I_{RMS} by $\sqrt{2}$ in Equation (B-1) yields the peak value of the AC waveform (Reichenstein and Gomez 1985).

f = Power system frequency (60 Hz in the United States),

t = Time, and

ϕ = Incidence angle (closing angle or phase angle of the voltage at fault initiation).

This phase angle, which continuously changes prior to the fault, affects the level of asymmetry of the fault current. It is common to assume that the fault occurs at an incidence angle that results in the maximum peak fault force (IEEE 2008, IEC 2011). This is a significant assumption. As can be seen in Figure B-2a, b, if the fault occurs at a time slightly different from the worst case, the fault current and resulting short-circuit condition can be substantially different.

θ = Impedance angle

$$\theta = \tan^{-1}\left(\frac{X}{R}\right)$$

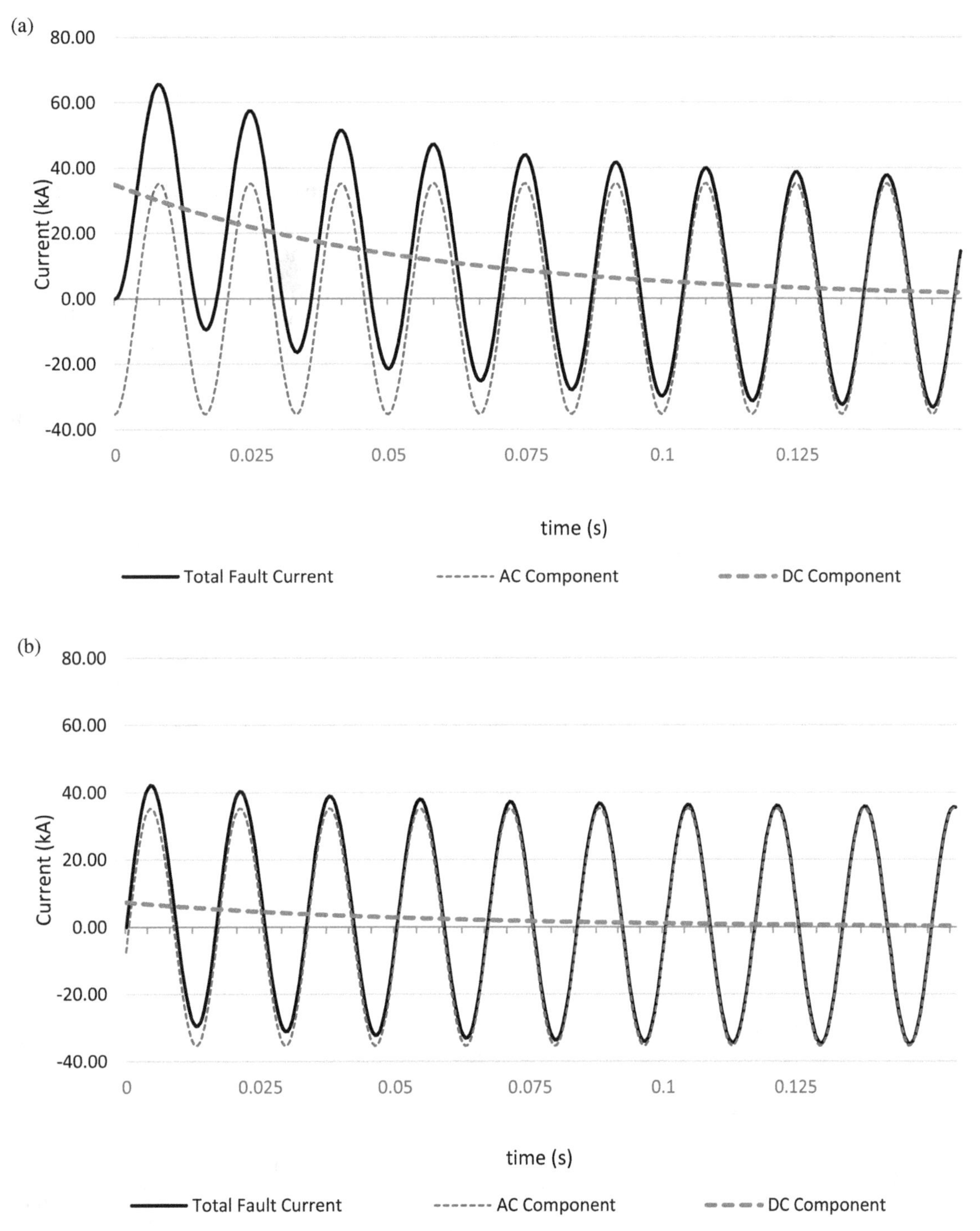

Figure B-2. (a) Effect of varying incidence angle on fault current ($I_{RMS} = 25$ kA, X/R = 20, $\phi = 0$ degree), (b) effect of varying incidence angle on fault current ($I_{RMS} = 25$ kA, X/R = 20, $\phi = 75$ degree).

X = System reactance,

R = System resistance, and

c = Time constant of the circuit (also often given as T_a).

$$c = \frac{X}{R * 2\pi f}$$

As seen in Figures B-1 and B-2a,b, the peak fault current occurs at approximately the first half cycle of the fault, although the precise time of the peak will vary slightly on the basis of the X/R ratio. The fault current decays to a steady state in a fraction of a second.

Various types of faults can occur on an electrical system. The most common type is a single line to ground fault, accounting for approximately 80% of faults. However, during a single line to ground fault, the current increase occurs only in a single phase. Because the electromagnetic forces are primarily caused by an interaction of magnetic fields between adjacent conductors, a lone single conductor with a large current does not create the large short-circuit forces of greatest concern.

The fault types of most concern with regard to short-circuit forces are phase-to-phase faults (15% of faults) and three-phase faults (5% of faults). Phase-to-phase faults often degenerate into three-phase faults (De Metz-Noblat et al. 2005).

Phase-to-phase faults (also referred to as two-phase or line-to-line faults) have phase angles that are 180 degrees out of phase (e.g., if $\phi_A = 0$ degree, then $\phi_B = 180$ degrees). An example plot of fault current versus time for the two phases in a phase-to-phase fault is shown in Figure B-3.

Three-phase balanced faults have phase angles that are 120 degrees out of phase (e.g., if $\phi_A = 0$ degree, then $\phi_B = 120$ degrees and $\phi_C = 240$ degrees). An example plot of fault current versus time for all three phases in a three-phase fault is shown in Figure B-4. Note that the incidence angle sign convention may vary depending on the utility. The sign convention typically used in the United States is as follows: $\phi_B = \phi_A + 120$ degrees and $\phi_C = \phi_A + 240$ degrees. However, in Europe and at certain US utility companies, it is common to use the following sign conventions: $\phi_B = \phi_A - 120$ degrees and $\phi_C = \phi_A = -240$ degrees. The former is used in this MOP.

Again, note that the voltage in the conductors will vary depending on the phase angle of the current at the time of fault closing (incidence angle). This effect can be seen in Figure B-5a, b.

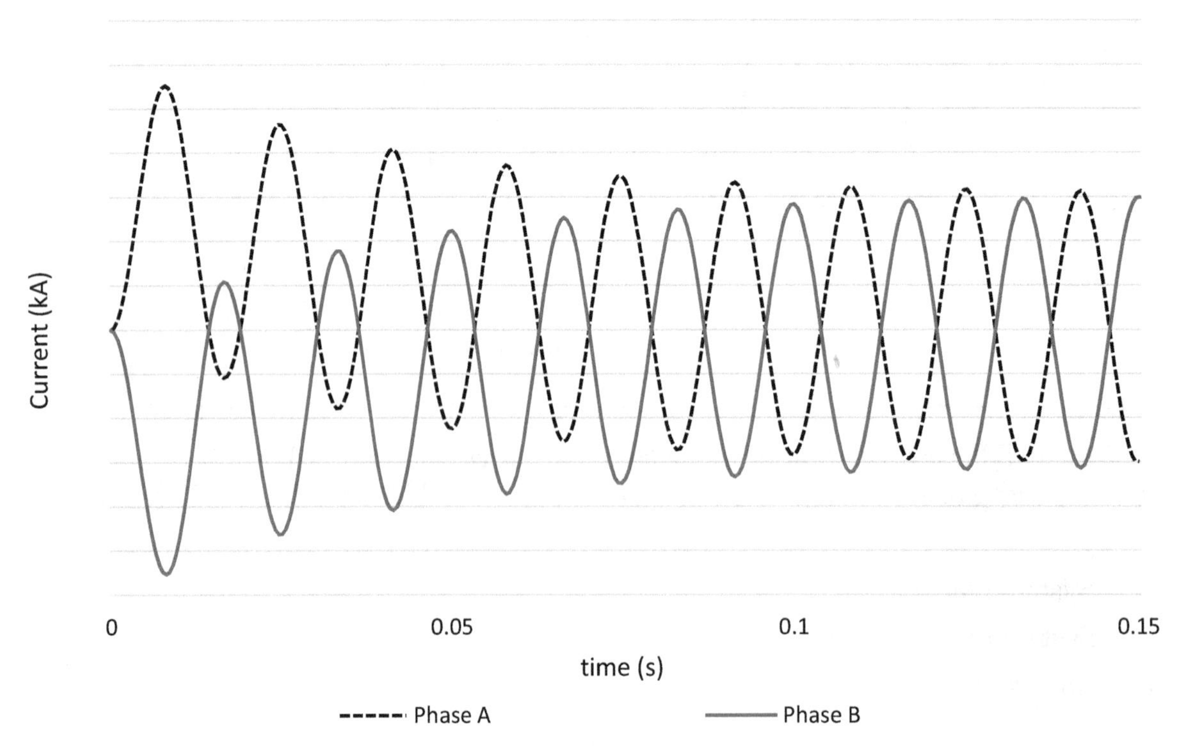

Figure B-3. Current in conductors under a phase-to-phase balanced fault on A and B phases (assuming that an X/R ratio of 14.3 and the fault is initiated at the following phase angles: $\phi_A = 0$ degree, $\phi_B = 180$ degrees).

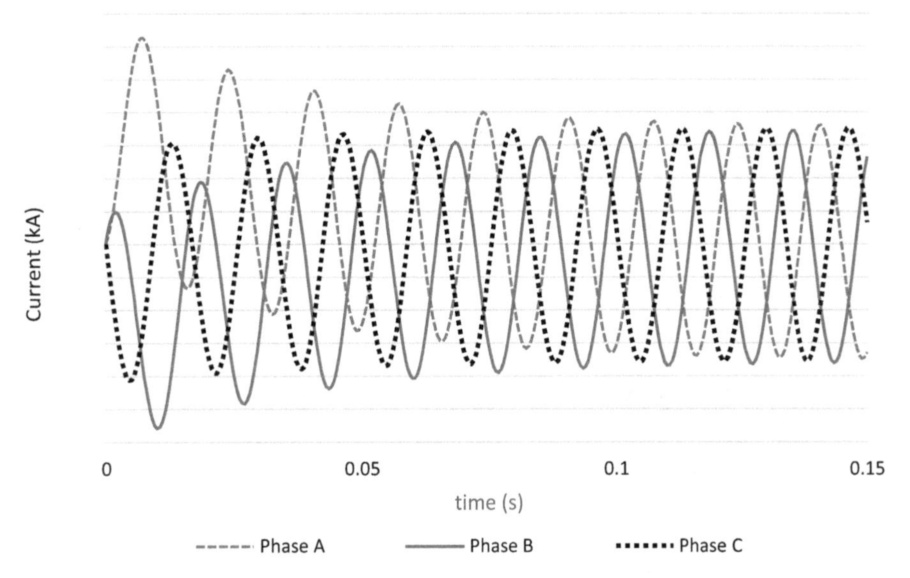

Figure B-4. Current in conductors under a three-phase balanced fault (assuming that an X/R ratio of 14.3 and the fault is initiated at the following phase angles: $\phi_A = 15$ degrees, $\phi_B = 135$ degrees, $\phi_C = 255$ degrees).

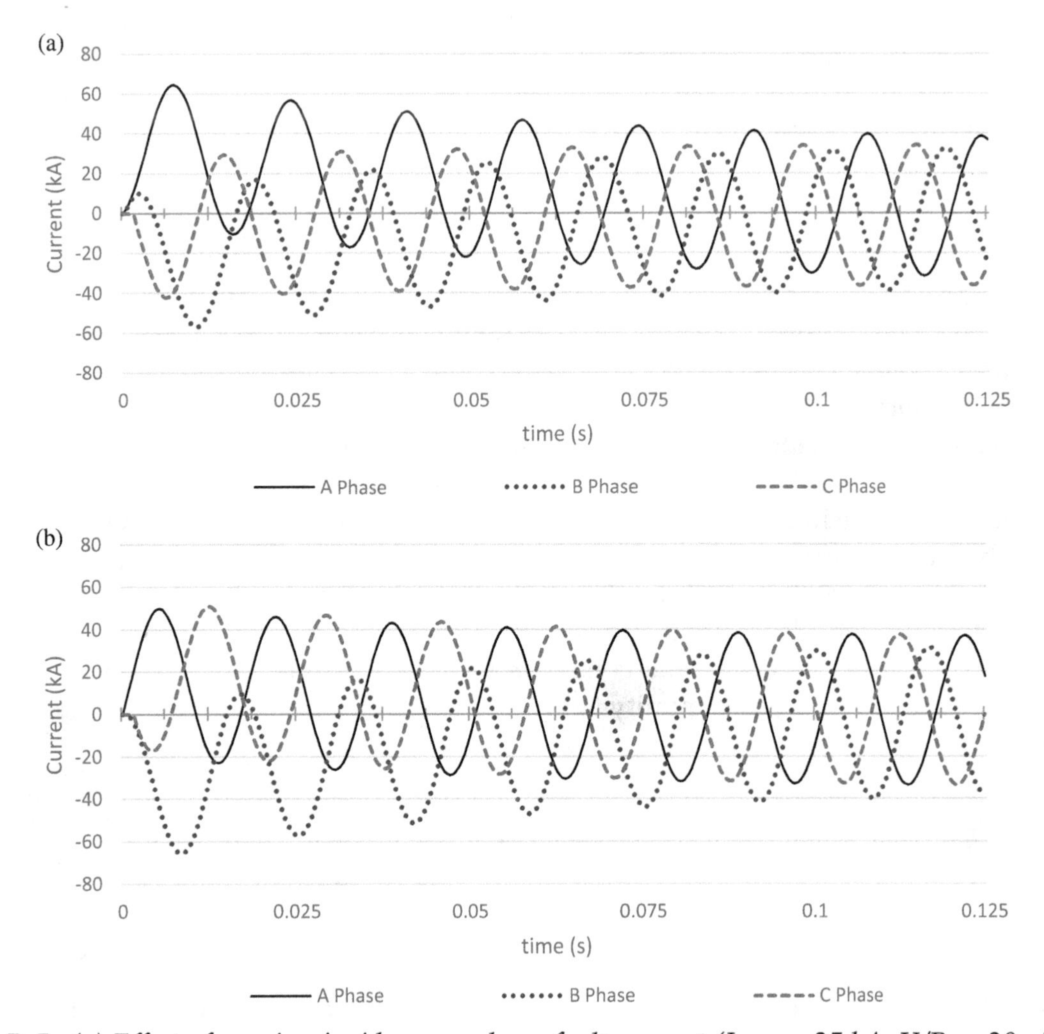

Figure B-5. (a) Effect of varying incidence angle on fault current ($I_{RMS} = 25$ kA, X/R = 20, $\phi_A = 15$ degrees, $\phi_B = 135$ degrees, and $\phi_C = 255$ degrees), (b) effect of varying incidence angle on fault current ($I_{RMS} = 25$ kA, X/R = 20, $\phi_A = 60$ degrees, $\phi_B = 180$ degrees, $\phi_C = 300$ degrees).

Referring to Chapter 3, Equations (3-38) to (3-40), it would appear that phase-to-phase faults cause the worst-case short-circuit loading because the Γ factor equals 1.0 for phase-to-phase faults and 0.866 for three-phase faults. However, phase-to-phase fault current is related to three-phase fault current by the following equation:

$$I_{\text{RMS\_2Ph}} = I_{\text{RMS\_3Ph}} \frac{\sqrt{3}}{2} \tag{B-2}$$

Therefore, for a given system, three-phase faults will result in larger short-circuit forces.

B.2 ELECTROMAGNETIC FORCE VARIATION WITH TIME

The force in a conductor caused by the interaction of electromagnetic fields with an adjacent conductor is described by the Biot–Savart law as follows (IEEE 2008):

$$F_{sc}(t) = \frac{\mu_0}{4\pi r^2} i_1(t) i_2(t) \left[d_1 \otimes (u_r \otimes d_2) \right] \tag{B-3}$$

where

$\mu_0 = 2.825 \times 10^{-7}$ lb/A^2, which is $4\pi \times 10^{-7}$ N/A^2 in SI units, magnetic constant, magnetic permeability in a classical vacuum;
r = Distance between the two conductor segments;
u_r = Unit directional vector in the direction r;
d_1 = Vector of length d_1 in the direction of the current flow in conductor Segment 1, and
d_2 = Vector of length d_2 in the direction of the current flow in conductor Segment 2.

A three-dimensional analysis could be performed using this general equation. This would lead to the most accurate forces on the conductors and would accommodate all possible arrangements. However, a general analysis would require discretizing the conductors, and for a particular point on a conductor, evaluating this equation for the said point and its fault current and all other points on adjacent conductors and their fault currents. This could be duplicated for all other discrete points. This would be a laborious process and as such would require software to complete for all but the simplest of conductor arrangements. Software capable of this general analysis has not gained widespread use in the United States at the time of publication of this MOP.

As such, the simplifying assumption of parallel, infinitely long conductors lying in a flat plane is typically made (CIGRE 1996, IEC 2011, IEEE 2008). If conductors are assumed parallel and a point is selected along a length of bus, a general expression can be formulated for a conductor point and a point on an adjacent parallel conductor. One can then integrate this general expression from negative infinity to positive infinity to obtain a closed-form solution for the fault force.

This parallel, infinitely long conductor simplification yields the following fault force equation (CIGRE 1996):

$$F_{sc}(t) = \frac{\mu_0}{2\pi} * i(t) \sum_n \frac{i_n(t)}{D_n} \tag{B-4}$$

where

$F_{sc}(t)$ = Fault force (μ_0 and D_n in US customary units lb/ft, and in SI units N/m);
$\mu_0 = 2.825 \times 10^{-7}$ lb/A^2, which is $4\pi \times 10^{-7}$ N/A^2 in SI units, magnetic constant, magnetic permeability in a classical vacuum;

$i(t)$ = Current in the conductor under consideration [Equation (B-1)];
$i_n(t)$ = Current in adjacent conductor(s) [Equation (B-1)]; and
D_n = Center-to-center perpendicular distance to adjacent conductor(s).

Although this arrangement of parallel, infinitely long conductors simplifies the calculations, it does not exist in real substations. Substations contain conductors of finite length and frequently contain staggered termination points, jogs in direction, and crossing conductors. The corner effects, end effects, and crossing member effects can cause short-circuit forces significantly different from those determined from Equation (B-4).

Ignoring end effects means that the tapering of forces at the ends of conductors is not accounted for. This is generally a conservative assumption. Conversely, ignoring corner effects can lead to a significant underestimation of forces, such as forces longitudinal to a support structure. At conductor crossings, such as high bus to low bus transitions, the direction of forces can be reversed.

In addition, Equation (B-4) and Equations (B-5) to (B-11) are valid only for rigid conductors. The short-circuit forces are dependent on the distance between conductors. The electromagnetic force on the strain bus (flexible conductors) is much more complex because the distance between conductors varies, significantly in many cases, throughout the fault event.

The designer should consult with an electrical engineer knowledgeable in faults and electromagnetism when significant corner, end, 3D, or conductor displacement effects are anticipated.

For a three-phase fault occurring on an arrangement of three conductors in a flat plane labeled A, B, and C, Equation (B-4) yields the following functions for the fault force versus time, assuming that the B phase is the center conductor.

$$F_{A\_3\mathrm{Ph}}(t) = \frac{\mu_0}{2\pi} i_A(t)\left(\frac{i_B(t)}{D} + \frac{i_C(t)}{2D}\right) \tag{B-5}$$

$$F_{B\_3\mathrm{Ph}}(t) = \frac{\mu_0}{2\pi} i_B(t)\left(-\frac{i_A(t)}{D} + \frac{i_C(t)}{D}\right) \tag{B-6}$$

$$F_{C\_3\mathrm{Ph}}(t) = \frac{\mu_0}{2\pi} i_C(t)\left(-\frac{i_B(t)}{D} - \frac{i_A(t)}{2D}\right) \tag{B-7}$$

Similarly, for a phase-to-phase fault occurring between Phases A and B, the functions for fault force versus time are

$$F_{A\_2\mathrm{Ph}}(t) = \frac{\mu_0}{2\pi} i_A(t)\left(\frac{i_B(t)}{D}\right) \tag{B-8}$$

$$F_{B\_2\mathrm{Ph}}(t) = \frac{\mu_0}{2\pi} i_B(t)\left(\frac{-i_A(t)}{D}\right) \tag{B-9}$$

For a fault occurring between Phases B and C, the functions are

$$F_{B\_2\mathrm{Ph}}(t) = \frac{\mu_0}{2\pi} i_B(t)\left(\frac{i_C(t)}{D}\right) \tag{B-10}$$

$$F_{C\_2\mathrm{Ph}}(t) = \frac{\mu_0}{2\pi} i_C(t)\left(\frac{-i_B(t)}{D}\right) \tag{B-11}$$

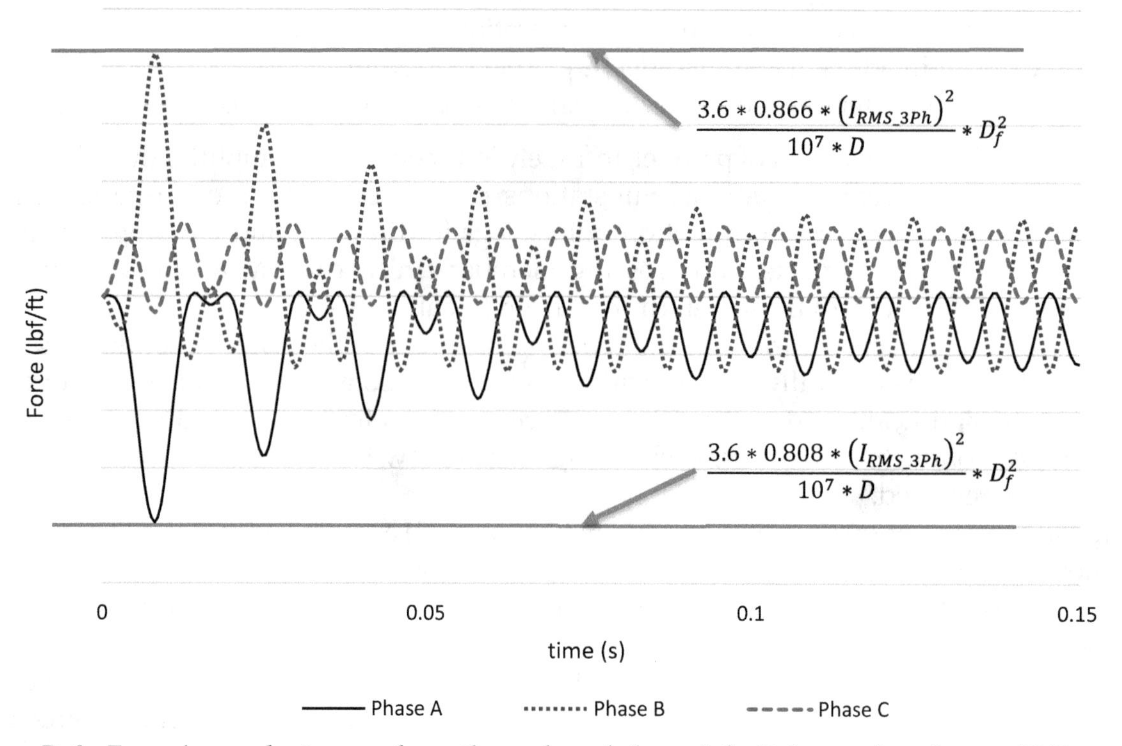

Figure B-6. Force in conductors under a three-phase balanced fault (assuming that an X/R ratio of 14.3 and the fault is initiated at the following closing phase angles: $\phi_A = 15$ degrees, $\phi_B = 135$ degrees, and $\phi_C = 255$ degrees).

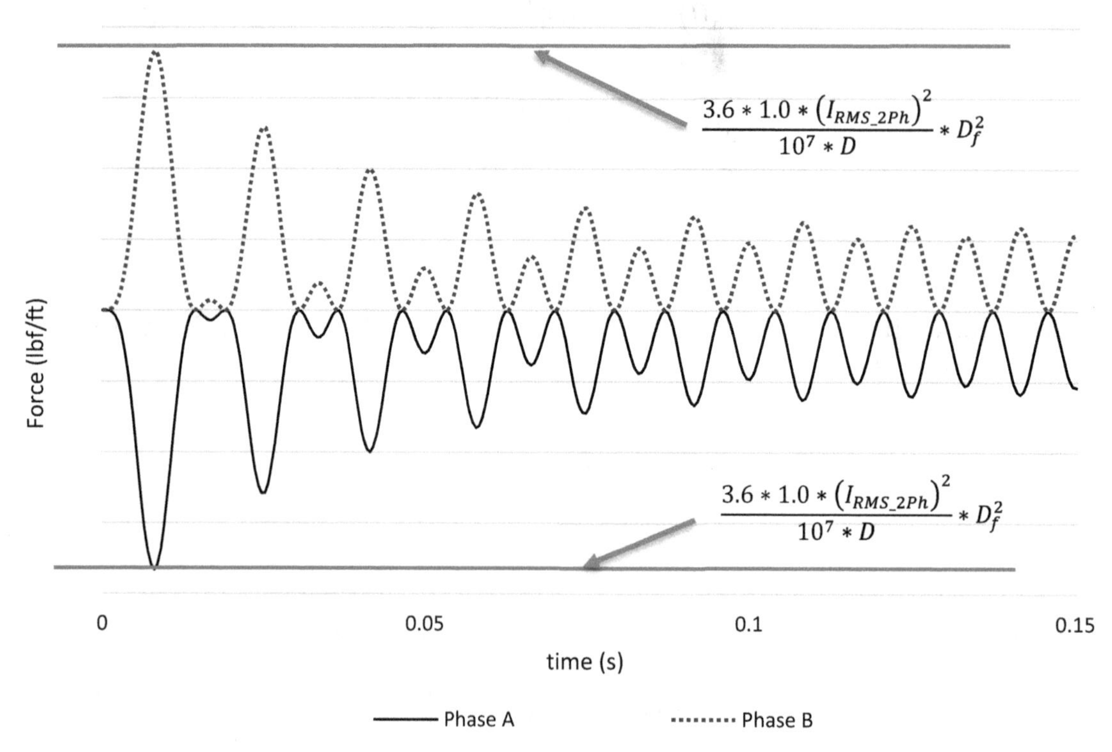

Figure B-7. Forces in conductors under a phase-to-phase balanced fault on A and B phases (assuming that an X/R ratio of 20 and the fault is initiated at the following phase angles: $\phi_A = 0$ degree, $\phi_B = 180$ degrees).

Figure B-6 presents the forces in Conductors A, B, and C resulting from the three-phase fault currents presented in Figure B-4. Figure B-7 presents the forces in Conductors A and B resulting from the phase-to-phase fault currents presented in Figure B-3.

Note the horizontal lines in Figures B-6 and B-7. These represent the magnitude of force obtained from Equation (3-38) (reproduced from IEEE 605-2008) (IEEE 2008) to be used in a simplified static analysis. There is no variation with time because these are static forces. The 0.866 and 0.808 coefficients in Figure B-9 refer to the Γ factor in Section 3.1.8.1 and IEEE 605-2008 (IEEE 2008). A Γ factor of 0.866 is used for the center conductor in a three-phase fault and 0.808 factor for outer conductors in a three-phase fault. Note that the peak force achieved in approximately the first half cycle of the short-circuit force varying with time is the same force that is obtained using Equation (3-38) and multiplying the basic fault force, F_{sc}, by the square of the half cycle decrement factor, D_f.

As discussed previously, the fault currents are dependent on the assumed closing angles and thus the forces are also dependent on the assumed closing angles. This effect can be seen in Figure B-8a, b. The closing angles causing the maximum peak force for a phase-to-phase

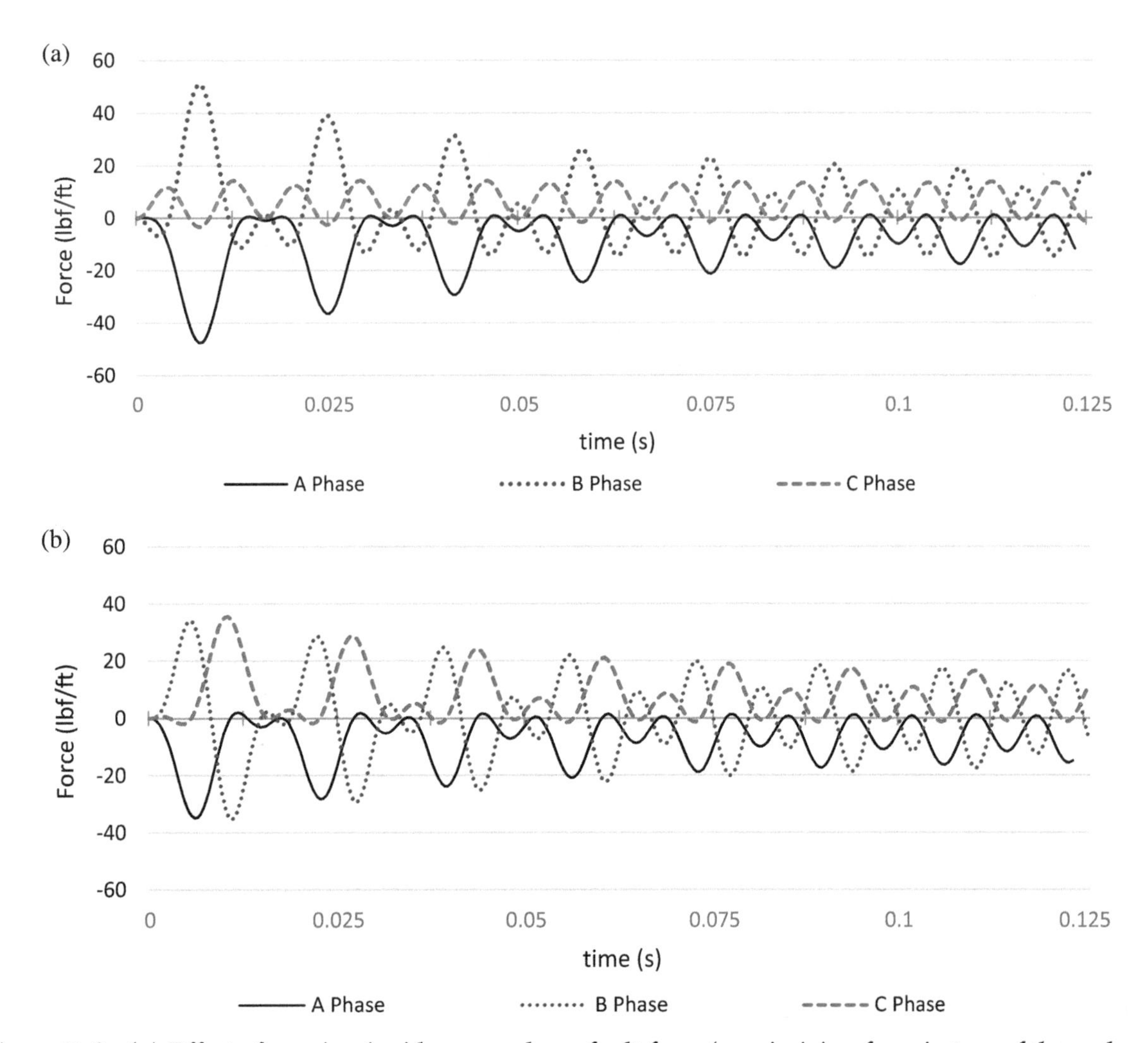

Figure B-8. (a) Effect of varying incidence angle on fault force (maximizing force in two of three phases) $I_{RMS} = 25$ kA, X/R = 20, $\phi_A = 15$ degrees, $\phi_B = 135$ degrees, $\phi_C = 255$ degrees, (b) effect of varying incidence angle on fault force (minimizing peak magnitudes) $I_{RMS} = 25$ kA, X/R = 20, $\phi_A = 60$ degrees, $\phi_B = 180$ degrees, $\phi_C = 300$ degrees.

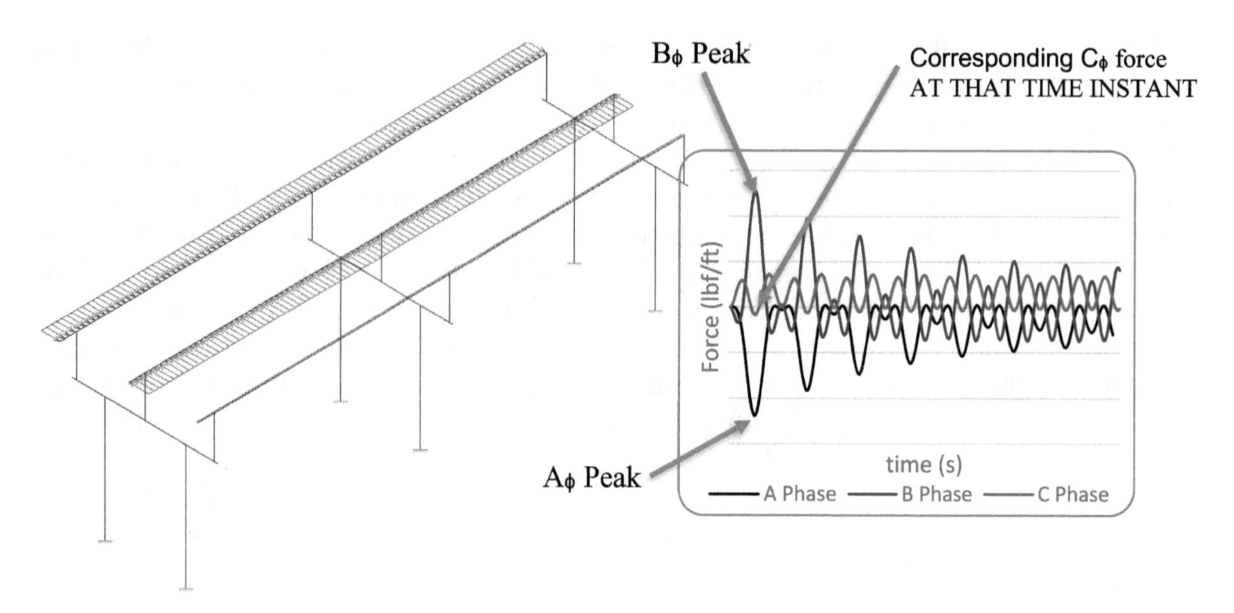

Figure B-9. Peak forces in conductors from Figure B-6 applied to conductors supported by three-phase bus supports.

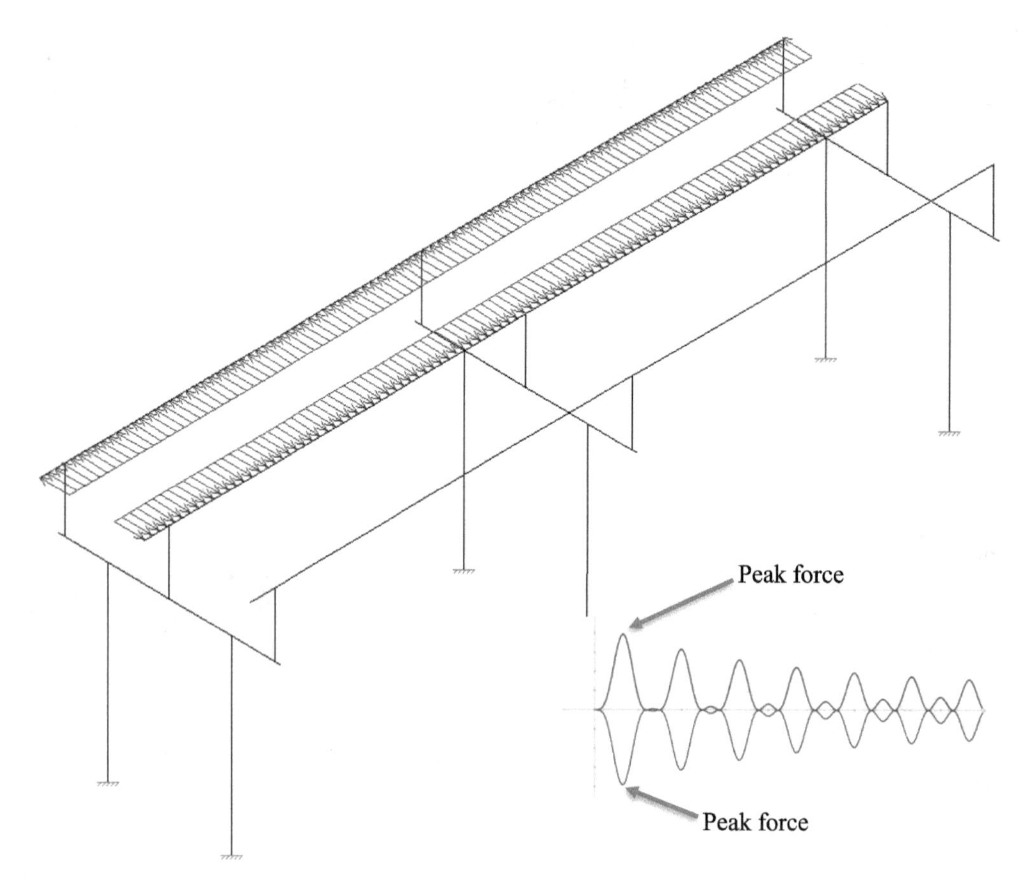

Figure B-10. Peak forces in conductors from Figure B-7 applied to conductors supported by three-phase bus supports.

fault are approximately 0 degree and 180 degrees. For a three-phase fault, the maximum peak force on the A phase occurs at approximately $\phi_A = 15$ degrees, $\phi_B = 135$ degrees, $\phi_C = 255$ degrees, $\phi_A = 195$ degrees, $\phi_B = 315$ degrees, and $\phi_C = 75$ degrees. B-phase maximums and C-phase half-cycle minimums will occur at approximately these values as well. However, the precise closing angles to cause these maximums/minimums will be shifted slightly depending on the X/R ratio. Therefore, the electrical engineer should investigate the different closing angles to find those that cause the worst-case forces in the conductors of interest.

B.3 APPLICATION OF SHORT-CIRCUIT FORCES TO STRUCTURES

The loads in Section B.2 are typically assumed to be imparted on the conductors as uniformly distributed loads. Figures B-9 and B-10 show the peak loading from Figures B-6 and B-7. As discussed in Section B.2, member ends, corners, conductor crossings, and angles in conductors relative to adjacent conductors can cause forces that are nonuniform or in directions that differ from those shown in the figures subsequently.

The peak force (Figure B-9) is valid for the particular closing angles used in Figure B-6 and at the particular instant of time when the forces on Phases A and B are at their peak. The force shown in Phase C is not the maximum force that will be exerted on the conductor during this particular fault.

It is important to note that at any instant of time during the balanced three-phase fault or phase-to-phase fault with parallel, infinitely long conductors, the sum of the forces on the conductors equals zero.

$$F_{A\_3\mathrm{Ph}}(t) + F_{B\_3\mathrm{Ph}}(t) + F_{C\_3\mathrm{Ph}}(t) = 0 \tag{B-12}$$

$$F_{A\_2\mathrm{Ph}}(t) + F_{B\_2\mathrm{Ph}}(t) = 0 \tag{B-13}$$

This is particularly important for structures that support more than one phase. For a three-phase fault, the electrical engineer should investigate which phase angles at fault initiation lead to the greatest effect on the structure.

REFERENCES

CIGRE (International Council on Large Electric Systems). 1996. *The mechanical effects of short-circuit currents in open air substations (rigid and flexible busbars).* Brochure 105. Paris: CIGRE.

De Metz-Noblat, B., F. Dumas, and C. Poulain. 2005. *Calculation of short-circuit currents.* Cahier Technique No. 158. Rueil-Malmaison, France: Schneider Electric.

Hartman, C. N. 1985. "Understanding asymmetry." *IEEE Trans. Ind. Appl.* 1A-21 (4): 842–848.

IEC (International Electrotechnical Commission). 2011. *Short-circuit currents—Calculation of effects—Part 1: Definitions and calculation methods.* IEC 60865-1. 3.0 edition. Geneva: IEC.

IEEE (Institute of Electrical and Electronics Engineers). 2008. *Guide for bus design in air insulated substations.* IEEE 605-2008. New York: IEEE.

Reichenstein, H. W., and J. Gomez. 1985. "Relationship of X/R, IP, and IRMS' to asymmetry in resistance/reactance circuits." *IEEE Trans. Ind. Appl.* IA-21 (2): 481–492.

Wang, R. 2018. "Impedance and generalized Ohm's law." *E84 Lecture Notes, Ch. 3: AC Circuit Analysis.* Claremont, CA: Harvey Mudd College.

APPENDIX C
SEISMIC DESIGN PARAMETERS

This appendix provides additional information on the seismic design of wire-supporting substation structures. In some cases, these structures can support substation equipment. The design requirements for structural members supporting substation equipment (intermediate support) are provided in IEEE 693 (IEEE 2018). An example of these structures supporting substation equipment can be rack structures (Figure C-1) or substation wire-termination and dead-end structures with a lower equipment support level (Figure C-2). Table 3-15 in Chapter 3 contains additional examples. Appendix C discusses only the Response Modification Factor (*R*). Future appendixes may provide addition information on other seismic design parameters.

Figure C-1. Substation rack structure.
Source: Courtesy of US Department of Energy.

Figure C-2. Substation dead-end structure (supporting equipment).
Source: Courtesy of CenterPoint Energy.

The Response Modification Factor approach, also referred to as the *R* factor, was originally developed for building codes on the basis of historical earthquake performance (SEAOC 2008, Uang and Bruneau 2018). The *R* factor approach allows for inelastic behavior of selected structural elements (individual member, connections, and supports) using linear elastic design and linear elastic response spectrum analysis. The design ground motion spectra are reduced from the anticipated ground motion, with an *R* factor that will achieve acceptable building earthquake performance with selected structural element inelastic behavior. This outcome is dependent on using suitable structural systems and structure detailing with appropriate levels of ductility, regularity, and continuity. To obtain this structural performance, the element forces and deformations produced in most building structures will substantially exceed the point at which structural elements start to yield and buckle and therefore behave in an inelastic manner. Inelastic behavior is further defined were plastic hinges develop, in at least, the most critical regions of the structure. The formulation of plastic hinges and redistribution of the member forces, in a properly designed and redundant structure, provides the overstrength capacity for the structure to resist the higher ground motions, which is above the design ground motion level used to obtain the member forces.

The Response Modification Factor (*R*) is the ratio of the force that would develop under the specified ground motion if the structure had an entirely linearly elastic response to the prescribed strength design forces. This is shown in Figure C-3 by the ratio of points *E*/*S* or Seismic Response Coefficients C_e/C_s. The ductility reduction from point *E* to point *S* is possible for a number of reasons. When the structure yields and deforms inelastically, the effective period of response of the structure tends to lengthen, which for many structures results in a reduction of the strength demand. In addition, the inelastic action results in a

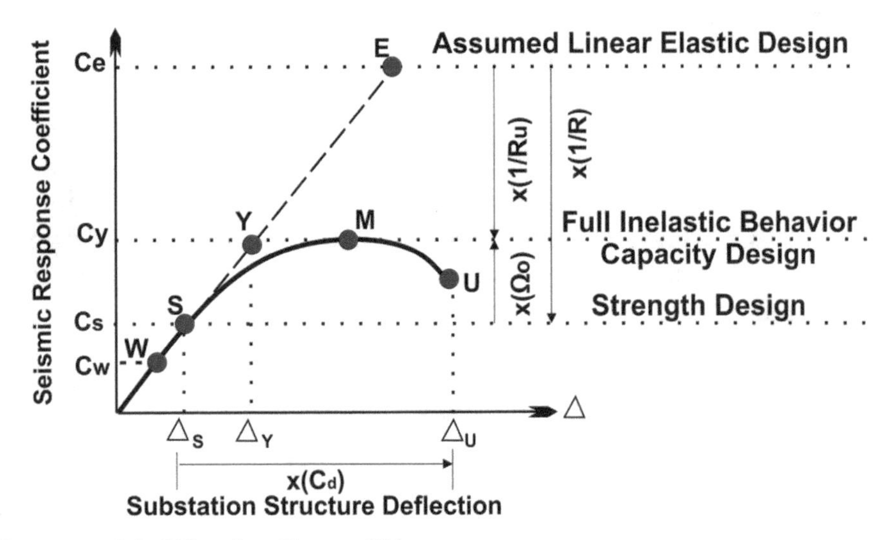

Figure C-3. Response Modification Factor (R).

significant amount of energy dissipation (hysteretic damping) as well as viscous damping. The combined effect (*R*) explains why a properly designed structure at point *S* with full yield strength, which is significantly lower than the elastic seismic force demand (point *E*), is expected to provide satisfactory performance under the specified design ground motions. If the substation structure designer does not expect this type of performance level of the structure, there is an option to use *R* equal to 1.0, which is given in the notes of Table 3-13, Chapter 3 of this MOP.

The Response Modification Factor (*R*) can be defined by two design parameters: Structure System Ductility Reduction Factor (R_u) and a Structural Overstrength Factor (Ω_0). The Overstrength factor provides additional capacity in critical structural components for inelastic performance. The Structure System Ductility Reduction Factor, R_u, is the ratio between point E over point *M* capacity design level, (*E*/*M*), or C_e over C_y, (C_e/C_y) in Table C-1. The Overstrength Factor, Ω_0, is the ratio of the full yield strength level (point *M* capacity design level) where all plastic hinges form and the final structural capacity is reached, over the strength design level, point *S* (*M*/*S*), or C_y over C_s, (C_y/C_s). The Response Modification Factor, *R*, is the product of $R_u \times \Omega_0$ (System Ductility Reduction and the Overstrength Factor). The *x*-axis represents the deflections at the difference design levels (Figure C-3).

Industry documents [ASCE 7 (ASCE 2022) FEMA 303 (FEMA 1998) and others] provide a range of *R* factors (1 to 9) depending on the structural and nonstructural systems/components. The higher the R value, the more significant is the required inelastic structure/component behavior. Inelastic behavior includes member performance beyond yield and/or buckling, a redundant structure/component with the ability to redistribute member forces as members yield and buckle, and members and connections that can develop significant hysteretic and viscous damping (energy absorption capacity).

This MOP selected a maximum *R* value of 3 and associated parameters (Ω_0 and C_d shown in Table 3-13 in Chapter 3) to eliminate the need for special connection detailing requirements [where permitted according to ASCE 7 (ASCE 2022) and AISC 341, AISC 2016]. If *R* factors are selected from ASCE 7 (ASCE 2022), then all provisions associated with the selection should be considered to apply, including the detailing requirements of AISC 341 (AISC 2016) and the limitations related to the structure height or seismic hazard at the substation site. If the designer determines that the substation structure does not have the capability to develop the necessary

Table C-1. Response Modification Factor (R).

***R* Factor Approach**
For high force levels, > 1 g times substation structure seismic mass, codes accept that damage is allowed for economic considerations. $R = R_u \times \Omega_o$; $R_u = C_e/C_y$ (system-level ductility reduction factor); $\Omega_o = C_y/C_s$ (system overstrength factor); C_y = capacity design (elastic no energy dissipation); C_s = ductility design (inelastic-energy dissipation). Based on the building code concept to maintain a linear elastic analysis approach.
Point *E* represents the required seismic force level if the structure remains linear-elastic.
To maintain elastic design (equivalent lateral load or modal response spectrum analysis),
Point *E* is reduced to **Point *S*** by the response modification factor R, for strength design.
Point *S* is represented when the substation structure leaves the linear elastic region into the inelastic range; for example, members yield, buckle, or develop plastic hinges, with damage accepted.
The redundancy (ductility of members/connections, post-buckling behaviour, load redistribution, P-Δ effects) allows the deflection to continue to increase to **Point *M*** (ultimate lateral strength and capacity design), and degrade to **Point *U***.
Ductility, inelastic deformation capacity, is required of structural components to experience inelasticity.
R-Factors are associated with a targeted plastic mechanism, while maintaining the gravity load-carrying capacity for each seismic force–resisting system given in ASCE 7, but the final structure performance relies on the AISC Seismic Provisions to ensure sufficient ductility capacity.
Point *W* represents allowable strength design. The R_w values to be used with ASD are 1.33 × strength design R values: $R_w = 1.33 \times R$.

inelastic behavior, and/or energy dissipation mechanisms, and/or has limited to no redundancy to redistribute member forces as members yield and buckle, the designer can select an R factor equal to 1, with $\Omega_0 = 1$ and $C_d = 1$; see Note 4 in Table 3-13, Chapter 3 of this MOP.

REFERENCES

AISC (American Institute of Steel Construction). 2016. *Seismic provisions for structural steel buildings.* AISC 341. New York.

ASCE. 2022. *Minimum design loads and associated criteria for buildings and other structures.* ASCE 7-22. Reston, VA: ASCE.

FEMA (Federal Emergency Management Agency). 1998. *NEHRP recommended provisions for seismic regulations for buildings and other structures, part 2—Commentary.* FEMA 303. Washington, DC: FEMA.

IEEE (Institute of Electrical and Electronics Engineers). 2018. *Recommended practice for seismic design of substations.* IEEE 693. Piscataway, NJ: IEEE.

SEAOC (Structural Engineers Association of California). 2008. "A brief guide to seismic design factors." *Structure* (September): 30–32. Accessed September 6, 2023. https://www.structuremag.org/?p=5522.

Uang, C. M., and M. Bruneau. 2018. "State-of-the-art review on seismic design of steel structures." *Struct. Eng. J.* 144 (4): 03118002.

APPENDIX D

DRAFT PRE-STANDARD SUBSTATION CIVIL/ STRUCTURAL DESIGN STANDARD

D.1 PURPOSE

The ASCE Task Committee on Substation Structure Design Guide envisions the need for a design standard for substation facilities. This appendix presents the recommendations of this committee written in a prescriptive form and is presented as a draft pre-standard for public review and comment. Comments should be directed to the ASCE Committee on Electrical Transmission Structures.

D.2 SCOPE

This draft *Pre-Standard for Substation Structure Design* specifies requirements for the loading, design, testing, and construction and maintenance of outdoor electrical substation structures and their foundations. Appropriate loading criteria, deflection criteria, strength reduction factors, and methods of analysis are provided for use in design using Ultimate Strength Design (USD) methodology utilizing steel, concrete, wood, and aluminum materials (see USD definition in Section D.7.2). This standard does not address control enclosures or other structures not associated with direct functioning of the electrical power substation and its equipment. These items include security fencing, lighting, storage buildings, perimeter fencing, blast walls, site development structures, and emergency facilities, barriers, and so on.

This draft pre-standard can be used by the facility owner or their authorized agent responsible for developing the design of the substation structures for the facility.

Note 1: This draft pre-standard is intended for designing new substation facilities. It may be applied to the assessment of existing substation facilities. Further, the principles contained herein may be applied to temporary or emergency facilities with adjustments cited herein.

Note 2: The principles contained herein may be applied to substation and switchyard facilities.

D.3 APPLICABLE DOCUMENTS

The following standards, codes, and guidelines are referenced in this draft pre-standard; the latest revisions apply unless noted:

Aluminum Association:

- *Aluminum Construction Manual* (AA 1981)
- *Aluminum Design Manual, Including Specification for Aluminum Structures* (AA 2020)

American Concrete Institute:

- ACI PRC-201.2, *Guide to Durable Concrete*
- ACI PRC-302.1, *Guide to Concrete Floor and Slab Construction*
- ACI CODE-318, *Building Code Requirements for Structural Concrete*
- ACI PRC-440.1, *Guide for the Design and Construction of Concrete Reinforced with FRP Bars*

American Galvanizers' Association:

- *Recommended Details for Galvanized Structures*
- *The Design of Products to Be Hot-Dip Galvanized after Fabrication*

American Institute of Steel Construction:

- AISC 303, *Code of Standard Practice for Steel Buildings and Bridges*
- AISC 341, *Seismic Provisions for Structural Steel Buildings*
- AISC 360, *Specification for Structural Steel Buildings*

American National Standards Institute:

- ANSI/NEMA C29.19, *American National Standard for Composite Insulators—Station Post Type*
- ANSI/NEMA C29.9, *American National Standard for Wet-Process Porcelain Insulators-Apparatus, Post Type*
- ANSI O5.1, *Wood Poles—Specifications and Dimensions*

American Society of Civil Engineers:

- ASCE 7, *Minimum Design Loads for Buildings and Other Structures*
- ASCE Standard 10 (ASCE 2015), *Design of Latticed Steel Transmission Structures*
- ASCE MOP 74, *Guidelines for Electrical Transmission Line Structural Loading*
- ASCE Standard 48, *Design of Steel Transmission Pole Structures*
- ASCE MOP 123, *Prestressed Concrete Transmission Pole Structures: Recommended Practice for Design and Installation*

American Society for Testing and Materials:

- ASTM A36/A36M (ASTM 2019a), *Standard Specification for Carbon Structural Steel*

- ASTM A143/A143M (ASTM 2020b), *Standard Practice for Safeguarding Against Embrittlement of Hot-Dip Galvanized Structural Steel Products and Procedure for Detecting Embrittlement*
- ASTM A153/A153M (ASTM 2016), *Standard Specification for Zinc (Hot-Dip) on Iron and Steel Hardware*
- ASTM A370 (ASTM 2022a), *Standard Test Methods and Definitions for Mechanical Testing of Steel Products*
- ASTM A615/A615M (ASTM 2022b), *Standard Specification for Deformed and Plain Carbon-Steel Bars for Concrete Reinforcement*
- ASTM A673/A673M (ASTM 2017), *Standard Specification for Sampling Procedure for Impact Testing of Structural Steel*
- ASTM A706/A706M (ASTM 2022c), *Standard Specification for Low-Alloy Steel Deformed and Plain Bars for Concrete Reinforcement*
- ASTM A767/A767M (ASTM 2019b), *Standard Specification for Zinc-Coated (Galvanized) Steel Bars for Concrete Reinforcement*
- ASTM F1554 (ASTM 2020a), *Standard Specification for Anchor Bolts, Steel, 36, 55, and 105 ksi Yield Strength*

American Welding Society:

- AWS D1.1/D1.1M, *Structural Welding Code—Steel*
- AWS D1.2/D1.2M (AWS 2014), *Structural Welding Code—Aluminum*

International Electrotechnical Commission:

- IEC 60865 (IEC 2011), *Short-Circuit Currents—Calculation of Effects—Part 1: Definitions and Calculation Methods*

Institute of Electrical and Electronics Engineers:

- IEEE C37.04, *IEEE Standard for Ratings and Requirements for AC High-Voltage Circuit Breakers with Rated Maximum Voltage Above 1,000 V*
- IEEE C37.30.1, *Standard Requirements for AC High-Voltage Air Switches Rated Above 1,000 V*
- IEEE C57.19.01, *Standard Performance Characteristics and Dimensions for Outdoor Apparatus Bushings*
- IEEE C57.19.100, *Guide for Application of Power Apparatus Bushings*
- IEEE 605, *Guide for Design of Substation Rigid-Bus Structures*
- IEEE 691, *Guide for Transmission Structure Foundation Design and Testing*
- IEEE 693, *Recommended Practice for Seismic Design of Substations*
- IEEE 1527, *Recommended Practice for the Design of Flexible Buswork Located in Seismically Active Areas*
- IEEE, *National Electrical Safety Code* (ANSI C2)

Precast/Prestressed Concrete Institute:

- PCI MNL-116 (PCI 2021), *Manual for Quality Control for Plants and Production of Structural Precast Concrete Products*
- PCI MNL-120 (PCI 2017), *Design Handbook*

D.4 DEFINITIONS

Substation: An assemblage of equipment (such as switches, power circuit breakers, bus, transformers) and auxiliary equipment through which electrical energy in bulk is passed for the purpose of switching or modifying its voltage levels.

Switchyard: An assemblage of equipment (such as switches, power circuit breakers, bus, and auxiliary equipment) that move power between different transmission lines. Switchyards do not include the use of power transformers to modify voltage levels.

Unit Substation: Lower-voltage metal-enclosed or open-air unit substations, often referred to as a metal clad switchgear, to house switches, fuses, circuit breakers, power transformers, instrument transformers, and controls.

Transmission Line: Transmission lines are electrical power lines, typically with voltages at 69 kV and above and may be installed overhead or underground.

Air-Insulated Substation and Switchyard: An air-insulated substation or switchyard has the insulating medium of air.

Gas-Insulated Substation and Switchyard: A substation unit enclosed inside a metallic sheath filled with SF_6 gas under pressure used as the insulating medium.

Electrical Clearance: Electrical clearances provide the physical separation needed for a phase-to-phase, phase-to-structure, and phase-to-ground air gaps to provide safe working areas and to prevent flashovers. Minimum electrical clearances are specified in the *National Electrical Safety Code* (IEEE 2023).

Bus-Work System: The network of conductors that interconnects transmission lines, power transformers, circuit breakers, disconnect switches, and other equipment.

Rigid Bus-Work (Bus) System: A rigid bus conductor is an extruded metallic conductor.

Strain Bus-Work (Bus) System: A strain bus conductor is a stranded wire conductor installed under tension.

Cable Bus-Work (Bus) System: Cable bus conductors are low-tension, stranded conductors supported on station post insulators.

Short-Circuit Force: Short-circuit forces are structure loads that are caused by short-circuit currents created as the result of electrical faults from equipment or material failure, lightning, or other weather-related causes, and electrical contact accidents.

Dead-End Structures: Structures designed to resist line termination loads from phase conductors and shielding wires. Dead-end structures are referred to as take-off structures, pull-off structures, termination structures, anchor structures, gantry frames, or strain structures.

A-Frame Dead-End: A type of dead-end structure that will convert large overturning moments, created by the wire tensions, into axial force couples.

H-Frame Dead-End: A type of dead-end structure typically used in conjunction with reduced tensions.

Shielding Mast: Structures with or without wires used to shield equipment in the substation from direct lightning strikes.

D.5 LOAD CASES AND COMBINATIONS FOR STRENGTH DESIGN

D.5.1 Introduction

The substation owner or authorized agent shall design substation facilities with sufficient strength for the basic load cases defined in Section D.5.2. Consideration shall also be given to the supplemental and serviceability load cases defined in Section D.5.3. Load cases shall be multiplied by the applicable load factors defined in Section D.5.4. Where substation facilities warrant a reliability level different from that defined in this draft pre-standard (e.g., MRI_{300} or Mean Recurrence Interval of 300 years) because of a site-specific application, the provisions of Section D.5.5 shall be followed.

The cumulative load demand requirements of this section are represented by the following:

$$\sum(\gamma \cdot LC \cdot Q_{MRI}) \quad \text{(D-1)}$$

where

LC = Load cases defined in Section D.5.2,
γ = Load factors defined in Tables D-1 and D-2, and
Q_{MRI} = Reliability adjustment factor defined in Section D.5.5.

Table D-1. Ultimate Strength Design Load Combinations.

Case	Combinations (all eight cases may not apply)
1*	$1.1\,D + 1.0\,W_{300} + 0.75\,SC + 1.1\,T_{W-300}$
2*	$1.1\,D + (1.0\,I_{100} + 1.0\,W_{WI\text{-}100}) + 0.75\,SC + 1.1\,T_{WI\text{-}100}$
3	$1.1\,D + 1.0\,SC + 1.1\,T_{APP}$
4*	$1.1\,D + 1.0\,E + 0.75\,SC + 1.1\,T_{EA}$
5*	$0.9\,D + 1.0\,W_{300} + 0.75\,SC + 1.1\,T_{W\text{-}300}$
6*	$0.9\,D + (1.0\,I_{100} + 1.0\,W_{WI-100}) + 0.75\,SC + 1.1\,T_{WI-100}$
7	$0.9\,D + 1.0\,SC + 1.1\,T_{APP}$
8*	$0.9\,D + 1.0\,E + 0.75\,SC + 1.1\,T_{EA}$

*The combination of SC loads with extreme events listed previously shall be determined by the owner.

Notes:

D = Structure and equipment dead load, see Section D.6.1.

W_{300} = Extreme wind load (F) obtained from Equation (D-3) or Equation (D-4) using a 300-year MRI wind map from Figure D-5; see Section D.6.5.

I_{100} = Extreme ice load from a 100-year MRI ice map (Figure D-6).

W_{WI-100} = Concurrent wind speed (Figure D-8) in combination with ice from the 100-year MRI ice map (Figure D-6); see Section D.6.4.

T_{WI-100} = Wire tension resulting from the following loads acting simultaneously: weight of the wire; weight of ice corresponding to ice thickness in the 100-year MRI ice map (Figure D-6); wind load on the iced wire corresponding to wind speed in the 100-year MRI ice map per Figure D-7 and wire temperature per Figure D-8; see Section D.6.4.

E = Seismic load as defined in Section D.6.9.8.1.

T_{W-300} = Wire tension caused by wire weight acting simultaneously with the wind force corresponding to the wind speed from the 300-year MRI wind map at an ambient temperature determined by the owner; see Section D.6.4.

SC = Short circuit load; see Section D.6.10.

T_{APP} = Wire tension caused by the wire weight acting simultaneously with any appropriate ice weight and temperature as determined by the owner (every-day or normal operational conditions); see Section D.6.4.

T_{EA} = Wire tension corresponding to wire dead load acting simultaneously with the seismic loading in accordance with Section D.6.9.12, at an ambient temperature determined by the owner; see Section D.6.4.

Table D-2. Allowable Strength Design Load Combinations (Service Level Loads).

Case	Combinations (all eight cases may not apply)
1[a]	$1.0\,D + 1.0\,W_{100} + 0.5\,\text{SC} + 1.0\,T_{W-100}$
2[a]	$1.0\,D + (1.0\,I_{50} + 1.0\,W_{WI-50}) + 0.5\,\text{SC} + 1.0\,T_{WI-50}$
3	$1.0\,D + 0.7\,\text{SC} + 1.0\,T_{\text{APP}}$
4[a]	$1.0\,D + 0.7\,E + 0.5\,\text{SC} + 1.0\,T_{EB}$
5[a,b]	$0.6\,D + 1.0\,W_{100} + 0.5\,\text{SC} + 1.0\,T_{W-100}$
6[a,b]	$0.6\,D + (1.0\,I_{50} + 1.0\,W_{WI-50}) + 0.5\,\text{SC} + 1.0\,T_{WI-50}$
7[b]	$0.6\,D + 0.7\,\text{SC} + 1.0\,T_{\text{APP}}$
8[a,b]	$0.6\,D + 0.7\,E + 0.5\,\text{SC} + 1.0\,T_{EB}$

[a]The combination of SC loads with extreme events listed previously shall be determined by the owner.

[b]Reference ASCE 7-22, Section C2.4.1 for an explanation of 0.6 dead load factor. These load combinations are also intended for foundation design.

Notes:

D = Structure and equipment dead load.

W_{100} = Extreme wind load (F_{WD}) obtained from Equation (D-3) or Equation (D-4), using the ASCE 7 Hazard Tool website.

I_{50} = Extreme ice thickness for a 50-year MRI ice event using the ASCE 7 Hazard Tool website and the conversion value from Figure D-9.

W_{WI-50} = Concurrent wind speed from Figure D-7.

T_{WI-50} = Wire tension resulting from the following loads acting simultaneously: weight of the wire; weight of ice corresponding to ice thickness using Figure D-6 and the conversion factor in Figure D-9; wind load on the iced wire corresponding to wind speed using the ASCE 7 Hazard Tool website and wire temperature per Figure D-8.

E = Seismic load as defined in Section D.6.9.8.1, "Basic Seismic Load Effect."

T_{W-100} = Wire tension caused by wire weight acting simultaneously with the wind force corresponding to the wind speed using the ASCE 7 Hazard Tool website at an ambient temperature determined by the owner,

SC = Short-circuit load.

T_{APP} = Wire tension caused by the wire weight acting simultaneously with any appropriate ice weight and temperature as determined by the owner (every-day or normal operational conditions).

T_{EB} = Wire tension corresponding to wire dead load acting simultaneously with 70% of the seismic loading per Section D.6.9.12, at an ambient temperature determined by the owner.

The structural capacity of a substation facility shall exceed the effects of the prescribed loads (demands) in this standard, as described by the following formula:

$$\phi \cdot R_n \geq \Sigma(\gamma \cdot \text{LC} \cdot Q_{\text{MRI}}) \qquad \text{(D-2)}$$

where ϕR_n is the design structural capacity or deflection restriction of the substation facility as defined by the appropriate design standard for the applicable structure type or an appropriate serviceability restriction.

The substation owner or authorized agent is referred to the following documents for the applicable material strength factors and design requirements:

- ACI CODE-318 Building Code Requirements for Structural Concrete and Commentary
- AISC 360, Specification for Structural Steel Buildings

- ANSI O5.1, Specifications and Dimensions for Wood Poles
- ASCE Standard No. 10, Design of Latticed Steel Transmission Structures
- ASCE Standard No. 48, Design of Steel Transmission Pole Structures
- ASCE Manual No. 91, Design of Guyed Electrical Transmission Structures
- ASCE Manual No. 123, Prestressed Concrete Transmission Pole Structures

For strengths of other substation facilities such as insulators and conductors, refer to the specifications of the applicable supplier.

D.5.2 Basic Load Cases

Substation facilities shall be designed such that their design strength equals or exceeds the effects of the following load cases. Determination of the magnitude of each load shall be in accordance with the applicable section of this draft pre-standard.

- Dead loads (D)
- Equipment-operating loads
- Wire tension loads
- Extreme wind loads
- Combined ice and concurrent wind loads
- Seismic loads
- Construction and maintenance loads
- Legislated loads
- Short-circuit loads

D.5.3 Supplemental and Serviceability Load Cases

Where site-specific circumstances warrant, the following load cases shall be considered in the design of substation facilities. These loads do not typically have an associated mean recurrence interval MRI.

- Load criteria associated with deflection limitations, see Section D.6.2;
- Dynamic wire loading with associated weather and short-circuit events. Dynamic loading must be evaluated relative to appropriate component resistance considering the nature of the loading events;
- Other high-consequence events.

D.5.4 Load Combination and Load Factors

This standard allows for the use of USD and Allowable Strength Design (ASD) concepts (see Section D.7.2 for the definitions of USD and ASD).

Unless otherwise specified within this draft pre-standard, the loads developed in accordance with this draft pre-standard shall be factored in accordance with Table D-1 (USD) and Table D-2 (ASD). All legislative loads shall be factored and combined on the basis of the load factors within the respective documents, and structures shall be designed using the material strength provision in the corresponding standard.

Operational loads shall be included with load combination Tables D-1 and D-2 using load factors of 1.0 and 0.75, respectively, where these loads can occur in conjunction with the other load cases according to the operating requirements of individual utilities.

Tables D-1 and D-2 list the required design load cases, combinations, and minimum load factors to be used for substation structures. For the load conditions that include ice, the effect of icing on the wire dead load and concurrent wind load with the ice formation shall be included.

D.5.5 Reliability Adjustment

The MRI used in the calculation of climatic loads may be adjusted for substation facilities requiring a reliability level different from that defined within this draft pre-standard (e.g., MRI_{700}). This may be due to site-specific applications or studies, or operating circumstances such as those described subsequently:

- Unique public safety,
- Challenging site access and restoration circumstances,
- Temporary construction, and
- Emergency restoration.

The substation owner or authorized agent shall determine the appropriate MRI. Wind speed and ice accretion maps for other MRIs can be found in ASCE *Substation Structure Design Guide*, MOP 113 or ASCE 7-22 (ASCE 2022), Table 1.3-1.

Note: Climatic load cases, as defined in this draft pre-standard, are determined by statistical modeling techniques as part of ASCE 7-22. As such, the structural reliability level can be adjusted using the same methodology.

D.6 LOADS

D.6.1 Dead Loads

The weight of the structure and components such as insulators, hardware, and electrical equipment shall be included and multiplied by the applicable load factor in Table D-1 or Table D-2 in the design of substation structures.

D.6.2 Equipment Operating Loads

Mechanical loads resulting from operation of equipment such as switches and circuit-interrupting devices shall be included with load combination Tables D-1 and D-2 using load factors of 1.0 and 0.75, respectively, where these loads can occur in conjunction with the other load cases according to the operating requirements of individual utility companies. The equipment manufacturer shall be consulted regarding the magnitude and application of such loads and any resulting support structure deflection limits.

D.6.3 Terminal Connection Loads for Electrical Equipment

Connectors to equipment shall be designed to accommodate sufficient movement or slack between equipment. Slack requirements for seismic loads are covered in IEEE 1527 (IEEE 2018c).

Terminal pad connection capacities for disconnect switches are specified in IEEE C37.30.1 (IEEE 2011), for circuit breakers in IEEE C37.04 (IEEE 2018a), and for transformer bushings in IEEE C57.19.01 (IEEE 2000) and IEEE C57.19.100 (IEEE 2012). Moment capacities are typically not provided by these documents.

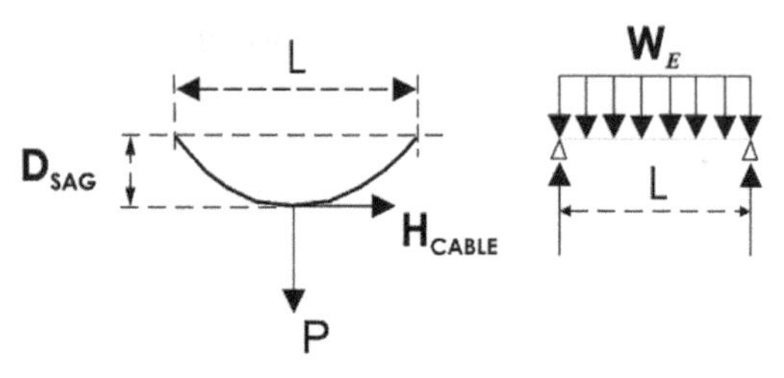

Figure D-1. General cable theorem.
Source: Norris and Wilbur (1960).

D.6.4 Wire Loads

The loads induced by all attached wires shall be included in the design of the structure. In the absence of specific cable data, Figure D-1 shows the relationship between sag and tension. Wire tension loads shall be calculated on the basis of tensions, span lengths, line angles appropriate for the site, and for the temperature, ice, and wind loadings specified in this draft pre-standard. The effects of wind and ice on non-structural attachments attached to wires shall be used in the calculation of design wire tensions. Further, wire loads shall be multiplied by the appropriate load factor in Table D-1 or Table D-2 and applied in the various combinations defined in Section D.5. Climatic, construction, and legislated loads shall be included to determine the maximum load effect.

Dynamic wire loads, such as those resulting from galloping, ice shedding, and Aeolian vibration that are caused or enhanced by wind and ice, or flexible structures and supports, shall be considered for wire-supporting structures such as dead-end and rack structures.

D.6.5 Extreme Wind Loads

Wind loads on substation structures, equipment, and conductors (bus and wire) shall be applied in the direction that generates the maximum loading. This section provides two equations for determining the wind loads on substation structures supporting equipment and/or overhead wires. Equations (D-3) and (D-4) are applicable to structures and electric equipment. The wind equation parameters are represented in Figure D-2. Equation (D-3) is applicable for use on non-lattice type structures (such as solid tubular, square, rectangular, and structural shapes for equipment and support pedestals). Equation (D-4) is applicable on lattice-type structures (such as line dead-end and rack structures). For substation structures supporting wire loads, the longitudinal winds (in the direction of the wires) also produce structure wind loading and shall be considered in the load calculation.

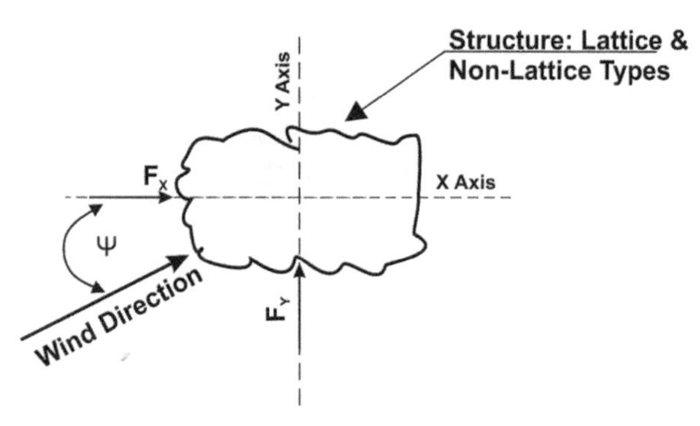

Figure D-2. Wind equation parameters.

The wind force in the direction of the wind (WD) shall be determined using the following equations:

Non-Lattice-Type Structures:

$$F_{WD} = Q\, K_z K_d K_{zt} (V_{MRI})^2 G_{RF} C_{fWD} A_{fWD} \tag{D-3}$$

Lattice-Type Structures:

$$F_{WD} = Q\, K_z K_d K_{zt} (V_{MRI})^2 G_{RF} (1 + 0.2\sin^2 2\Psi)\, (C_{fx} A_{fx} \cos^2\Psi + C_{fy} A_{fy} \sin^2\Psi) \tag{D-4}$$

where

F_{WD} = Wind force in the direction of wind (WD), pounds (N);
Q = Air density factor, default value = 0.00256 (0.613 SI) (Section D.6.5.1);
K_z = Terrain exposure coefficient defined in Section D.6.5.2;
K_d = Wind directionality factor = 1.0 for substation structures;
K_{zt} = Topographical factor, as defined in ASCE MOP 74 (ASCE 2020), 4th edition, or ASCE 7-22;
V_{MRI} = Basic wind speed [mph (m/s)] defined in Sections D.6.5.3 and D.6.5.4;
G_{RF} = Gust response factor (structure G_{SRF} and wire G_{WRF}) (Section D.6.5.5);
C_{fWD} = Force coefficient for face in the wind direction (WD), defined in Section D.6.5.6;
C_{fx} = Force coefficient in the x-axis, defined in Section D.6.5.6;
C_{fy} = Force coefficient in the y-axis, defined in Section D.6.5.6;
A_{fWD} = Projected wind surface area normal to the wind direction WD [ft² (m²)] for an entire structure or parts of the structure if warranted;
A_{fx} = Projected wind surface area normal to the x-direction [ft² (m²)] for an entire structure or parts of the structure if warranted;
A_{fy} = Projected wind surface area normal to the y-direction [ft² (m²)] for an entire structure or parts of the structure if warranted; and
Ψ = Wind direction angle measured from the x axis.

The wind force in the direction of the wind is then resolved into its x-axis and y-axis components, respectively, for the design of each direction:

$$F_x = F_{WD} \cos\Psi \tag{D-5a}$$

$$F_y = F_{WD} \sin\Psi \tag{D-5b}$$

For simple equipment support structures, such as instrument transformers, disconnect switches, etc., and overhead shield wire supports, the application of Equation (D-3) is through the principal axes of the support structure ($\Psi = 0$ degrees or 90 degrees). Where a skewed wind is applied to simple equipment supports, the appropriate force coefficient ($C_{f\Psi}$) and effective wind area (the area projection normal to the wind) in the direction of wind shall be used. For $C_{f\Psi}$ values, see ASCE MOP 74, 4th edition, and ASCE 7-22.

For structures supporting wire(s), the wire tension corresponding to the wind loading shall be calculated using the temperature that usually occurs at the time of the extreme wind-loading events.

The topographical factor, K_{zt}, shall be calculated using ASCE MOP 74, 4th edition, or ASCE 7-22. K_{zt} = 1.0 is permitted for normal terrains.

The wind directionality factor K_d (accounts for the directional variation of the maximum wind direction relative to the position of a structure ASCE 7-22) shall be 1.0 for substation structures.

D.6.5.1 Air Density Factor. For a standard atmosphere, the air density factor (Q) is 0.00256 (or 0.613 SI). The standard atmosphere is defined as a sea-level pressure of 29.92 in. of mercury (101.32 kPa) with a temperature of 59 °F (15 °C). The wind speed, V_{MRI}, in Equations (D-3) and (D-4), shall be expressed in terms of miles per hour when using the constant 0.00256 and meters per second when the constant 0.613 is used.

The air density factor of 0.00256 shall be used, except where sufficient elevation data are available to justify a different value. The air density factor may be reduced at higher altitudes or higher temperatures, but it is permitted to use 0.00256 for all elevations and temperatures. Variations of the air density, for other air temperatures and elevations that are different from the standard atmosphere, are given in ASCE MOP 74, 4th edition, or ASCE 7-22.

D.6.5.2 Terrain Exposure Coefficient. The terrain exposure coefficient (K_z) accounts for the wind speed variation with height as a function of terrain type (Table D-3). Wind speed varies with height because of ground friction and the amount of friction varies with ground roughness. The ground roughness is characterized by the various exposure categories described in Section D.6.5.2.1.

D.6.5.2.1 Exposure Categories. Three exposure categories, B, C, and D, are required for use with this standard. The maximum substation structure height (H), used as a reference for identifying exposure categories, includes all structures within the substation that contribute to the delivery of electric power and excludes lightning masts or free-standing and guyed microwave communication structures. One exposure category is intended for all structures within the substation.

Exposure B This exposure is classified as urban and suburban areas, well-wooded areas, or terrain with numerous closely spaced obstructions having the size of single-family dwellings or larger. Use of this exposure category shall be limited to those areas for which terrain representative of Exposure B roughness prevails in the upwind direction for a distance of at least 2600 ft (792 m) or 20 times H, whichever is greater, for an H of more than 30 ft (9.1 m). For an H of less than 30 ft (9.1 m), Exposure B roughness prevails upwind for a distance of 1,500 ft (457 m).

Table D-3. Terrain Exposure Coefficient, K_z.

Height above ground z (ft)	K_z Exposure B	K_z Exposure C	K_z Exposure D
0–15	0.57	0.85	1.04
30	0.69	0.98	1.17
40	0.74	1.04	1.23
50	0.79	1.09	1.28
60	0.83	1.13	1.32
70	0.86	1.17	1.35
80	0.90	1.20	1.38
90	0.92	1.23	1.41
100	0.95	1.25	1.44

Note: 1 ft = 0.3048 m.

Exposure C Open terrain with scattered obstructions having height typically less than 33 ft (10 m). Category C includes a flat open country and grasslands but excludes shorelines in hurricane-prone regions. Exposure C is the default for this standard.

Exposure D Flat, unobstructed areas and water surfaces. This category includes smooth mud flats, salt flats, and unbroken ice. Exposure D applies to winds flowing across a distance greater than 5,000 ft (1,524 m) or 20 times *H*. Shorelines in Exposure D include inland waterways, the Great Lakes, and coastal areas of California, Oregon, Washington, Alaska, and shorelines in hurricane-prone regions. This exposure shall apply to those structures exposed to the wind blowing from over the water or other smooth surfaces. Exposure D extends a distance of 600 ft (183 m) or 20 times *H*, whichever is greater. *Note*: Hurricane-prone regions were moved from Exposure C in prior editions of ASCE 7 to Exposure D in ASCE 7-22.

Figures D-3 and D-4 show examples of exposure category transition zones. Also shown in the figures is the requirement that the tallest substation structure (*H*) be used to determine the exposure transition distance for all structures in the substation.

Values of the terrain exposure coefficient K_z are listed in Table D-3 for heights up to 100 ft (30.5 m) above ground. Values for K_z are not listed in Table D-3, and for heights greater than 100 ft (30.5 m), they can be determined using Equation (D-6).

$$K_z = 2.41\left(\frac{z}{z_g}\right)^{2/\alpha} \qquad \text{for } 15\text{ ft}(4.6\text{ m}) \le z < z_g \tag{D-6}$$

where

z = Effective height at which the wind is being evaluated,
z_g = Gradient height (Table D-4), and
α = Power law coefficient for a 3-second gust wind (Table D-4).

If the structure height is less than 15 ft (4.6 m), K_z is taken as the value at 15 ft (4.6 m).

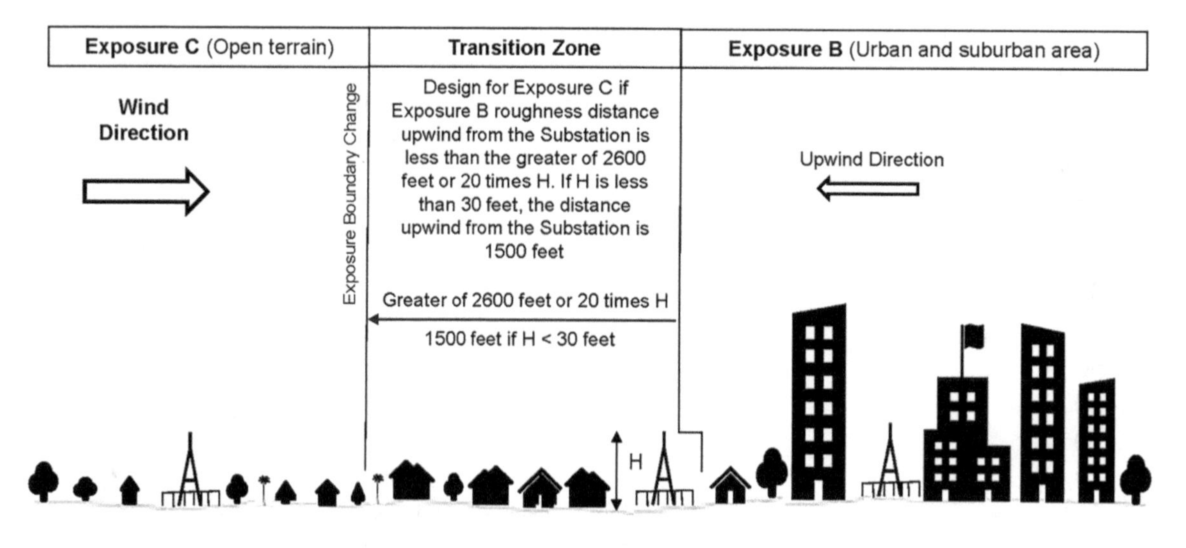

Figure D-3. Surface conditions required for the use of Exposure Category B and the transition from Exposure C to Exposure B.

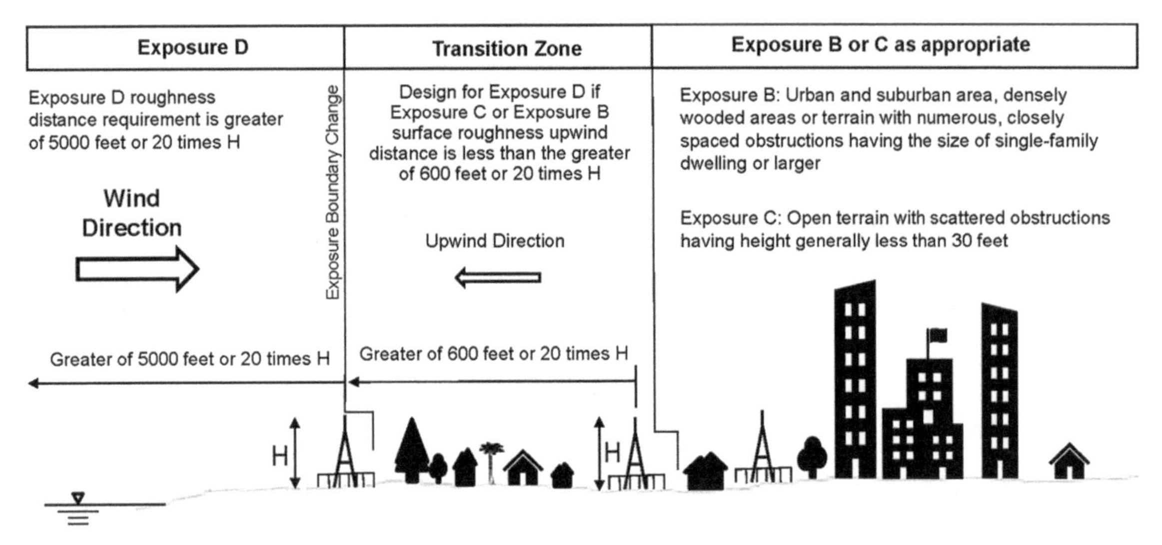

Figure D-4. Surface conditions required for the use of Exposure Category D and the transition from Exposure D to Exposure C or Exposure B.

Table D-4. Power Law Constants.

Exposure category	α Power law coefficient For 3 s gust wind	z_g Gradient Height feet (m)
B	7.5	3,280 (1,000)
C	9.8	2,460 (750)
D	11.5	1,935 (590)

The effects of terrain roughness on the wind force are significant. Exposure C shall be used, unless the engineer has determined with good engineering judgment that other exposures are more appropriate. The power law constants α and z_g for the different terrain categories are given in Table D-4.

D.6.5.2.2 Effective Height. An effective height, z, is used for the selection of the terrain exposure coefficient, K_z, the structure gust response factor, G_{SRF}, and the wire gust response factor, G_{WRF}. Sections D.6.5.5.1 and D.6.5.5.2 define the location of the effective height for the structure and wire.

D.6.5.3 Basic Wind Speed. The USD basic wind speed (V_{MRI}) used in Equations (D-3) and (D-4) shall be the 300-year MRI 3-second gust wind speed shown in Figure D-5. This wind speed is referenced at 33 ft (10 m) above ground in a flat and open country terrain (Exposure Category C). The basic wind speed for ASD loads is the 100-year MRI, Exposure Category C, at 33 ft (10 m) above ground. The wind speed values for the 100-year MRI can be obtained from the online ASCE 7 Hazard Tool application.

In regions where topographical characteristics may cause significant variations of wind speed over short distances (such as in mountainous terrain), local meteorological data shall be collected to establish the design wind speed.

The following ice density shall be used for calculating ice loads used in the design of substation structures: glaze ice is not less than 56 lb/ft^3 (900 kg/m^3) and rime ice is 25 pcf (400 kg/m^3).

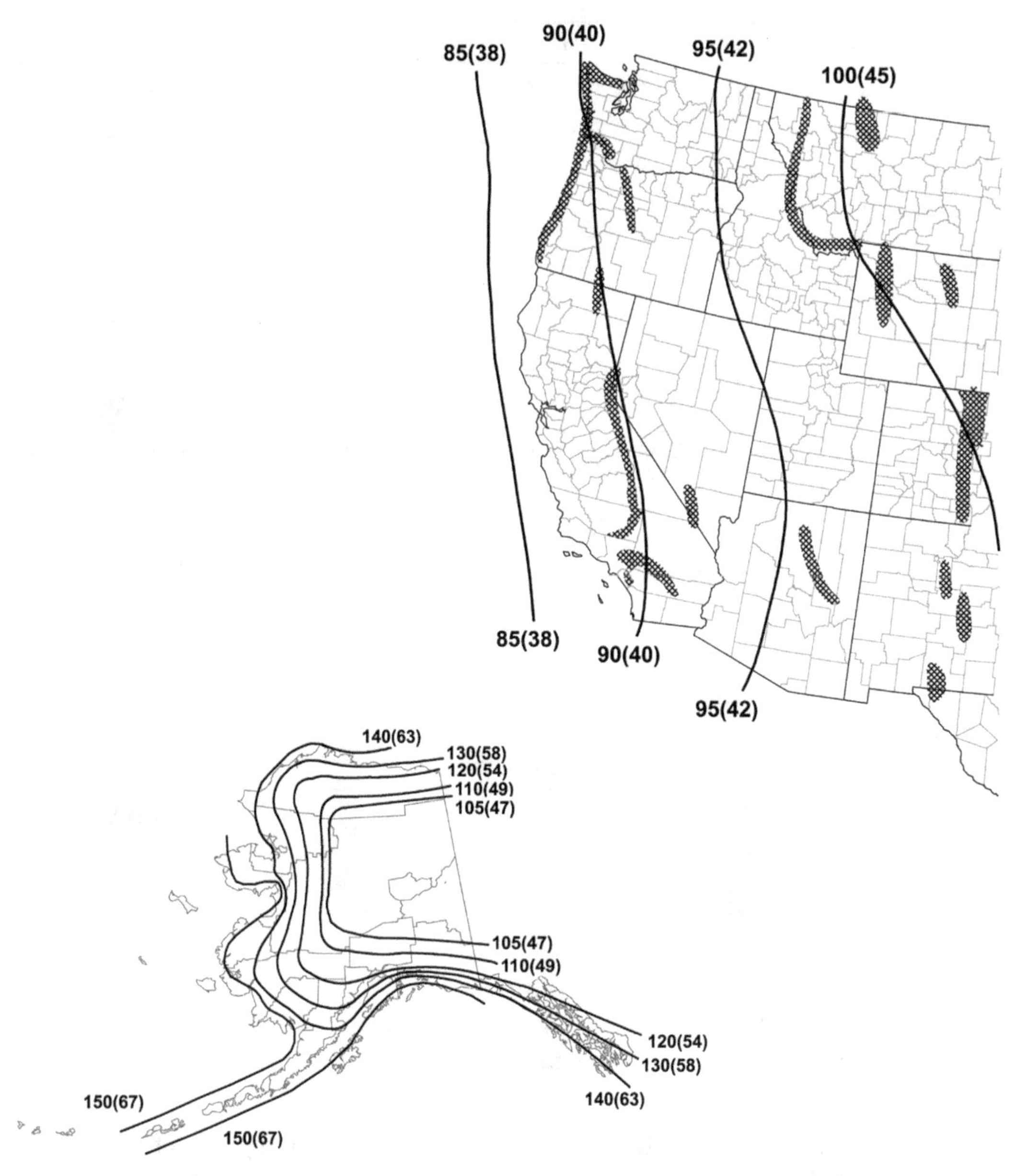

Figure D-5. 300-year MRI 3 s gust wind speed map in mph (m/s) at 33 ft (10 m) above ground in Exposure C, ASCE 7-22, for USD Load Combinations in Table D-1.

Notes:

1. Values are nominal design 3-second gust wind speeds in mph (m/s) at 33 ft (10 m) above ground for Exposure Category C.
2. Linear interpolation is permitted between contours. Point values (dots on map) are provided to aid with interpolation.
3. Islands, coastal areas, and land boundaries outside the last contour shall use wind speed contour.
4. Mountainous terrain, gorges, ocean promontories, and special wind regions shall be examined for unusual wind conditions.
5. Wind speeds correspond to approximately a 15% probability of exceedance in 50 years (annual exceedance probability = 0.00333, MRI = 300 years).
6. Location-specific basic wind speeds shall be permitted to be determined using the ASCE 7 Hazard Tool website.

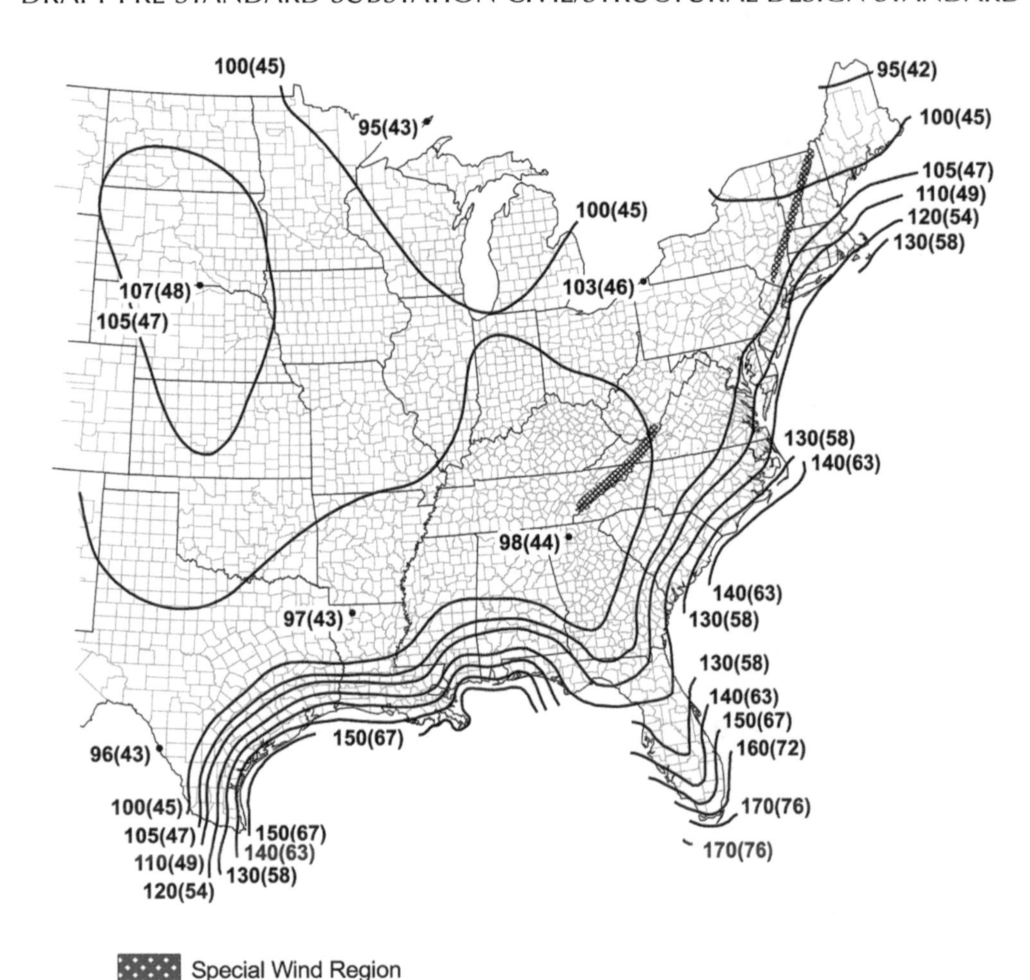

Figure D-5. (Continued)

Table D-5. Exceedance Probability for Various MRIs.

Typical conditions	MRI (Years)	Probability that the MRI Load is exceeded in any one year (%)	Probability that the MRI Load is exceeded at least once in 50 years (%)
Reduced reliability	100	1.00	40
Recommended reliability	300	0.33	15
Enhanced reliability	700	0.14	7

D.6.5.4 Mean Recurrence Interval Wind Speed. The wind calculations in this standard are based on a 300-year mean recurrence interval (MRI). It is the responsibility of the owner's engineer of record to select the level of wind speed reliability. Extreme wind speed MRIs referenced in this standard are given in Table D-5. Additional wind speed MRIs (corresponding higher wind speeds) are available in ASCE 7-22, Section 26.5.1.

D.6.5.5 Gust Response Factor. The gust response factors, G_{SRF} and G_{WRF}, account for the dynamic wind speed component (gust) on the wind response of structures and wires, respectively. Structure and wire gust response factors, G_{SRF} and G_{WRF}, respectively, shall be determined using Equations (D-7) and (D-8) or Tables D-8 and D-9.

D.6.5.5.1 Structure Gust Response Factor, G_{SRF}. The structure gust response factor, G_{SRF}, accounts for the response of the substation structure to the wind gust. It is used for computing the wind loads acting on substation structures and, on the insulator, and hardware assemblies attached to the structures.

For rigid structures supporting substation equipment, but not overhead wires, a structure gust response factor, G_{SRF}, of 0.85 shall be used. Rigid structures are structure/equipment combinations that have a fundamental frequency of 1 Hz and greater.

For wire-supporting structures, the gust response factor shall be calculated using Equation (D-7). A factor of 0.67 is included in Equation (D-7) (I_{ZS}) to simplify the calculation by assuming that the average wind force is equal to the force calculated at two-third of the structure height, defined as the structure effective height

$$G_{SRF} = \left[\frac{1+6.1\,\varepsilon\,(I_{ZS})B_S}{1+6.1\,(I_{ZS})}\right] \quad \text{(D-7)}$$

where

$I_{ZS} = c_{exp}(33/0.67h_s)^{1/6}$, turbulence intensity at effective height of the structure in ft;
c_{exp} = Turbulence intensity constant, based on exposure (Table D-7);
$\varepsilon = 0.75$, Wire-supporting structures (such as dead-end and rack structures);
$\varepsilon = 1.00$, Flexible structures (non-overhead wire supporting, < 1 Hz);
$B_S = \sqrt{1/(1+0.375(h_s/L_s))}$ quasi-static background response, does not include a dynamic component;
h_s = Total above ground height of the structure; and
L_s = Integral length scale of turbulence (Table D-7).
Note: 1 ft = 0.3048 m

An approximate range of structure natural frequencies, and damping, is provided in Table D-6.

Table D-7 gives parameters for Terrain Exposure Categories B, C, and D that are used to calculate the gust response factors.

Table D-8 provides structure gust response factors for structures up to 100 ft (30 m) in height, measured from ground elevation. Equation (D-7) shall be used to calculate G_{SRF} values in the table and for heights above 100 ft (30 m).

Table D-6. Approximate Structure Fundamental Natural Frequencies.

Type of structure	Fundamental frequency f_t (Hz)	Damping ratio ξ_t
Lattice tower	2.0–4.0	0.04
H-frame	1.0–2.0	0.02
Pole	0.5–1.0	0.02

Table D-7. Parameters for Calculation of the Gust Response Factor by Exposure.

Exposure	c_{exp}	L_s (ft)	L_s (m)	h_s min (ft)	h_s min (m)
B	0.3	320	97.54	30	9.14
C	0.2	500	152.4	15	4.57
D	0.15	650	198.12	7	2.13

Table D-8. Structure Response Factor (G_{SRF}) Structure Heights $\leq$ 100 ft. (30 m) for equipment support structures and wire-support structures.

	Exposure Category B		Exposure Category C		Exposure Category D	
Height (ft)	G_{SRF} equip. supports	G_{SRF} wire supports	G_{SRF} equip. supports	G_{SRF} wire supports	G_{SRF} equip. supports	G_{SRF} wire supports
$10 < 45$	1.00	0.83	1.00	0.85	1.00	0.87
$46 < 100$	1.00	0.82	1.00	0.86	1.00	0.88
> 100			Use Equation (D-7)			

Notes:

1. Equation (D-7) can be used to calculate G_{SRF} values [in the table and above 100 ft (30 m)].
2. $G_{SRF} = 1.00$ for equipment support structures.
3. G_{SRF} for wire-support structures, where $\varepsilon = 0.75$.

D.6.5.5.2 Wire and Strain Bus Gust Response Factor (G_{WRF}). Substation structures supporting overhead wire and strain bus (both will be referred to as wires) shall include the wire gust response factor (G_{WRF}) for determining the wind loads acting on the wire.

The wire gust response factor, G_{WRF}, shall be calculated using Equation (D-8) or Table D-9

$$G_{WRF} = \left[\frac{1+4.6(I_{ZW})B_W}{1+6.1(I_{ZW})}\right] \quad \text{(D-8)}$$

where

$I_{ZW} = c_{exp}(33/h_w)^{1/6}$, Turbulence intensity at effective height of the wire in feet;
c_{exp} = Turbulence intensity constant, based on exposure (Table D-7);

$B_W = \sqrt{1/(1+0.8(L/L_s))}$, quasi-static background response (does not include a dynamic component);
L = Wire horizontal span length;
h_w = Wire and strain bus effective height on the structure (height above the ground to the wire attachment point at the structure); and
L_s = Integral length scale of turbulence (Table D-7).
Note: 1 ft = 0.3048 m

Table D-9. Wire Gust Response Factors (G_{WRF}), Effective Heights up to 100 ft.

Wind exposure category	Span length $L < 100$ ft	Span length $100 \leq L < 250$ ft	Span length $250 \leq L < 500$ ft	Span length $500 \leq L \leq 750$ ft
B	0.85	0.80	0.75	0.70
C	0.88	0.85	0.82	0.78
D	0.89	0.88	0.85	0.82

Notes:

1. Equation (D-8) can be used to calculate G_{WRF} values in the table and outside the table parameter ranges.
2. Values in the table were generated for a 100-ft wire height and the maximum span listed in the range.
3. 1 ft = 0.3048 m.

Table D-9 gives the wire gust response factor for Terrain Exposure Categories B, C, and D with a span length range of up to 750 ft (229 m) and a wire effective height of up to 100 ft (30 m). Equation (D-8) with the appropriate exposure category constants shown in Table D-7 shall be used to determine G_{WRF} values in the table and outside the table parameter ranges.

D.6.5.6 Force Coefficient. The force coefficient, C_f (also referred to as drag or pressure coefficient or shape factor), in the wind force Equations (D-3) and (D-4) is the amplification factor of the resulting wind force per unit area in the direction of the wind.

The ratio of a member's length to its diameter (or width) of the windward face is known as the aspect ratio. The force coefficients given in Table D-11 are applicable to members with aspect ratios greater than 40. Correction factors for members with an aspect ratio of less than 40 shall be used to determine a correction C_f' that is substituted for C_f in Equations (D-3) and (D-4), as follows:

$$C_f' = (c)C_f \tag{D-9}$$

where c is the correction factor from Table D-11, and C_f is the force coefficient from Table D-10.

D.6.5.6.1 Lattice Structure Force Coefficients. Wind forces shall be applied in the directions resulting in maximum member forces and reactions. For lattice structures in substations, the force coefficient shall be calculated according to Table D-12 and adjusted from a flat to round section, as needed, according to Equations (D-10a) and (D-10b).

Table D-10. Force Coefficients (C_f) for Structural Shapes, Bus, and Surfaces Commonly Used in Substation Structures.

Member shape	Force coefficient (C_f)
Structural shapes (average value)*	1.6
Structural circular pipes and round tubes	0.9
Hexadecagonal (16-sided polygonal)	0.9
Dodecagonal (12-sided polygonal)	1.0
Octagonal (8-sided polygonal)	1.4
Hexagonal (6-sided polygonal)	1.4
Square and rectangle structural tubes, square shapes	2.0
Round rigid bus, strain or cable bus, and stranded conductor	1.0
Insulators (station post, suspension, and strain)	1.0

*The 1.6 value has been commonly used for structural shapes such as angles, channels, and wide flange-type shapes. See ASCE MOP 74, 4th edition (Appendix G) (ASCE 2020) for these and other structural shapes.

Table D-11. Aspect Ratio Correction Factors.

Aspect ratio	Correction factor (c)
0–4	0.6
4–8	0.7
8–40	0.8
> 40	1

Table D-12. Force Coefficients, C_f, for Normal Wind on Latticed Structures Having Flat-Sided Members.

Tower cross section	C_f
Square	$4.0\Phi^2 - 5.9\Phi + 4.0$
Triangle	$3.4\Phi^2 - 4.7\Phi + 3.4$

Note: Φ is the solidity ratio.

Equation (D-10b) gives a correction factor for converting the C_f values in Table D-12 to C_f values for round-section members

$$C_c = 0.51\Phi^2 + 0.57 \text{ but not greater than } 1.0 \quad \text{(D-10a)}$$

$$C_f(\text{round}) = C_f(\text{flat-sided members}) \times C_c(\text{correction factor}) \quad \text{(D-10b)}$$

where Φ is the solidity ratio, the ratio of the area of all members in the windward face to the outline area of the windward face of a latticed structure.

D.6.5.7 Application of Wind Forces to Structures. The wind forces on any square or triangular lattice substation structure shall be calculated according to the directions provided in Section D.6.5.6.

D.6.6 Ice Loads with Concurrent Wind

Maps of glaze ice thickness to be used with this standard for USD loads are shown in Figure D-6. Figure D-7 shows the wind speed that shall be used with USD and ASD ice thicknesses. Figure D-8 shows the ambient temperature that shall be used for USD and ASD ice events. Figure D-9 shows the conversion factor (0.50) to be used with the 500-year MRI ice thickness provided in the ASCE 7 Hazard Tool website to obtain the ASD 50-year MRI ice thickness.

This standard is based on a minimum of a 100-year MRI for glaze ice design. A longer MRI can be selected on the basis of site conditions. The design engineer is responsible for selecting, from available data, the most appropriate ice thickness to use for the location of the facility being designed. Utilities can conduct icing studies, with the assistance of a consulting meteorologist with ice expertise, to develop more accurate ice loading for substation sites in their service areas.

A 500-year MRI ice, with concurrent wind, map, is available in ASCE 7-22. Conversion tables to obtain additional MRIs for design are provided in ASCE 7-22. Using the ASCE 7-22 ice MRI factors to modify the return period does not take into account the changes in contours with changing MRI values.

The amount of ice that accretes on a wire or structure depends on the temperature and wind speed at the wire or structure height. Design thicknesses of glaze ice t_z for heights z above ground shall be obtained from Equations (D-10c) and (D-10d):

$$t_z = t_{\text{MRI}} \left(\frac{z}{33} \right)^{0.10} \text{ for } 0 < z < 900 \text{ ft} \quad \text{(D-10c)}$$

$$t_z = t_{\text{MRI}} \left(\frac{z}{10} \right)^{0.10} \text{ for } 0 < z < 275 \text{ m} \quad \text{(D-10d)}$$

where

t_z = Design ice thickness at height z above ground,
t_{MRI} = Nominal ice thickness (e.g., t_{100} for 100-year MRI), and
z = Height above ground (feet or meters).

The design ice thickness t_z shall be used for Exposure Category B, C, or D.

D.6.7 Legislated Loads

Substation structures shall be designed to resist the loading specified by all applicable federal, state, and municipal codes, and legislative or administrative acts.

D.6.8 Other High-Consequence Events

Events with the potential to cause devastating damage to substation facilities, resulting in extended outages, shall be considered, where applicable. Examples of these events include tornadoes, floods, landslides, avalanches, and sabotage.

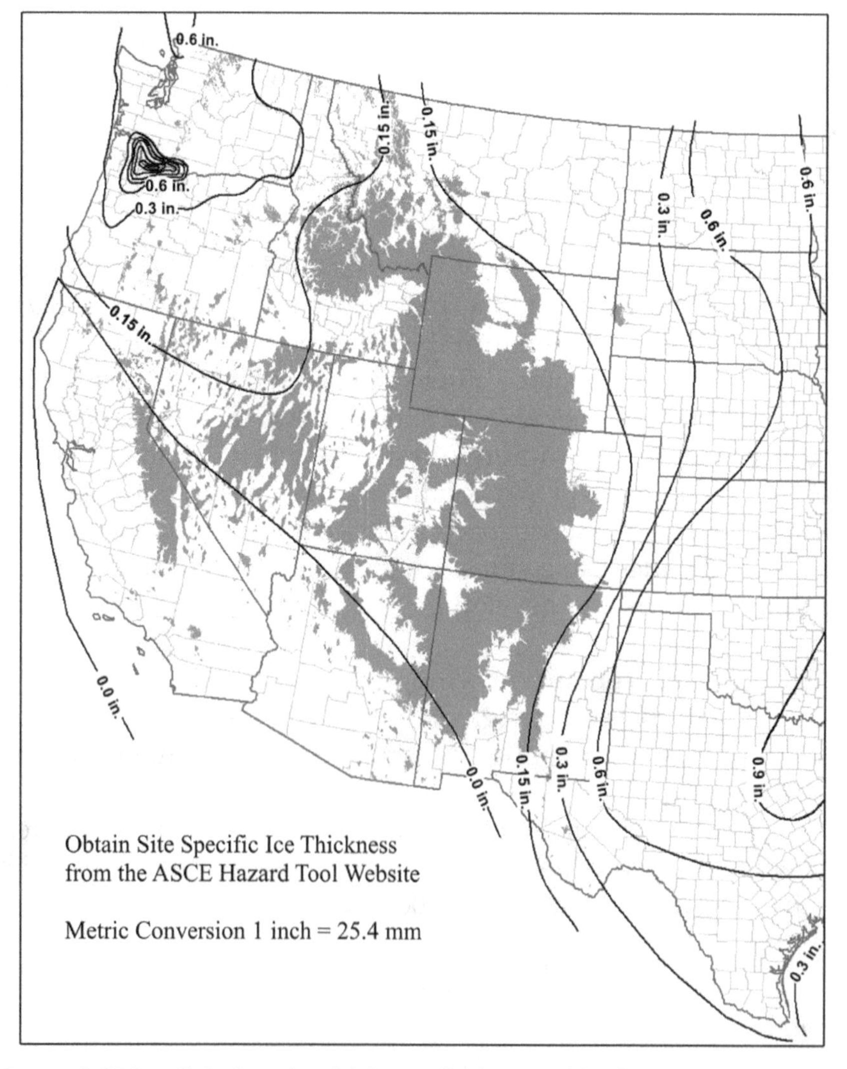

Figure D-6. 100-year MRI radial glaze ice thickness (in.) caused by freezing rain to be used with Figure D-7 concurrent wind, Western United States, for USD load combinations in Table D-1.
Source: Table D-1 map from ASCE MOP 74, 4th edition (ASCE 2020).

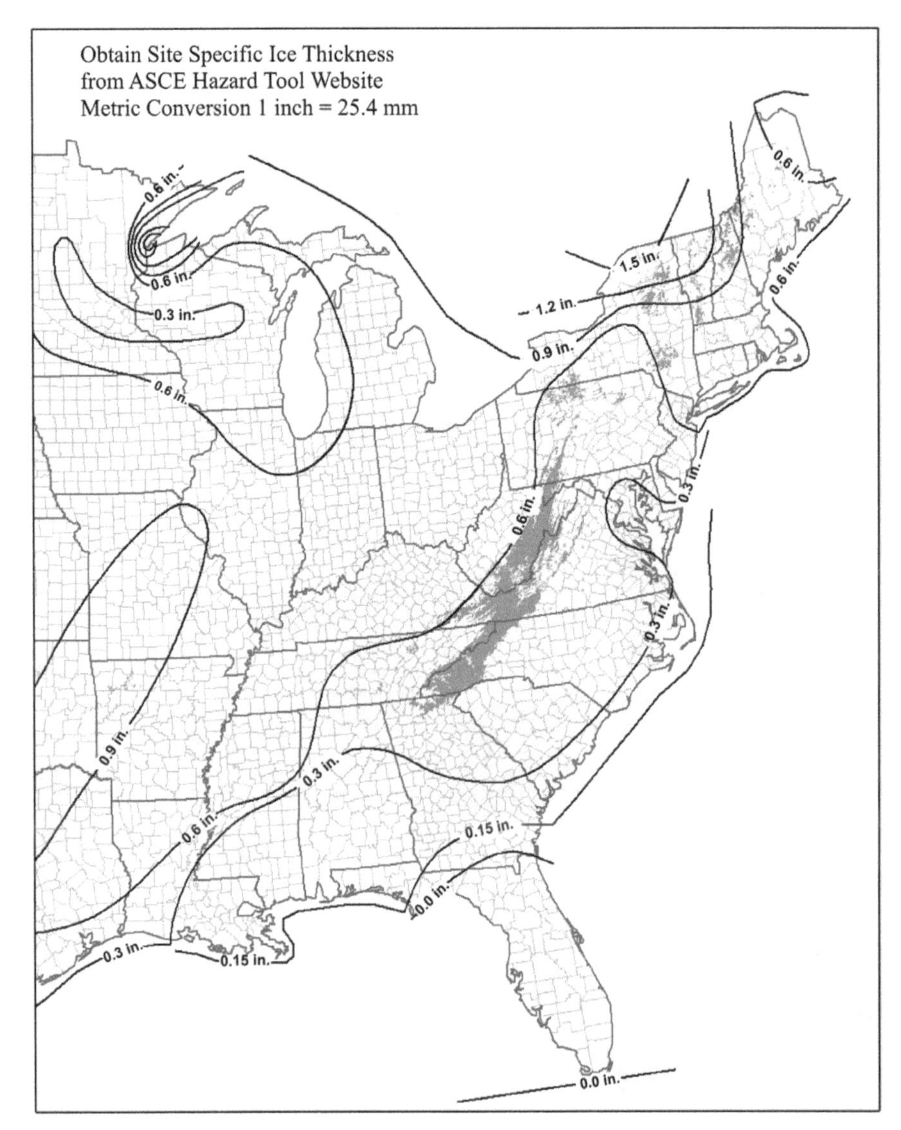

Figure D-6. (Continued) 100-year MRI radial glaze ice thickness (in.) caused by freezing rain to be used with Figure D-7 concurrent wind, Eastern United States, for USD load combinations in Table D-1. Source: Table D-1 map from ASCE MOP 74, 4th edition (ASCE 2020).

D.6.9 Seismic Loads

Seismic hazards shall be determined from data provided by the United States Geological Survey (USGS).

D.6.9.1 Purpose and Scope. This section covers seismic loads for new substation structures and foundations whose function is not solely for the support of a particular substation equipment type, such as line dead-end, rack, and lightning protection structures. Substation structures with the sole purpose of supporting a particular substation equipment type, such as instrument transformer, capacitors, disconnect switches, are covered by IEEE 693 (IEEE 2018b). Seismic loads for the design of foundations for equipment and their supports shall be as specified in IEEE 693 (IEEE 2018b). In addition, the design of anchor rods for equipment and their supports shall be as specified by IEEE 693 (IEEE 2018b) and combined using load combinations including overstrength factors provided by this standard.

The seismic design loads provided in this standard are obtained by factoring the Maximum Considered Earthquake (MCE) (Section D.6.9.2) loads by two-third [Equations (D-12a) and

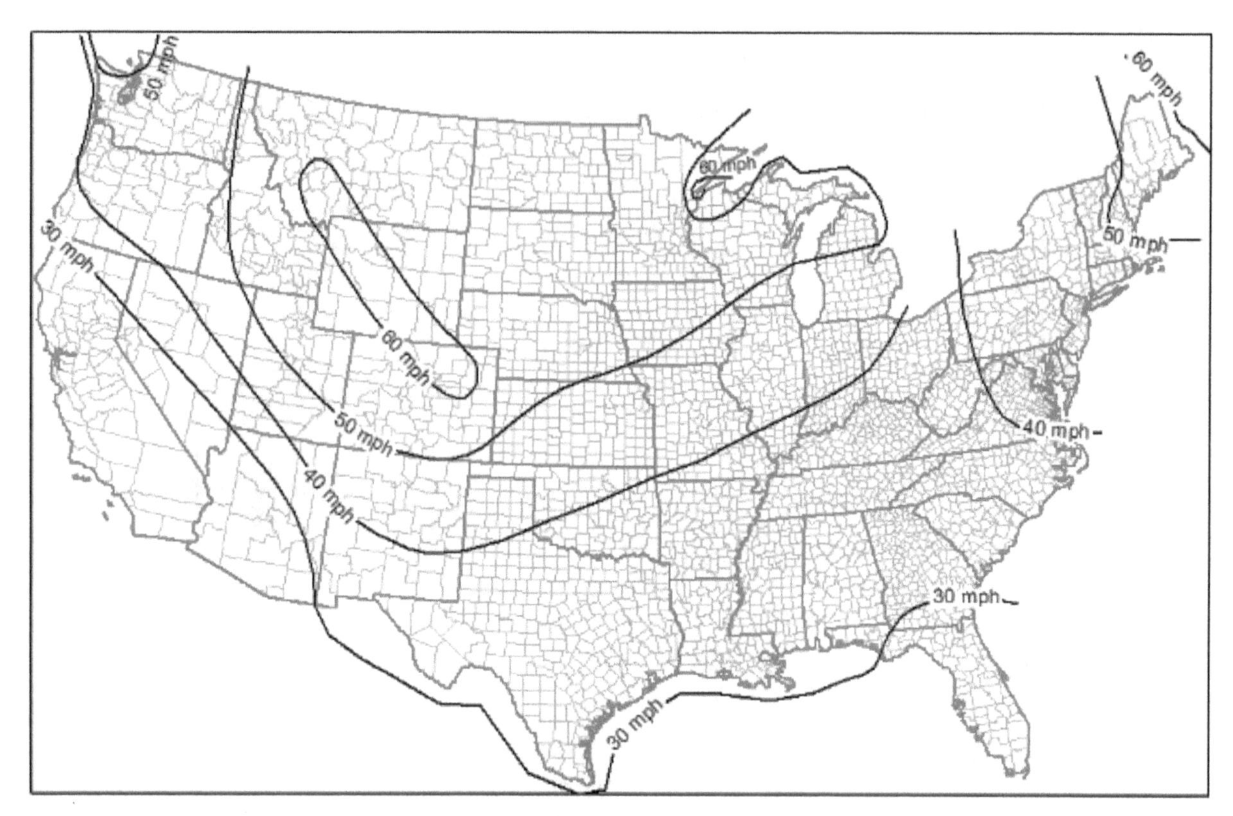

Figure D-7. Concurrent wind speed at 33 ft (10 m) above ground for Exposure C, for USD load combinations in Table D-1 and ASD load combinations in Table D-2.

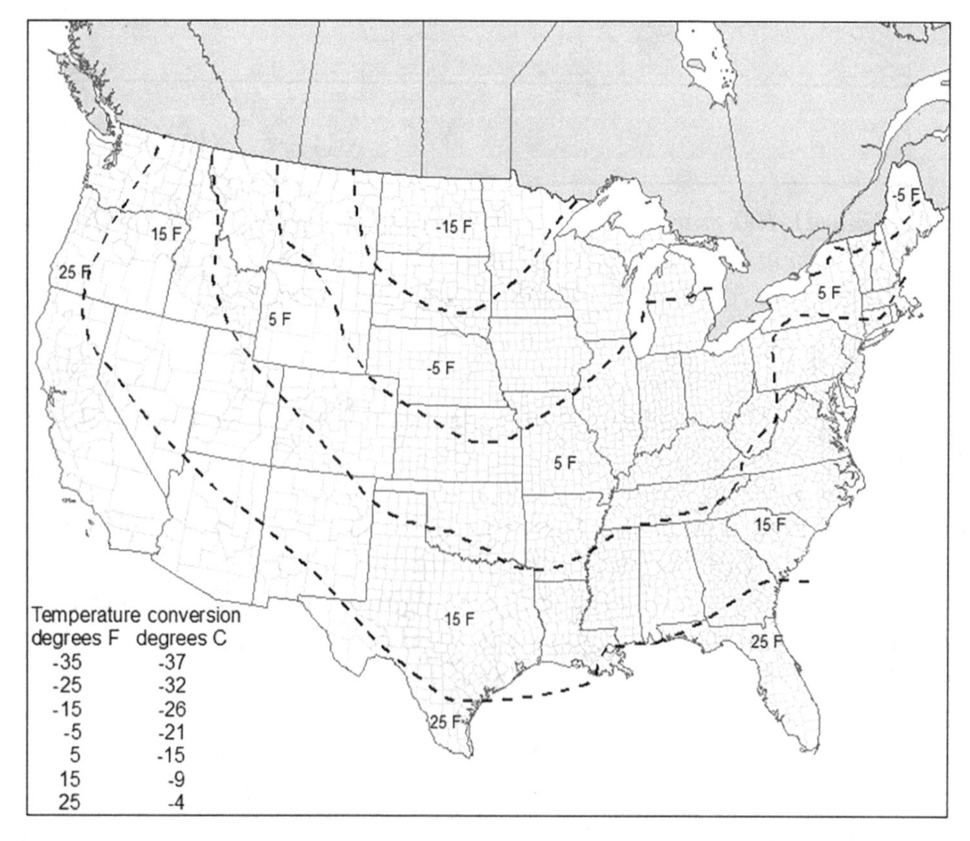

Figure D-8. Temperatures concurrent with ice thicknesses caused by freezing rain. Source: ASCE (2020).

Mean Recurrence Interval (MRI)	Ice Thickness Conversion Factor
50 year	0.5
100 year	0.6
500 year	1.0

Figure D-9. Ice thickness conversion factors for return periods other than a 500-year MRI.

(D-12b)] and further dividing by the response modification factor R ($R \geq 1$) [Table D-15, Equation (D-14)]. Connections shall be designed to higher demand by designing to the prescribed seismic forces amplified by the overstrength factor Ω_0 ($\Omega_0 \geq 1.0$) (Table D-15). Substation structures shall be designed in accordance with Table D-15 where $R \leq 3$ are not required to meet the provisions in AISC 341 (AISC 2016a).

Structures located in low seismic hazard areas with $S_{DS} < 0.167$ and $S_{D1} < 0.067$ need not be designed to the seismic provisions in Section D.6.9, where S_{DS} and S_{D1} are determined in accordance with Section D.6.9.2.

Figure D-10 shows the US Geological Survey (USGS) Relative Seismic Hazard map.

D.6.9.2 Seismic Ground Motion Acceleration Parameters. The site seismic ground motion acceleration parameters shall be determined from the ASCE 7 Hazard Tool website. The ground motion spectral response acceleration for 5% critical damping corresponding to a short period is referred to as S_{MS}, and the 1.0 second period value is referred to as S_{M1} (long period). The USGS maps are based on Risk Targeted MCE for 1% probability of collapse within a 50-year period, which is acceptable for substation structures.

The ground motion spectral response acceleration values (S_{MS} and S_{M1}) shall be adjusted on the basis of site classes. The soil site class is based on different soil types and associated shear wave velocity (Vs_{30}) measured in the upper 100 ft (30 m) of the soil profile. Table D-13 shows the different soil site classes based on the soil description. Table D-14 shows the shear wave velocity for associated soil site classes.

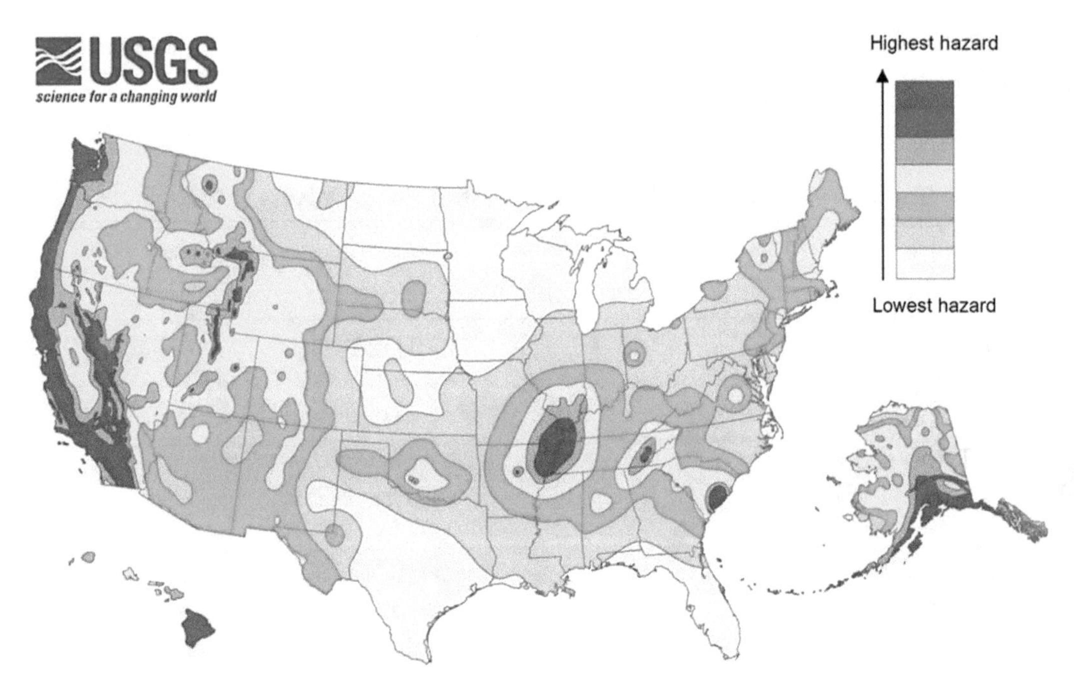

Figure D-10. Relative seismic hazard map.
Source: USGS (2018).

Table D-13. Soil Site Class and Associated Soil Type.

Soil site class	Soil description
A	Hard rock
B	Medium hard rock
BC	Soft rock
C	Very dense sand or hard clay
CD	Dense sand or very stiff clay
D	Medium dense sand or very stiff clay
DE	Loose sand or medium stiff clay
E	Very loose sand or soft clay
F	Very poor soil (determined by a geotechnical engineer, in accordance with ASCE 7-22, Chapters 20.2.1 and 21.1)

Where the soil properties are not known in sufficient detail to determine the soil site class, spectral response accelerations shall be based on the most critical spectral response acceleration at each period of Site Class C, Site Class CD, Site Class D, and Site Class DE, unless the geotechnical data determine that Site Class E or F soils are present at the site, subject to the limitations described in Section D.6.9.2.1. A geotechnical engineer shall be consulted for determining site-specific soil conditions.

D.6.9.2.1 Site-Specific Ground Motion Procedures. The definitions of near-fault sites given in ASCE 7-22 Chapter 11.4.1 shall be used where applicable: (1) Within 9.5 mi (15 km) or less from the surface projection of a known active fault capable of producing M_w7 or larger events, or (2) Within 6.25 mi (10 km) or less from the surface projections of a known active fault capable of producing M_w6 or larger events, but smaller than M_w7.

Two exceptions are applied to identify near faults: (1) Faults with estimated slip rates less than 0.04 in. (1 mm) per year shall not be used to determine whether a site is a near-fault site, and (2) The surface projection used to determine a near-fault site classification shall not include portions of the fault at depths of 6.25 mi (10 km) or greater.

Table D-14. Soil Site Class–Associated Shear Wave Velocity.

Soil site class	Shear wave velocity Vs_{30} (ft/s)	Shear wave velocity Vs_{30} (m/s)
A	$>$ 5,000	$>$ 1,524
B	$>$ 3,000–5,000	$>$ 914–1,524
BC	$>$ 2,000–3,000	$>$ 610–914
C	$>$ 1,450–2,000	$>$ 442–610
CD	$>$ 1,000–1,450	$>$ 305–442
D	$>$ 700–1,000	$>$ 213–305
DE	$>$ 500–700	$>$ 152–213
E	$\leq$ 500	$\leq$ 152
F	Determined by a geotechnical engineer in accordance with ASCE 7-22 Chapter 20.2.1 and 21.1.	Determined by a geotechnical engineer in accordance with ASCE 7-22 Chapter 20.2.1 and 21.1.

Site-specific ground motions hazard should be performed for structures on Site Class F in accordance with ASCE 7-22, Chapter 21.1, unless exempted in accordance with Chapter 20.3.1. A ground motion hazard analysis in accordance with ASCE 7-22, Chapter 21.2, can be used to determine ground motions for any structure. When the procedures of either ASCE 7-22, Section 21.1 or 21.2, are used, the design response spectrum shall be determined in accordance with ASCE 7-22, Section 21.3, the design acceleration parameters shall be determined in accordance with ASCE 7-22, Section 21.4, and, if required, the MCE_G peak ground acceleration parameter PGA_M shall be determined in accordance with Section 21.5.

D.6.9.2.2 Design Spectral Response Accelerations. The design spectral accelerations S_{DS} and S_{D1} are determined using Equations (D-11) and (D-12) or obtained directly from the ASCE 7 Hazard Tool website. S_{DS} and S_{D1} are adjusted for different soil site classes on the basis of shear wave velocity (Table D-14). ASCE 7-22 and the ASCE 7 Hazard Tool websites now include the site coefficients (soil class) used in the seismic map S_{DS} and S_{D1} values.

Short Period Design Spectral Acceleration:

$$S_{DS} = \frac{2}{3} S_{MS} \tag{D-11}$$

where S_{MS} is 5% damped, spectral response acceleration parameter at short period.

The S_{DS} value shall be obtained directly from the ASCE 7 Hazard Tool website.

Long Period (1.0 second) Design Spectral Acceleration:

$$S_{D1} = \frac{2}{3} S_{M1} \tag{D-12}$$

where S_{M1} is 5% damped, spectral response acceleration parameter at a period of 1 second.

The S_{D1} value shall be obtained directly from the ASCE 7 Hazard Tool website (for 5% damping). When damping values other than 5% are desired, IEEE 693 (IEEE 2018b) shall be used to develop the damped design spectral accelerations. Equations (D-11) and (D-12) shall be used to obtain the MCE_R spectral response accelerations parameter.

D.6.9.3 Design Response Spectra. Where a design response spectrum is required (Figure D-11), the design response spectrum shall be developed, as described in ASCE 7-22, Chapter 11.

D.6.9.4 Seismic Design Coefficients and Factors. Seismic coefficients, R, Ω_0, and C_d, for substation structures whose function is not solely for the support of a particular substation equipment type, such as line dead-end, rack, and lightning protection structures, are given in Table D-15 and defined as follows:

- R = Response modification coefficient,
- Ω_0 = Overstrength factor, and
- C_d = Deflection amplification factor.

For combinations of different types of structural systems along the same loading axis (horizontal and vertical), the R value (Table D-15) used for design in that direction shall not be greater than the least value of any of the systems used in that same direction. R values for structural systems and materials not given in Table D-15 can be found in ASCE 7-22.

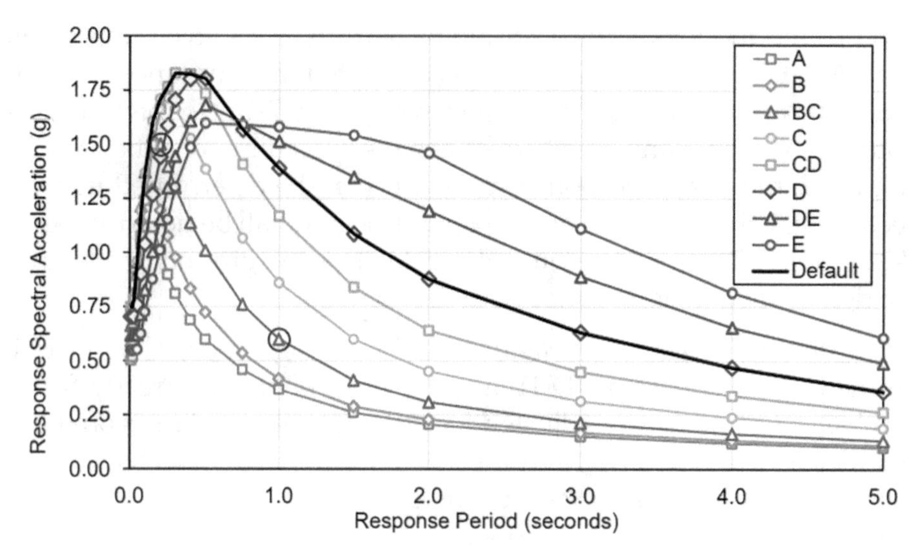

Figure D-11. Example of a multiperiod response spectrum.

Table D-15. Seismic Design Coefficients.

Structure or component type	R	Ω_0	C_d	Photograph Figure Number	Remarks/notes
Steel moment frame–supporting wires	2	2	2	D-12	Note 4
Steel lattice portal frame/truss-supporting wires	3	1.5	3	D-13	Notes 1 and 4
Steel truss–supporting wires	1.5	1	1.5	D-14	Notes 1 and 4
Frame-type rigid bus support	2	2	2	D-15	Notes 2 and 4
Cantilever rigid bus support	1	1	1	D-16	Note 5
Single-pole-supporting wire: steel, wood, and concrete	1.5	1.5	1.5	D-17	Note 4
Single poles without wires: steel, wood, and concrete	1.5	1.5	1.5	D-18	Note 4
Fire or sound barrier walls	1.25	2	2.5	D-19	Note 4
Components (rigid bus conductor and insulators)	1.0	1.0	1.0	D-15 D-16	Note 2, see Section D.6.9.11
Structures not otherwise covered herein:	1.25	2	2.5		Notes 3 and 4

Notes:

1. Latticed and braced structures composed of axial members that resist lateral loads primarily in compression and tension.
2. When more than 50% of the structure mass is concentrated at the top of a single-column rigid-bus support, the moment capacity of the column/connection at the top shall be at least one-half the moment capacity at the base of the column.
3. The user of this standard can select an R Factor using their engineering judgment for structures or components not listed in Table D-15. The use of R factors in Table D-15 shall follow the recommendation of this section.
4. It is permitted to use $R = 1$, $\Omega_0 = 1$, and $C_d = 1$ for all structures and components.
5. Typical single column bus support structures or three phases supported by a single column are permitted to use $R = 1$, $\Omega_0 = 1$, and $C_d = 1$.

Figure D-12. Dead-end structure.
Source: Courtesy of US Department of Energy.

This standard for substation structures limits the R factor ≤ 3. For the structures and components listed in Table D-15 and Section D.6.9.8.2, all structural connections in the seismic load path, including column-to-base plate connections and anchorage to foundations, shall be designed for seismic forces factored by Ω_0. Exceptions to this criterion are

Figure D-13. Box-type structure.
Source: Courtesy of US Department of Energy.

Figure D-14. Substation dead-end structure with line traps underhung.
Source: Courtesy of US Department of Energy.

Figure D-15. Frame-type rigid bus support.
Source: Courtesy of Brian Low.

Figure D-16. Insulator and rigid bus conductors support.
Source: Courtesy of US Department of Energy.

1. Structures with $0.167 < S_{DS} < 0.33$ and $0.067 < S_{D1} < 0.133$, unless the structure is a cantilever column system such as a cantilever support, or pole type; or
2. Ductile anchor elements designed to the ductile anchorage provisions in ACI CODE-318 (ACI 2019); or
3. IEEE 693 (IEEE 2018b), *Qualified Equipment and Support Anchorages*, see Section D.6.9.10.1 for applicable provisions.

D.6.9.5 Importance Factor. The importance factors for seismic loads, I_e, provided by this standard are given in Table D-16. The owner shall designate each electrical installation (substation or circuit pathway) as either essential (vital to power delivery and cannot be bypassed in the system or are undesirable to lose because of economic or operational considerations) or non-essential on the basis of its relative criticality to the power system.

The selection of the appropriate importance factor is the responsibility of the owner. The importance factors, I_e, specified in this section are the same as the recommended values for I_p, used in IEEE 693 (IEEE 2018b) for foundation design ($I_e = I_p$).

D.6.9.6 Seismic Analysis. Table D-17 provides examples of structure configurations common to substations with mass, stiffness, and torsional irregularities that shall consider dynamic analysis.

Table D-16. Importance Factor for Seismic Loads (I_e).

Structures and equipment essential to operation	1.25
All other structures and equipment	1.0

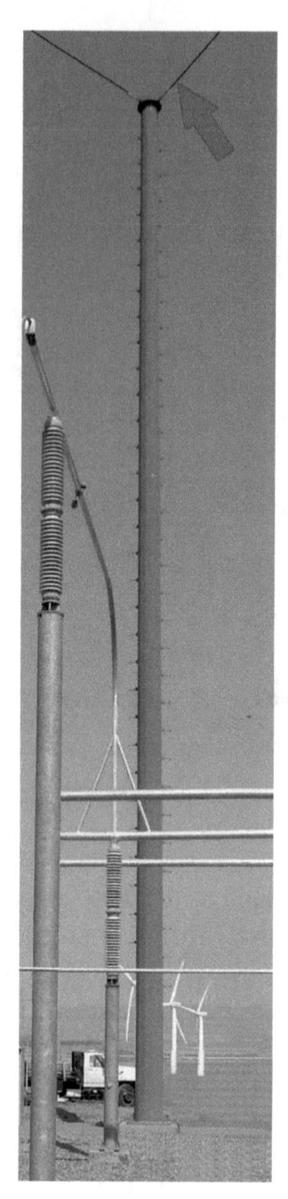

Figure D-17. Shielding mast with wires.
Source: Courtesy of US Department of Energy.

D.6.9.6.1 Selection of Analysis Procedures. When performing a seismic analysis of substation structures, the following methods are acceptable: (a) Equivalent Lateral Force Procedure and (b) Dynamic Analysis.

D.6.9.6.2 Equivalent Lateral Force Procedure. The ELF procedure described in this section is suitable for determining seismic loads for most substation structures that respond as single-degree-of-freedom systems. In cases where the contribution of higher modes of vibration may be significant, the engineer shall apply I_{mv}, as described in Equation (D-13).

The seismic base shear, V, in a given direction shall be determined by

$$V = C_s W(I_{mv}) \tag{D-13}$$

where C_s is seismic response coefficient, and W is total effective seismic weight, which includes the dead load of permanent equipment above the base of the structure.

Figure D-18. Static or shielding mast without wires.
Source: Courtesy of Michael Miller.

Figure D-19. Transformer protection firewall.
Source: Courtesy of US Department of Energy.

Table D-17. Substation Structure Configurations for Dynamic Analysis.

Example photo	Sketch of a substation structure	Description
Figure D-20	Figure D-21	Type 1: Rack structure with geometric or torsional irregularity caused by a nonsymmetrical lateral force–resisting system.
Figure D-22	Figure D-23	Type 2: Steel-braced frame/truss with stiffness irregularity caused by an unbraced panel
Figure D-24	Figure D-25	Type 3: Distribution rack structure with mass irregularity caused by a difference in the mass of equipment. Mass irregularity exists if the total equipment weight at any elevation W_i is > 1.5 times any other elevation (W_j) and/or distance to the center of mass from the centerline is > $\frac{B}{6}$

Figure D-20. Lattice substation structure.
Source: Courtesy of Brian Low.

I_{mv} = 1.5 if multimode contributions are significant; otherwise, I_{mv} = 1.0

$$C_s = \frac{S_{DS}}{\frac{R}{I_e}} \tag{D-14}$$

S_{DS} = Design spectral response acceleration from Equation (D-11)

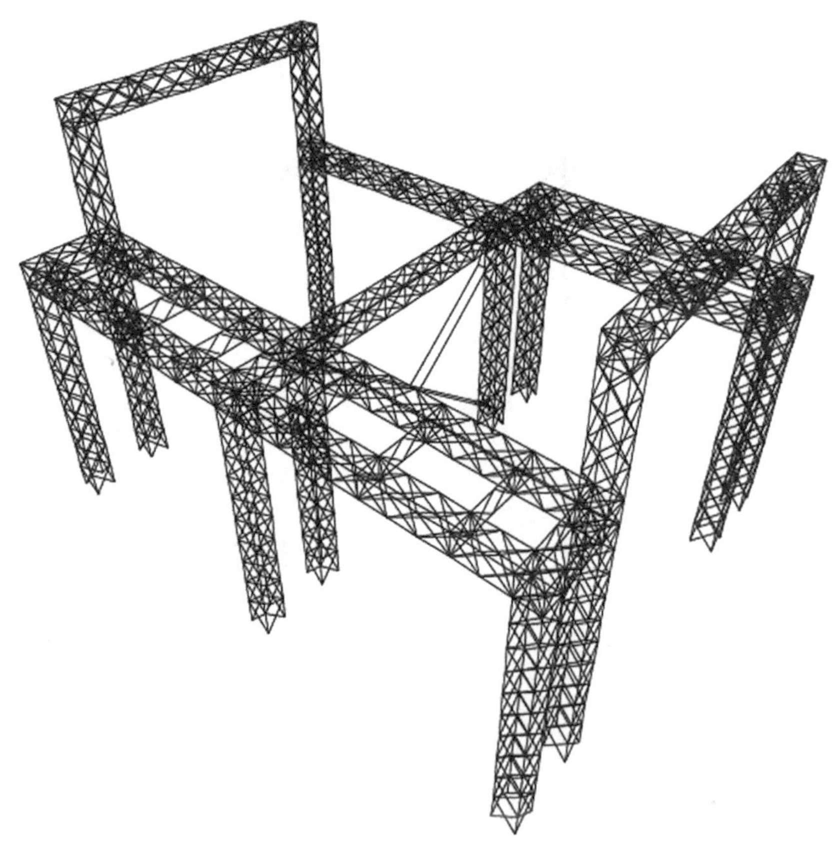

Figure D-21. Computer dynamic model of Figure D-20.

Figure D-22. Braced Frame.
Source: Courtesy of Brian Low.

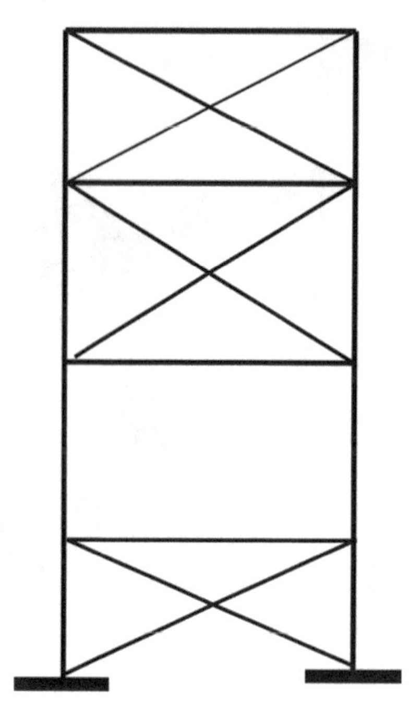

Figure D-23. Sketch for Figure D-22.

Figure D-24. Rack structure.
Source: Courtesy of US Department of Energy.

R = Response modification factor listed in Table D-15
I_e = Importance factor listed in Table D-16
C_s shall not exceed the following:

$$C_s \leq \frac{S_{D1}}{T\left(\frac{R}{I_e}\right)} \tag{D-15}$$

where T is calculated fundamental period of vibration of the structure.

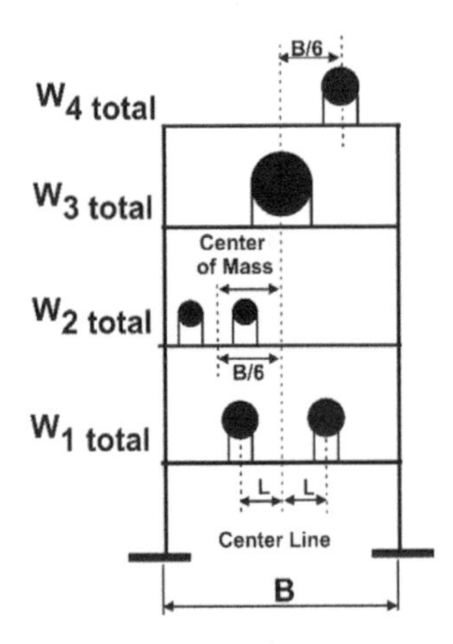

Figure D-25. Computer model of Figure D-24 (the terms equipment mass and weight are interchangeable).

$C_{s,\min}$ shall not be less than 0.03

$$C_{s,\min} = 0.044 S_{DS} I_e \geq 0.03 \tag{D-16}$$

In addition, for $S_1 \geq 0.6$ g, C_s shall not be less than

$$C_{s,\min} \leq \frac{0.8 S_1}{\frac{R}{I_e}} \tag{D-17}$$

If T is not calculated, the coefficient $C_s = \dfrac{S_{DS}}{\frac{R}{I_e}}$

D.6.9.6.2.1 Vertical Distribution of Seismic Forces. The vertical distribution of the total lateral seismic force shall be proportional to the height and effective seismic weight of the structure and computed with the following equations:

$$F_x = C_{vx} V \tag{D-18}$$

and

$$C_{vx} = \frac{w_x h_x^k}{\sum_{i=1}^{n} w_i h_i^k} \tag{D-19}$$

where

C_{vx} = Vertical distribution factor,
V = Total design lateral force at the base of the structure per Equation (D-13),
w_i and w_x = effective seismic weight at the level i or x being considered (typically locations of significant concentrated mass, points of connection of horizontal members or substructures, or as appropriate for discretization of the structure model),
h_i and h_x = height above the base to level i or x,
k = Exponent dependent on structure natural frequency (f) as follows:
$k = 1$ Structure frequency $(f) \geq 2$ Hz,
$k = 2$ Structure frequency $(f) \leq 0.4$ Hz,
$k = 2.25–0.625f$ when structure frequency (f) is between 0.4 and 2.0 Hz, and
n = Number of mass vertical locations (elevations) selected.

D.6.9.6.2.2 Horizontal Distribution of Seismic Forces. For structures with horizontal bracing (horizontal truss) connecting vertical trusses or frames of the lateral load–resisting system, the horizontal distribution of lateral seismic force at any elevation shall be based on the relative lateral stiffness of the vertical trusses or frames of the lateral load–resisting system. For structures without horizontal bracing, the horizontal distribution of seismic force to the vertical trusses or frames of the lateral load–resisting system shall be based on the tributary weights to each vertical truss or frame. In Table D-17, mass irregularity exists if any weight W_1, W_2, W_3, or W_4 varies by more than a factor of 1.5 of the weight of any one elevation, or if the horizontal distribution of weight on any one elevation varies from the structure support centerline by more than $B/6$, where B is the support width.

D.6.9.6.2.3 Substation Rigid support Structures. For substation rigid support structures that have a fundamental period, T, less than 0.03 s (frequency $\geq$ 33 Hz), the lateral force may be computed as

$$V = 0.30 S_{DS} W (I_e) \qquad \text{(D-20)}$$

where

S_{DS} = Design spectral response acceleration from Equation (D-11),
W = Total effective seismic weight, and
I_e = Importance factor listed in Table D-16.

D.6.9.6.3 Dynamic Analysis Procedure. The dynamic analysis procedure is applicable for all substation structures with or without irregularities. The analysis shall include a minimum number of modes to obtain a cumulative modal mass participation of at least 90% of the actual mass in each orthogonal horizontal direction of response considered in the model.

The value of each force-related design parameter of interest for each mode of response shall be computed using the properties of each mode and the design response spectrum defined in Section D.6.9.3 divided by the quantity R/I_e.

The value of each parameter of interest calculated for the various modes shall be combined using the square root of the sum of the squares (SRSS) or the complete quadratic combination (CQC) methods. The SRSS method assumes that the vibration modes are independent and therefore does not account for cross-coupling vibration mode effects. For structures with coupled modes of vibration and structures with unsymmetrical stiffness or mass, this can result in an underestimation of the dynamic response. When using the SRSS method, the response of vibration modes whose frequencies differ by less than 10% shall be first summed using absolute values. The CQC method has the advantage of accounting for the cross-coupling effects of vibration modes. For both methods, a sufficient number of natural modes of vibration shall be used such that the total response does not increase by more than 10% with the addition of more modes. The acceleration, displacement, force, and moment response caused by motion in the three orthogonal directions shall be combined by using the SRSS or the procedure described in Section D.6.9.8. A base shear (V) shall be determined in each of the two orthogonal horizontal directions using the calculated fundamental period ($T = 1/f$) of the structure per Section D.6.9.6.2. When the combined response for the modal base shear (V_t) is less than 100% of the calculated base (V), the forces shall be multiplied by V/V_t, where V equals the equivalent lateral force procedure base shear, calculated in accordance with D.6.9.6.2, and V_t equals the base shear from the required modal combination.

D.6.9.7 Vertical Seismic Load Effect. The effective seismic design force for vertical seismic motions shall be determined according to the following:

$$E_v = (0.80)S_{DS}D \quad \text{(D-21a)}$$

where S_{DS} is the design spectral acceleration at short period, from Equation (D-11), and D is the dead load.

Or when the design spectral response vertical acceleration S_{av} is known, it can be used to determine E_v:

$$E_v = S_{av}D \quad \text{(D-21b)}$$

Either value of E_v is acceptable.

EXCEPTION:

The vertical seismic load effect, E_v, is permitted to be taken as zero when determining demands on the soil–structure interface of foundations.

D.6.9.8 Seismic Load Effects and Combinations. Structures, foundations, and anchorage shall be designed for the seismic load effects described in this section, including the combination of horizontal and vertical (as applicable) seismic components.

D.6.9.8.1 Basic Seismic Load Effect. Horizontal seismic loads effect, E_h, applied in two orthogonal horizontal directions, and vertical seismic load effect, E_v (acting upward or downward), shall be considered to determine the controlling member and foundation design forces where

$$E = E_h + E_v \quad \text{(D-21c)}$$

or

$$E = E_h - E_v \quad \text{(D-21d)}$$

Applicable structures shall be designed for the seismic component combinations described in Equations (D-22) to (D-24). The combination requiring the maximum component strength shall be used.

The seismic load effect, E, in the load combinations described in Section D.5.4 shall be determined as

$$E = E_{h1} + 0.4E_{h2} + 0.4E_v \quad \text{(D-22)}$$

$$E = E_{h2} + 0.4E_{h1} + 0.4E_v \quad \text{(D-23)}$$

$$E = E_v + 0.4E_{h1} + 0.4E_{h2} \quad \text{(D-24)}$$

If an SRSS seismic component combination is used, the seismic load effect E shall be determined from

$$E = \sqrt{{E_{h1}}^2 + {E_{h2}}^2 + {E_v}^2} \quad \text{(D-25)}$$

where E_{h1}, E_{h2} represent horizontal seismic load effect in mutually perpendicular directions from Section D.6.9.6.2 or D.6.9.6.3, and E_v represents vertical seismic load effect (acting upward or downward) from Section D.6.9.7.

If a CQC seismic component combination is used, the seismic load effect E is determined from

$$E = \sqrt{\sum_{i=1}^{2}\sum_{j=1}^{2} E_{hi}^2 \alpha_{ij} E_{hj}^2 + E_v^2} \quad \text{(D-26)}$$

$$\alpha_{ij} = \frac{8\xi^2(1+\beta)\beta^{3/2}}{(1+\beta^2)^2 + 4\xi^2(1+\beta)^2} \tag{D-27}$$

$$\beta = \frac{\omega_i}{\omega_j} \quad \text{for } (\omega_j > \omega_i) \tag{D-28}$$

where

E_{hi}, E_{hj} = Horizontal seismic load effect in mutually perpendicular directions from Section D.6.9.6.3,
E_v = Vertical seismic load effect (acting upward or downward) from Section D.6.9.7,
a_{ij} = Correlation coefficient with constant damping applied to all vibration modes,
ω_i, ω_j = Frequencies of the ith and jth modes of vibration,
ξ = Damping ratio, and
β = Vibration mode frequency ratios.

It is assumed in this standard that seismic loads are applied during the condition of no wind or ice and at 60 °F (15.6 °C). The owner shall determine whether it is appropriate to combine seismic loads with wind, ice, short-circuit, and operating loads. When appropriate, wire loads shall be applied to the structure; see Section D.6.9.12.

D.6.9.8.2 Seismic Load Effect with Overstrength Factor. The seismic load effect, E, described in Equations (D-22) to (D-25) shall be modified by using an overstrength factor (Ω_0) for designing connections, as described in Table D-15 and this section. The modified seismic load effect E_m is determined from

$$E_m = \Omega_0(E_{h1} + 0.4E_{h2}) + 0.4E_v \tag{D-29}$$

$$E_m = \Omega_0(E_{h2} + 0.4E_{h1}) + 0.4E_v \tag{D-30}$$

$$E_m = E_v + \Omega_0(0.4E_{h1} + 0.4E_{h2}) \tag{D-31}$$

If the SRSS seismic component combination is used, the modified seismic load effect E_m shall be determined from

$$E_m = \sqrt{(\Omega_0 E_{h1})^2 + (\Omega_0 E_{h2})^2 + E_v{}^2} \tag{D-32}$$

If the CQC seismic component combination is used, the modified seismic load effect E_m is determined from

$$E_m = \sqrt{\sum_{i=1}^{2}\sum_{j=1}^{2} \Omega_0^2 (E_{hi}^2 \alpha_{ij} E_{hj}^2) + E_v^2} \tag{D-33}$$

The overstrength factor (Ω_0) is intended to provide a level of assurance that the connections in the structural seismic load path have sufficient strength to resist demand forces that exceed the basic seismic load effects that may have been reduced by an R factor (Table D-15).

D.6.9.9 Seismic Deflection Considerations. The effects of structure displacements resulting from seismic loads shall be evaluated to ensure that structural stability is maintained and attached equipment and conductor slack are not adversely affected. The displacements obtained from elastic analysis shall be multiplied by the quantity C_d/I_e. Refer to Table D-15 for deflection amplification factor C_d for typical substation structures.

D.6.9.10 Anchorage Design Forces. Seismic load effects shall be considered in the design of substation structures and equipment anchorage, as recommended in Sections D.6.9.10.1 and D6.9.10.2 of this standard.

D.6.9.10.1 IEEE 693 (IEEE 2018b) Qualified Equipment and Supports Anchorage Design Forces. Anchors (anchor material, diameter, and welds) for IEEE 693 (IEEE 2018b) qualified equipment shall be designed in accordance with the provisions in IEEE 693 (IEEE 2018b). The anchorage in the concrete foundation (anchor embedment in concrete) shall be designed in accordance with the provisions in this standard for anchor loads corresponding to the IEEE 693 (IEEE 2018b) seismic qualification, amplified by the overstrength factor (Ω_{PL}) where appropriate, as specified in Equations (D.6-34) to (D.6-37).

D.6.9.10.1.1 Design-Level Qualifications. If a ductile anchorage design is provided, then the design-level (ASD) loads shall be factored by 1.4 and the anchorage designed to resist this magnitude of load using LRFD methods.

If a nonductile anchorage design is provided, then the design-level (ASD) seismic loads shall be factored by 1.4, and $\Omega_{PL} = 1.5$ shall be applied only to the horizontal components of the seismic loads, using the methods of qualification approved by IEEE 693 (IEEE 2018b). This maintains the intent of the design level being projected to the performance level by achieving a factor of about 2 (1.4 load factor $\times$ 1.5 overstrength factor $= 2.10$).

The following strength design load combinations shall be used for the design of anchorages:

$$1.2\,D + \Omega_{PL}(1.4\,E_{DL}) \tag{D-34}$$

$$0.9\,D + \Omega_{PL}(1.4\,E_{DL}) \tag{D-35}$$

where

D = Dead load effects;
E_{DL} = Seismic loads from IEEE 693 (IEEE 2018b) Design Level (ASD) qualification; and
Ω_{PL} = Overstrength factor for IEEE 693 (IEEE 2018b) qualified items, 1.0 for ductile anchorage designs, or 1.5 for nonductile anchorage designs, when forces from design-level qualifications are used.

D.6.9.10.1.2 Performance-Level Qualifications For performance-level forces, use $\Omega_{PL} = 1.0$. The following strength design load combinations shall be used for the design of anchorages:

$$1.2\,D + \Omega_{PL}(E_{PL}) \tag{D-36}$$

$$0.9\,D + \Omega_{PL}(E_{PL}) \tag{D-37}$$

where

D = Dead load effects;
E_{PL} = Seismic loads from IEEE 693 (IEEE 2018b) performance-level qualification;
Ω_{PL} = Overstrength factor for IEEE 693 (IEEE 2018b)-qualified items, 1.0 for ductile and nonductile anchorage designs when forces from performance-level qualification are used.

D.6.9.10.2 Equipment to Structure Anchorage Design Forces. If the equipment is seismically qualified in accordance with IEEE 693 (IEEE 2018b), and if this qualification shall be maintained for the as-installed position on the structure designed to the provisions of this standard, then the structure designer shall verify that the seismic forces experienced by the equipment on the structure will not exceed the IEEE 693 (IEEE 2018b) seismic qualification forces. If the IEEE 693

(IEEE 2018b) qualification is performed for 2.5 times the acceleration levels (equipment only tested), the anchorage determined from the shake table test needs to be used for designing the connections. If a dynamic analysis of the equipment/support structure system is performed, then the analysis anchor loads can be considered to design the connections.

When equipment qualified to the IEEE 693 (IEEE 2018b) provisions is installed on a structure designed to the provisions of this standard, the connection of the equipment to the structure (or intermediate support) shall be designed to resist the anchorage forces not less than those corresponding to the IEEE 693 (IEEE 2018b) seismic qualification, and as modified by the overstrength factors described in this standard.

When the equipment is not qualified to the IEEE 693 (IEEE 2018b) provisions, the equipment to structure anchorage design forces shall be determined in accordance with Section D.6.9.11.

D.6.9.11 Seismic Demand on Other Components. Other components include rigid bus conductors, insulators, appurtenances, and equipment that are not subject to seismic qualifications per IEEE 693 (IEEE 2018b). The seismic demand as determined in this section shall be used to design the component and anchorage to the support structure.

The horizontal seismic design force (F_p) shall be determined in accordance with Equation (D-38) and applied at the component's center of gravity and distributed relative to the component's mass distribution.

$$F_p = 1.6 S_{DS} I_e W_p \quad \text{(D-38)}$$

where

F_p = Seismic design force;
S_{DS} = Design Spectral Acceleration, short period, as determined from Section D.6.9.2;
I_e = Importance factor per Section D.6.9.5; and
W_p = Total effective seismic weight, which includes the dead load weight.

The seismic design force for vertical seismic motions shall be determined from Section D.6.9.7. The combination of seismic load effects from the horizontal and vertical components of seismic forces shall be determined from Section D.6.9.8.1.

For design of components such as cable trays, raceway, and cable support inside control enclosures, refer to ASCE 7-22, Chapter 13. $I_p = 1.5$.

D.6.9.12 Flexible Wire Seismic Forces (Strain Bus or Cable Bus). Forces applied to structures from strain bus seismic loading shall consider the following conditions:

- Taut strain bus condition
  The structure design shall provide sufficient slack not to allow for the development of taut condition tensions by using the following procedure:

$$\Delta_{\text{slack}} \geq \sqrt{(C_{d1}\Delta_{e1})^2 + (C_{d2}\Delta_{e2})^2} \quad \text{(D-39)}$$

  where

  Δ_{slack} = Conductor slack provided in the installation under normal temperature conditions, and no wind or ice;
  Δ_{ei} = Calculated elastic seismic displacement of the ith support structure parallel to the strain bus, with the structures unconnected; and
  $C_{di} = 2.0$.

- Slack strain bus condition

  To estimate the wire tension for the condition with seismic loading transverse to the strain bus support structures, the following process shall be followed:

  1. Determine the strain bus unit weight and profile (coordinates of end points, sag) for the ambient conditions of no wind, no ice, and a temperature of 60 °F. The utility may decide to use a different temperature if appropriate for consideration with the seismic loading.
  2. Determine the strain bus lateral seismic loading per unit length as S_{MS} multiplied by the strain bus unit weight, where S_{MS} is the maximum considered seismic spectral response accelerations parameter for a short period according to Equation (D-11).
  3. Determine the strain bus tension at the support structures for the combined effect of the vertical and lateral loads described in Steps 1 and 2. This may be determined from sag/tension calculations in a manner similar to transverse wind loading on conductors.
  4. 50% of the wire weight and 50% of the wire inertia load (Item 2 times span length) shall be applied at the seismic tension load points on the structure.

  The tension force determined from this procedure shall be used in conjunction with the structure seismic loading for designing the structure. Strain bus tension loads need not be factored by Ω_0.

D.6.10 Short-Circuit (Fault) Loads

Short-circuit (fault) loads shall be determined for designing substation structures. This section provides the equations to determine short-circuit forces. Current flowing in a conductor causes a magnetic field. When currents in adjacent conductors are large enough, their magnetic fields can interact, as shown in Figure D-26, which causes conductors to attract or repel. At normal operating currents, the magnetic fields are sufficiently small that these forces are generally insignificant. However, during a short-circuit event, the currents increase by orders of magnitude higher than normal operating conditions and can cause large forces in conductors.

Short-circuit forces shall be determined by either the simplified static method or the dynamic analysis method. The simplified method is presented in this standard. The dynamic

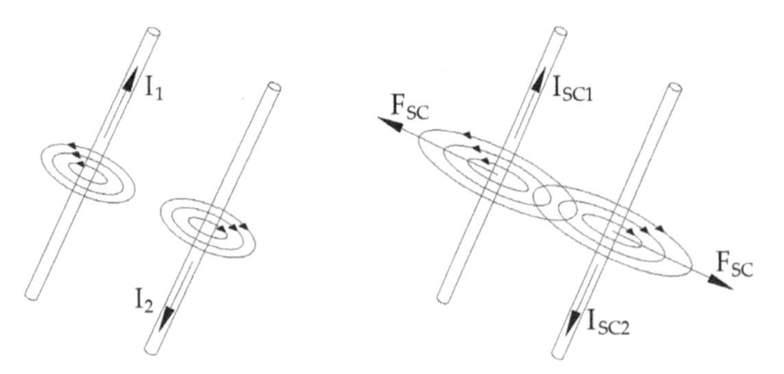

Figure D-26. Magnetic fields generated by current flowing in adjacent conductors. Left: Typical operating currents resulting in inconsequential magnetic fields and a lack of electromagnetic force. Right: Current during a fault resulting in large magnetic fields and electromagnetic force.

analysis method provides more accurate short-circuit forces, but it requires a more complex modeling procedure.

In the following sections, the simplified equations for a short-circuit force are presented. The corrected force given by Equation (D-40) is used in a simplified static analysis to design conductors, insulators, and supporting structures.

D.6.10.1 Simplified Static Short-Circuit (Fault) Force on Rigid Conductors. The equation for the peak fault force presented from IEEE 605 (IEEE 2008), Section 11.3, is reproduced as Equation (D-40). This force represents the peak force assuming full DC offset.

The equation for the basic maximum distributed force between two parallel, infinitely long conductors is

$$F_{sc} = \frac{\frac{\mu_0}{2\pi}\Gamma\left(2\sqrt{2}I_{sc}\right)^2}{D} \tag{D-40}$$

For SI units, this simplifies to

$$F_{sc} = \frac{16 * \Gamma\left(I_{sc}^2\right)}{10^7(D)} \tag{D-41}$$

For customary units, this simplifies to

$$F_{sc} = \frac{3.6\,\Gamma\left(I_{sc}^2\right)}{10^7(D)} \tag{D-42}$$

where

F_{sc} = Fault force in N/m or lb/ft;
I_{sc} = Symmetrical RMS fault current in Amps;
D = Conductor center-to-center spacing in meters or feet;
Γ is 1.0 for phase-to-phase faults and 0.866 (middle conductor) or 0.808 (outer conductors) for three-phase faults; and
$\mu_0 = 4\pi*10^{-7}$ Newton/Amp^2 = $2.825*10^{-7}$ pounds/Amp^2 (Magnetic Constant – magnetic permeability in a classical vacuum).

The basic maximum fault force from Equation (D-40) is modified to obtain the corrected force by multiplying by the square of the half-cycle decrement factor, D_f, as given in IEEE 605 (IEEE 2008):

$$F_{sc\_corrected} = K_f F_{sc}(D_f^2) \tag{D-43}$$

where

$D_f = \dfrac{1+e^{-(1/2fc)}}{2}$,
c = Time constant of the circuit (also often given as T_a),
$c = \dfrac{X}{R*2\pi f}$,
f = Power system frequency (60 Hz in the United States),
K_f = Mounting flexibility factor (IEEE 605, IEEE 2008, Figure 20),
X = System reactance (Ohms), and
R = System resistance (Ohms).

This corrected force given by Equation (D-43) is used in a simplified static analysis to design conductors, insulators, and supporting structures. The corrected peak force of Equation (D-43) is slightly less than the basic force of Equation (D-40).

D.6.10.2 Simplified Static Short-Circuit (Fault) Force on Strain Bus. Each utility, as part of its SCF design criteria, shall consider parameters for the application of short-circuit forces to a strain bus. If it is determined that an SCF on a strain bus is required, refer to either IEEE 605 (IEEE 2008) or IEC 60865 for the force applications without or with jumpers, respectively. IEC 60865 (IEC 2011) includes the stiffness of the attachment points, S, whereas IEEE 605 (IEEE 2008) does not include this parameter. IEC 60865 (IEC 2011) also includes an effective conductor Young's Modulus, whereas IEEE 605 (IEEE 2008) does not.

D.6.11 Construction and Maintenance Loads

The loads resulting from the construction and maintenance of electric substation structures, including any applicable OSHA requirements, shall be considered and incorporated with other loads as applicable in accordance with the installation and maintenance practices of each electric utility. Additional information on construction and maintenance considerations is discussed in Section D.12.0.

D.6.12 Loading Criteria for Deflection Limits

For the purposes of carrying out deflection checks, substation structures shall be classified according to the following three structure classes, A, B, and C. These three classes are defined subsequently.

Class A structures support equipment with mechanical devices such as operating rods or control linkages where structure deflection could impair or prevent proper operation.

Class B structures support equipment without mechanical devices such as operating rods or control linkages, but where excessive deflection could result in compromised phase-to-phase or phase-to-ground clearances or unpredicted stresses in equipment, fittings, or bus conductors, also referred to as Class B equipment.

Class C structures support equipment relatively insensitive to deflection or are stand-alone structures that do not support any equipment.

Structures can be designed to support several pieces of equipment that require different structure classifications. When evaluating the deflection of a multiple-use structure, the deflection limits applicable to any point on the structure shall be determined by using the classification of the structure from that location upward.

Where the owner has not developed specific loading conditions for deflection analysis, the load conditions from Section D.6.12.1 and Section D.6.12.2 shall be used with a load factor of 1.0. The owner shall determine whether additional loads need to be applied in combination with the recommended deflection load cases. The deflection limits in Table D-18 shall be used with the load case combinations listed in the following sections. The structure class deflection limits do not include the foundation displacement or rotation. If the foundation deflection can cause the malfunction of the equipment operation, then those deflections shall be included in the evaluation of the structure class limitations.

Horizontal Members.

For determination of maximum deflections, the span of a horizontal member is the clear distance between structural connections to vertical supporting members, or for cantilever

Table D-18. Summary of Structure Deflection Limits.

Maximum structure deflection as a ratio of span length				
		Structure class		
Member Type	Deflection Direction	Class A	Class B	Class C
Horizontal	Vertical	1/200	1/200	1/100
Horizontal	Horizontal	1/200	1/100	1/100
Vertical	Horizontal	1/100	1/100	1/50

Note: For loading criteria for deflection limits, see Section D.6.12.1.

members, the distance from the point of investigation to the vertical supporting member (Figure D-27).

For horizontal members, the deflection is the maximum net displacement, horizontal or vertical, of the member relative to the member connection points.

Net and gross displacement are illustrated in Figure D-28.

Vertical Members.

For determination of maximum deflections, the span of a vertical member is the vertical distance from the foundation support to the point of investigation on the structure. The deflection to be limited is the gross horizontal displacement of the member relative to the foundation support.

D.6.12.1 Wind Load for Deflection Calculations. A 70 mph (31.3 m/s) wind speed shall be used for the velocity (V) in Equation (D-3) or Equation (D-4) to calculate wind load associated with deflection criteria for substations located outside hurricane-prone regions. Wind speeds

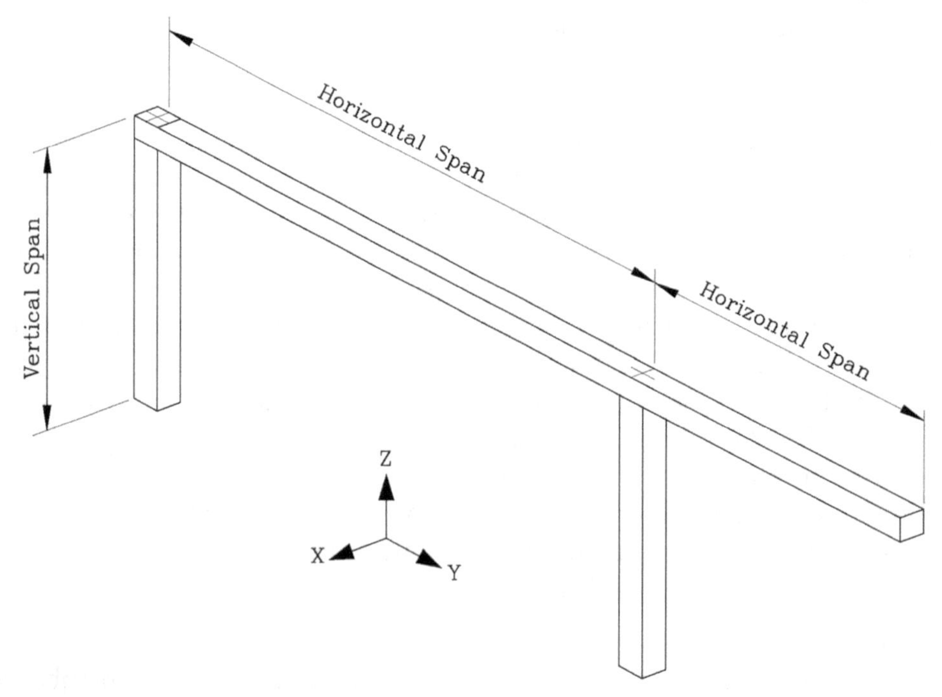

Figure D-27. Span definitions.

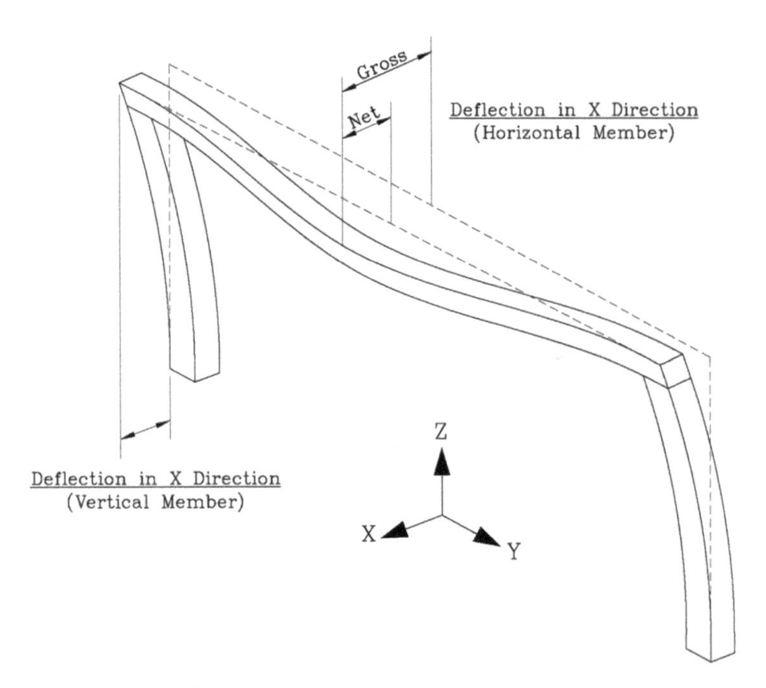

Figure D-28. Net versus gross deflection.

used to determine deflections in hurricane-prone regions shall be determined by the owner. Hurricane-prone regions are shown on the wind maps in Section D.6.5, along the East coastline, the Gulf of Mexico, and are also defined in Chapter 26 of ASCE 7-22. Without specific input from the owner, an 80 mph (35.8 m/s) wind speed for V in Equation (D-3) or Equation (D-4) shall be used to calculate wind load associated with deflection criteria for substations located in hurricane-prone regions. A load factor of 1.0 shall be used for all wind load deflection calculations.

D.6.12.2 Ice with Concurrent Wind Load for Deflection Calculations. Figure D-6 in this standard provides extreme ice thickness for a 100-year MRI. Figure D-7 provides concurrent wind speed for both a 100-year MRI and a 50-year MRI ice event. Figure D-8 provides the concurrent temperature for both 100-year MRI and 50-year MRI ice events. For deflection checks, the wind speeds shown on these maps shall be converted to a 5-year MRI. The conversion factors from 100-year to 5-year values are given in Table D-19.

D.6.12.3 Other Deflection Considerations. The owner shall determine the loading conditions for deflection checks and its associated clearances.

A decision whether to include loads resulting from short-circuit current or fault current in deflection analysis shall be made by the owner.

Table D-19. Conversion Factors for Ice Thickness and Concurrent Wind Load for Deflection Computations.

	100-year to 5-year mean recurrence
Ice thickness conversion factor	0.2
Concurrent Wind load conversion factor	1.0

D.6.13 National Electrical Safety Code Loads

The *National Electrical Safety Code* (IEEE 2023), Section 16, Paragraph 162.A, requires that substation structures supporting facilities (wires) that extend outside the substation fence shall comply with the loading and strength sections of IEEE C2-2023.

D.7 DESIGN OF MEMBERS

D.7.1 General Design Principles

The provisions of this section shall be used for designing structures and equipment support in substations. Specific guidelines for member design and fabrication are not included in this standard.

This standard refers the substation owner or authorized agent to the following standards, manuals of practice (MOPs), and notes any exceptions to the referenced documents:

- ASCE Standard 10, *Design of Latticed Steel Transmission Structures*
- ASCE Standard 48, *Design of Steel Transmission Pole Structures*
- ASCE MOP 91, *Design of Guyed Electrical Transmission Structures*
- ASCE MOP 123, *Prestressed Concrete Transmission Pole Structures*
- ASCE MOP 141, *Recommended Practice for the Design and Use of Wood Pole Structures for Electrical Transmission Lines*
- ACI CODE-318, *Building Code Requirements for Structural Concrete and Commentary*
- ANSI O5.1, *Specifications and Dimensions for Wood Poles*
- AISC 360, *Specification for Structural Steel Buildings*
- Aluminum Association's *Aluminum Design Manual*
- IEEE *National Electrical Safety Code* C2

For strengths of other substation facilities such as insulators and conductors, refer to the specifications of the applicable manufacturer.

Load factors, load combinations, load cases, and deflection criteria specified in Sections D.5 and D.6 shall be used with referenced design codes or documents.

There is no intention to exclude any material or section types. If the material or section type is not addressed in this draft pre-standard, the engineer shall use the appropriate design code or document.

D.7.2 Design Methods

Allowable Strength Design (ASD) is a method of proportioning structural members such that elastically computed stresses produced in the members by service loads do not exceed specified allowable stresses. ASD is also called *working stress design*. The base reactions developed through ASD are used in conjunction with foundation soil interaction design.

Ultimate Strength Design (USD) is the method of proportioning structural members such that the computed forces produced in the members by the factored loads do not exceed the member design strength. USD is also sometimes called *load and resistance factor design* (LRFD). USD is recommended for substation structures.

Plastic analysis or design shall not be used for substation structures.

Structures that support conductors and overhead ground wires that extend outside the boundaries of the substation shall also meet or exceed the load and strength requirements of the NESC C2 (IEEE 2023) and local regulatory codes. For structures that are required to meet NESC C2 (IEEE 2023) load criteria, the USD method shall be used because NESC C2 (IEEE 2023) specifies load factors and material strength factors.

D.7.3 Steel Structures

ASCE 10-15 shall be used for designing lattice structures using angles (Section D.7.3.1).

AISC 360 (AISC 2016b) shall be used for designing structures using standard structural shapes (Section D.7.3.2).

ASCE 48 (ASCE 2019a) shall be used for designing structures using hollow tubular member shapes (Section D.7.3.3).

D.7.3.1 Lattice Angle Structures. ASCE 10-15 shall be used for designing lattice structures. This standard uses factored design loads, linear material properties, and first- or second-order elastic analysis. It does not use strength reduction factors.

ASCE 10-15 has adjusted column equations for angles, which account for the effect of eccentricities in connections that are commonly used in lattice angle structures. The effects of flexural–torsional and torsional buckling are also included. For these reasons, ASCE 10-15 shall be used for designing lattice substation structures constructed using angle sections.

As an alternative to ASCE 10-15, steel lattice structures using angle sections shall be designed in accordance with AISC 360 (AISC 2016b). LRFD shall be used with factored design loads, linear material properties, LRFD member capacity reduction factors, and second-order elastic analysis.

When the loading is not through the centroidal axis, the combined stress equation for bending and axial loads shall be used and flexural–torsional and torsional buckling shall be considered for these members.

D.7.3.2 Standard Structural Shapes Other Than Angles. AISC 360 (AISC 2016b) shall be used for designing standard structural shapes other than angles. Examples of standard shapes are wide flanges, channels, HSS tubes, pipes, and tee sections. LRFD shall be used for designing these member shapes with factored design loads, linear material properties, LRFD member capacity reduction factors, and second-order elastic analysis.

D.7.3.3 Hollow Tubular Member Shapes. ASCE 48 (ASCE 2019a) or AISC 360 (AISC 2016b) shall be used for designing hollow structural shapes, as applicable.

Custom-fabricated hollow tubular member shapes include 4-, 6-, 8-, and 12-sided polygonal sections and circular sections. These members shall be designed and fabricated in accordance with ASCE 48 (ASCE 2019a).

ASCE 48 (ASCE 2019a) uses factored design loads, linear material properties, and a second-order geometric nonlinear analysis method (*P*-delta effects). ASCE 48 (ASCE 2019a) also allows for uniformly tapered members.

D.7.3.4 Local Buckling of Irregular Polygonal Shapes. For structures that use an irregular polygonal shape (unequal flats) and have high w/t ratios for the two flats that are on the neutral axis, the buckling stress of the long flats shall be calculated using Equation (D-44) (Bleich 1952)

$$\sigma_{cr} = \frac{\pi^2 E}{12(1-\upsilon^2)}\left(\frac{t}{w}\right)^2 k \text{ but less than } F_y \tag{D-44}$$

where

σ_{cr} = Critical buckling stress,
E = Young's modulus,
F_y = Yield strength stress,
υ = Poisson's ratio,
t = Plate thickness,
w = Plate width in compression, and
k = Local buckling factor.

For flats that have tension on one corner and compression on the other corner, the width w shall be taken as the distance from the corner in compression to the point along the flat that has zero compression. Then, k is 7.7 for this condition. For flats where both corners are in compression, k is 4.0 (ASCE 48) (ASCE 2019a), and the width w shall be taken as the length of the flat between the actual inside bend radii or four times the thickness (ASCE 48) (ASCE 2019a).

Using this method, the stress at each corner of the polygon shall be calculated for each load case and for the entire length of the member because the buckling strength is load case– and position-dependent.

D.7.4 Concrete Structures

D.7.4.1 Reinforced Concrete Structures. Reinforced concrete structures shall be designed and constructed in accordance with ACI CODE-318 (ACI 2019). This code uses the USD method with factored design loads, linear material properties, and second-order elastic analysis. Member strength reduction factors shall be used as specified in ACI CODE-318 (ACI 2019).

D.7.4.2 Prestressed Concrete Structures. Prestressed concrete structures shall be designed and constructed in accordance with ACI CODE-318 (ACI 2019). This code uses the USD method with factored design loads, linear material properties, and second-order elastic analysis. Member strength reduction factors shall be used as specified in ACI CODE-318 (ACI 2019).

D.7.4.3 Prestressed Concrete Poles. The prestressed concrete pole-type structures, either static cast or spun cast, shall be designed and constructed in accordance with ASCE MOP 123 (ASCE 2012). This guideline uses the USD method using the requirements of ACI and PCI recommendations.

D.7.5 Aluminum Structures

Aluminum structures shall be designed and fabricated in accordance with the Aluminum Association's *Aluminum Design Manual 2020,* using stresses for building-type structures. To design aluminum structures according to the requirements of IEEE 693 (IEEE 2018b), the safety factors shall be as specified for bridge-type structures for ASD in the *Aluminum Design Manual 2020.*

D.7.6 Wood Structures

Wood structures and poles shall be designed and constructed in accordance, MOP 141 (ASCE 2019b) and NESC C2 (IEEE 2023). ANSI-O5.1 (ANSI 2022) shall be used for wood pole stresses with the NESC C2 (IEEE 2023) 0.65 strength factor.

D.7.7 Base Plate Design

This section provides a method to determine the plate thickness for a base plate on leveling nuts. ASCE 48 (ASCE 2019a) shall be used for base plate design. The effective length of the bend line (b_{eff}) for bending planes depends on the size and shape of any galvanizing drain holes in the base plate.

D.7.7.1 Determination of Anchor Rod Loads. Assuming that the base plate behaves as an infinitely rigid body, the load in anchor rod i, BL_i, shall be calculated by using the following formula (Figure D-29):

$$BL_i = \left(\frac{P}{A_{BC}} + \frac{M_x y_i}{I_{BC_x}} + \frac{M_y x_i}{I_{BC_y}} \right) A_i \tag{D-45}$$

where

P = Total vertical load at the base of the column,
M_x = Base moment with respect to the x-axis,
M_y = Base moment with respect to the y-axis,
x_i, y_i = x and y distances of anchor rod i from reference axes,
A_i = Net area of anchor rod i,
$A_{BC} = \sum_{i=1}^{n} A_i$ (A_{BC} is the total anchor rod cage area),

$I_{BC_x} = \sum_{i=1}^{n} (A_i y_i^2 + I_i)$ (I_{BC_x} is the total anchor rod cage inertia with respect to the x-axis),

$I_{BC_y} = \sum_{i=1}^{n} (A_i x_i^2 + I_i)$ (I_{BC_y} is the total anchor rod cage inertia with respect to the y-axis),
n = Total number of anchor rods, and
I_i = Moment of inertia of anchor rod i.

Because I_i is often small, it can be omitted when calculating I_{BC_x} and I_{BC_y}.

Figure D-29 illustrates the application of Equation (D-45) to determine anchor rod loads.

D.7.7.2 Determination of Base Plate Thickness. The design procedure for base plates assumes that anchor rod loads produce uniform bending stress, F_y, along the effective portion of bend lines located at the face of the column. Each bend line is characterized by the following:

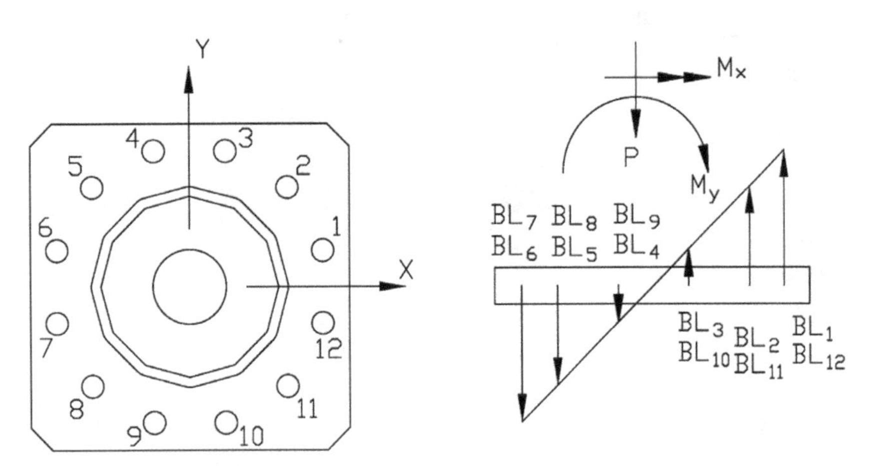

Figure D-29. Rigid plate free body diagram.

k = Number of anchor rod load BL_i's contributing moment along the bend line,

c_i = Shortest distance from the center of each anchor rod (i) to the bend line,

b_{eff} = Length of the bend line (depending on the shape of the column, the shape of the base plate, and k), and

BL = Anchor rod load.

ASCE 48 (ASCE 2019a), Appendix F, shall be used to design base plates for tubular steel pole columns.

The base plate bending stress F_{PL} for the assumed bend line shall be calculated by using Equation (D-46):

$$F_{PL} = \left(\frac{6}{b_{\text{eff}} t^2}\right)(BL_1 c_1 + BL_2 c_2 + \cdots + BL_k c_k) \quad \text{(D-46)}$$

where t is the base plate thickness, and BL_i is the effective anchor rod load for each anchor rod that causes a moment on the assumed bend line b_{eff}. The base plate thickness is determined by keeping F_{PL} below the yield stress F_y for anchor rod loads corresponding to the factored loads. To determine $t_{\min}$, Equation (D-46) shall be used to solve for t and rewritten as follows:

$$t_{\min} = \sqrt{\left(\frac{6}{b_{\text{eff}}\left(F_y\right)}\right)(BL_1 c_1 + BL_2 c_2 + \cdots + BL_k c_k)} \quad \text{(D-47)}$$

D.7.7.3 Anchor Rod Holes in Base Plates. Table D-20 provides hole diameters for base plates to meet oversize hole requirements of AISC 360 (AISC 2016b), Table J3.3, and also provides construction tolerances for cast-in-place anchor rods.

D.7.8 Rigid Bus Design

Rigid bus design shall be approached as a system requiring both an electrical engineer and a civil engineer.

D.7.8.1 Bus Materials. Copper: Copper found in legacy installations. Consult the manufacturer for design values.

Table D-20. Base Plate Anchor Rod Hole Diameters (inches).

Anchor rod diameter	Hole diameter	Anchor rod diameter	Hole diameter
0.625	0.8125	1.625	1.9375
0.750	0.9375	1.750	2.0625
0.875	1.0625	1.875	2.1875
1.000	1.2500	2.000	2.3125
1.125	1.4375	2.250	2.5625
1.250	1.5625	2.500	2.8125
1.375	1.6875	2.750	3.0625
1.500	1.8125	3.000	3.3125

Note: 1 in. = 25.4 mm.

Table D-21. Aluminum Alloy Properties.

Alloy and temper	ASTM standard	Minimum tensile strength (ksi)	Minimum yield strength (ksi)
6061-T6	B241 (ASTM 2022)	38	35
6063-T6	B241 (ASTM 2022)	30	25
6101-T6	B317 (ASTM 2015a)	29	25

Note: 1 ksi = 6.9 MPa.

Aluminum: Typical aluminum alloys and tempers used in a rigid bus design are shown in Table D-21.

The modulus of elasticity of aluminum is 10,100 ksi (69,600 Mpa).

D.7.8.2 Insulator Type. Porcelain Insulators: Porcelain station post insulators are assemblages generally consisting of end fittings, a bonding medium, and a porcelain body.

Consult manufacturers for tension, compression, torsion, bending, and other strength characteristics.

Composite Insulators: Composite station posts derive their strength from a fiberglass core that is covered by elastomeric weather sheds.

Consult manufacturers for tension, compression, torsion, bending, and other strength characteristics.

D.7.8.3 Insulator Mechanical Design

D.7.8.3.1 Insulator Nominal Strength. The nominal moment strength of a station post insulator shall be determined by

$$M_n = C_s * h_i \tag{D-48}$$

where C_s is the maximum rated cantilever value published by the manufacturer, as determined from testing in accordance with ANSI/NEMA C29.9 (ANSI 2017), and h_i is the total insulator height.

The nominal torsion, compression, and tension strengths of a station post insulator shall be taken as the value published by the manufacturer, as determined from testing in accordance with ANSI/NEMA C29.9 (ANSI 2017) or ANSI/NEMA C29.19 (ANSI 2020).

D.7.8.3.2 Insulator Factored Strength. For USD, the factored resistance of porcelain insulators shall be determined by multiplying the nominal insulator strength by the following resistance factors:

Bending: $\phi = 0.5$,

Torsion: $\phi = 0.6$,

Compression: $\phi = 0.6$, and

Tension: $\phi = 0.6$.

D.7.8.3.3 Insulator Strength Evaluation. For LRFD evaluation, insulator strength shall be evaluated for the individual moment, axial load, and torque caused by loading from a particular load case (extreme wind, short circuit, and so on). In the following equations M_{LC}, P_{LC}, and T_{LC} are the applied moment, axial load, and torque, respectively, for a given load case from Section D.5, and γ is the corresponding load factor for the case in question:

$$\varphi M_n \geq \Sigma(\gamma * M_{LC}) \quad \text{(D-49)}$$

$$\varphi P_n \geq \Sigma(\gamma * P_{LC}) \quad \text{(D-50)}$$

$$\varphi T_n \geq \Sigma(\gamma * T_{LC}) \quad \text{(D-51)}$$

D.7.8.3.4 Combined Forces. Insulators shall be evaluated for the effects of combined forces.

For LRFD evaluation of torsional forces in combination with bending moments and axial forces, the following equations shall be satisfied:

$$B = \frac{\Sigma(\gamma M_{LC})}{\varphi M_n} \quad \text{(D-52)}$$

$$P = \frac{\Sigma(\gamma P_{LC})}{\varphi P_n} \quad \text{(D-53)}$$

$$T = \frac{\Sigma(\gamma T_{LC})}{\varphi T_n} \quad \text{(D-54)}$$

These equations D-52, D-53, and D-54 shall then be used to evaluate Equation (D-55):

$$\frac{1}{2} * \left[B + P + \sqrt{(B+P)^2 + 4 * (T)^2}\right] \leq 1.0 \quad \text{(D-55)}$$

D.7.8.4 Bus System Design. IEEE Standard 605 (IEEE 2008) shall be used for bus and insulator design using the loads and factors from this standard.

D.7.8.5 Rigid Bus Seismic Design. Rigid bus shall be designed to withstand seismic forces and displacement demands as defined by this standard. Flexible connections shall be designed using the provisions of IEEE 1527 (IEEE 2018c) and IEEE 605 (IEEE 2008).

D.7.9 Special Design Considerations

D.7.9.1 Structures Design Support of Air Core Reactors. The following criteria shall be used in designing air core reactor installations:

- Use of fiberglass reinforcement within $D/2$ of the reactor (where D is the diameter of the reactor). Refer to ACI PRC-440.1 (ACI 2015b) for a design with fiberglass reinforcement.
- Anchor rods within $D/2$ distance of the reactor shall be nonmagnetic stainless steel. Anchor rods beyond $D/2$ distance of the reactor can be steel.
- No aluminum structures within $D/2$ distance of the reactor.
- No closed loops in metal allowed within D of the reactor.
- Closed loops can be broken by taping the crossing touching parts with electrical or other nonconductive tapes.
- No contact between dissimilar metals.

D.7.9.2 Wind-Induced Vortex Shedding. The potential of wind-induced vortex shedding shall be considered. The design of strain bus structures and ground masts shall consider the following:

- Stiffness of the member;
- Damping of the member;

- Natural frequencies of the structure, bus, and members;
- When known, the site wind conditions (dominate wind direction and exposure); and
- Addition of spiral strakes or other mitigating devices.

D.8 BARRIER STRUCTURE REQUIREMENTS: OIL SPILL CONTAINMENT, FIRE, SOUND, AND BALLISTIC WALLS

Oil spill and fire containment system design are a function of each utility's internal standards and are not covered here.

D.8.1 Structural Design Considerations

Structural design requirements: Fire, sound, and ballistic wall materials shall retain their strength when exposed to fire for the entire design duration.

The reinforced concrete containment wall shall be designed for a minimum overburden earth pressure of 250 pounds per linear foot per foot of depth and length (372 kg/m) for supporting construction and maintenance equipment.

The reinforce concrete containment floor shall be designed such that any bearing pressure resulting from "jack-and-skid" construction methods can be supported, where such methods are intended for use.

Any steel grating used for supporting fire-quenching rocks shall be designed for an additional maintenance load of 250 lb (113 kg) plus the weight of any other components (e.g., replacement oil containers) used in accordance with the utility's maintenance practices.

D.8.2 Barrier Walls

All barrier walls shall be designed such that they meet the wind, seismic, and dead loads as defined by this standard and any local governing jurisdictions, where these are applicable.

Design of barrier fire walls and connections shall consider exposure to the temperatures of resulting fire for a minimum of 2 h.

Design of barrier ballistic walls shall consider threats as defined by the owner's internal protection standards.

Design of barrier sound walls shall follow the provisions of this standard, local government jurisdictional code (where these are applicable), and proportioned or aligned to reduce sound travel in a specific direction.

D.9 DESIGN OF CONNECTIONS TO FOUNDATIONS

Substation structures and equipment that can overturn or slide during a seismic or extreme wind event shall be anchored.

ASCE 10-15 and ASCE 48 (ASCE 2019a), respectively, cover the anchorage of the substation structure through direct embedment or stub angles.

The approach for the design of anchor rods in this section is based on ultimate strength design loads (also referred to as factored loads).

D.9.1 Anchorage Systems for Given Foundation Types

Design of anchorage to concrete shall follow the provisions of ACI CODE-318 (ACI 2019), Chapter 17, except as noted in this section.

ACI CODE-318 (ACI 2019), Chapter 17, is most applicable to anchorage in structural slabs with or without a horizontally placed rebar. Anchorage consists of headed anchor rods, hooked anchor rods, headed studs, expansion anchors, undercut anchors, or adhesive anchors.

D.9.1.1 Spread Footing Foundation. A spread footing foundation (also referred to as a "pad and pedestal foundation") (Figure D-30) typically has a square or rectangular horizontal footing pad.

Transverse reinforcement shall not be counted upon to carry any side blow out loading for spread footing with pedestal or stem walls. The anchor force to the foundation shall be developed by using lap splices with vertical reinforcement or in accordance with ACI CODE-318 (ACI 2019), Chapter 17. See Figures D-31 and D-32 for shear pryout modes.

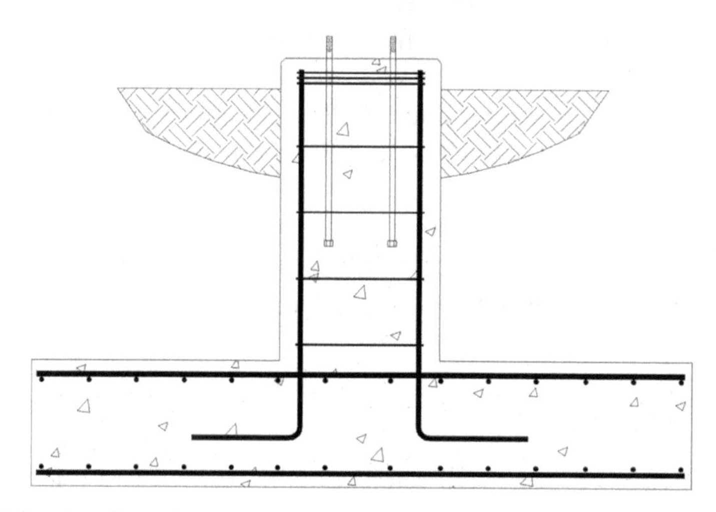

Figure D-30. Spread footing foundation.

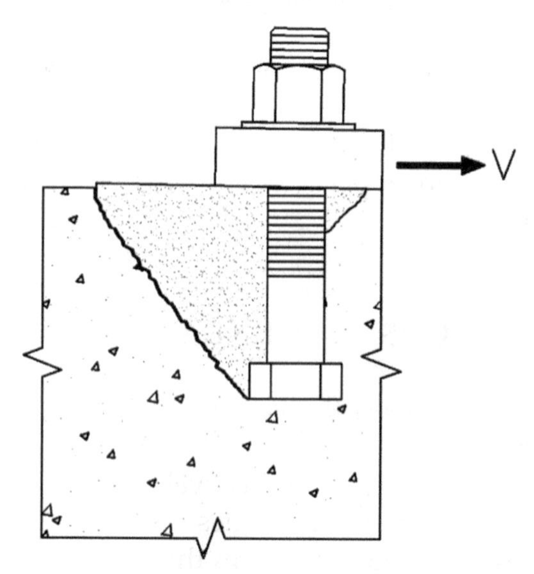

Figure D-31. Shear pryout with a short anchor rod far from a free edge.

D.9.1.2 Drilled Pier Foundation. Figure D-33 illustrates another type of common foundation in substations, a circular drilled pier. Concrete breakout area determination for round piers shall follow the pattern as shown in Figure D-34.

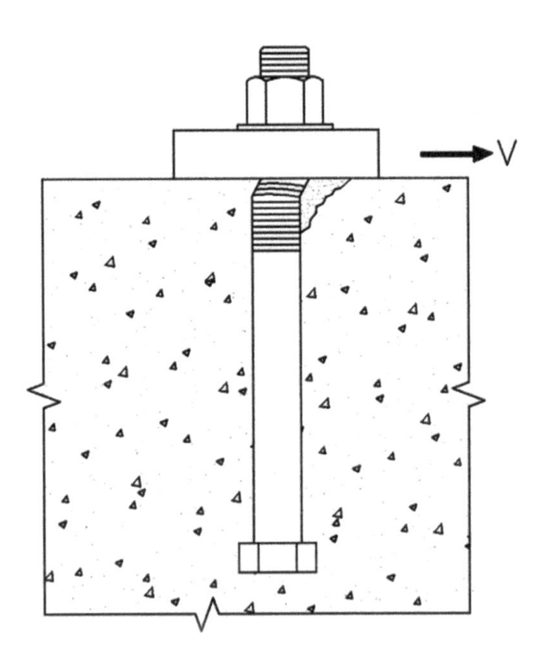

Figure D-32. Steel failure preceded by concrete spall for a long anchor rod.

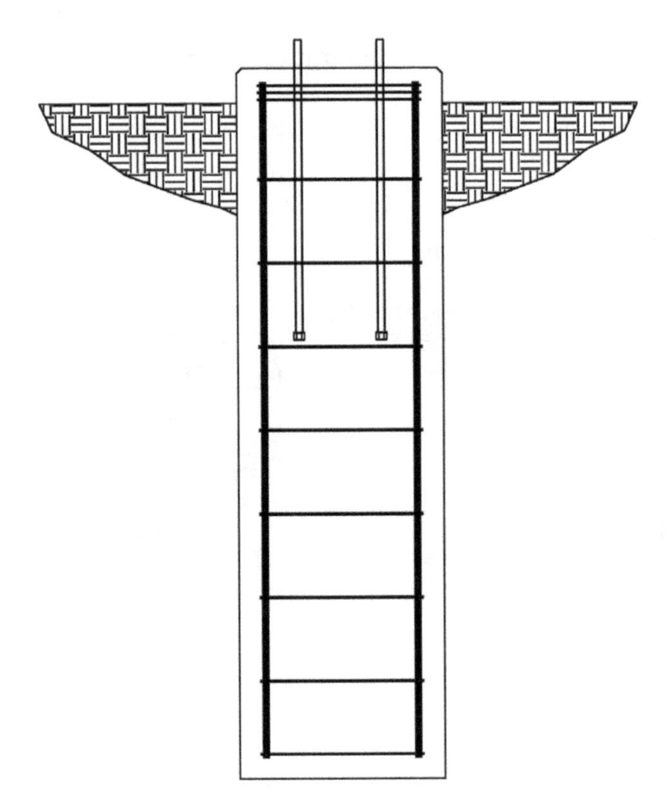

Figure D-33. Drilled pier foundation.

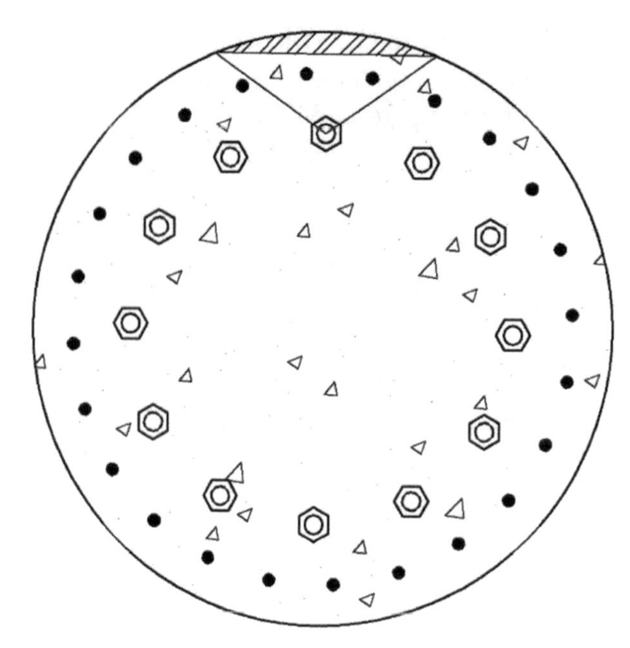

Figure D-34. Concrete shear breakout area of a round pier.

D.9.2 Anchor Materials

Table D-22 provides the properties of acceptable anchor rod materials. Anchor rods shall not be less than 0.75 in. (1.91 cm) in diameter, unless the design strength is greater than or equal to two times the required strength.

The utility shall determine whether the anticipated magnitude of low-service temperatures warrant the specification of Charpy V-Notch testing and range thereof. The owner shall determine whether a minimum Charpy V-Notch impact strength is required for anchor material used in areas with high seismic accelerations (ASCE 7-22, Seismic Design Categories C and D, Table 11-6).

Deformed reinforcing bar material used for anchor rods shall be in accordance with ASTM A615/A615M or ASTM A706/A706M. When deformed reinforcing bars are used, Charpy-V notch test in accordance with ASTM A673 and ASTM A370 shall be performed, and they shall

Table D-22. Anchor Material Properties.

Material ASTM standard	Yield strength, min F_y (kip/in.^{2})	Tensile strength F_u (kip/in.^{2})
ASTM A36/A36M	36	58–80
ASTM F1554, Grade 36	36	58–80
ASTM F1554, Grade 55	55	75–95
ASTM F1554, Grade 105	105	125–150
ASTM A615/A615M, Grade 60	60	90 min
ASTM A615/A615M, Grade 75	75	100 min
ASTM A615/A615M, Grade 80	80	105 min
ASTM A706/A706M, Grade 60 (weldable)	60–78	80 min
ASTM A706/A706M, Grade 80 (weldable)	80–98	100 min

Note: 1 kip/in.^{2} = 6.89 MPa.

have a minimum Charpy-V notch requirement of 15 ft·lb (20.3 N·m) at –20 °F (–28.9 °C) when tested in the longitudinal direction.

Shear stud connectors can be used to anchor plates to the concrete. Shear studs can be manufactured from ASTM A108 steel. Stud welding shall conform to AWS D1.1 (AWS 2020).

The anchor rod projection above the top of concrete plus a minimum of 6 in. (15.24 cm) into the concrete foundation shall be galvanized. Rods, nuts, washers, and steel hardware components shall be hot-dip-galvanized in accordance with ASTM A153 supplemented by ASTM F2329 (ASTM 2015b) as appropriate, and reinforcing steel in accordance with ASTM A767. Safeguard products against steel embrittlement in conformance with ASTM A143.

D.9.3 Anchor Arrangements and General Considerations

The two most commonly used arrangements to anchor a structure to the foundation are base plates supported by anchor rods with leveling nuts (Figure D-35), and anchor rods with the base plates on concrete or grout (Figure D-36).

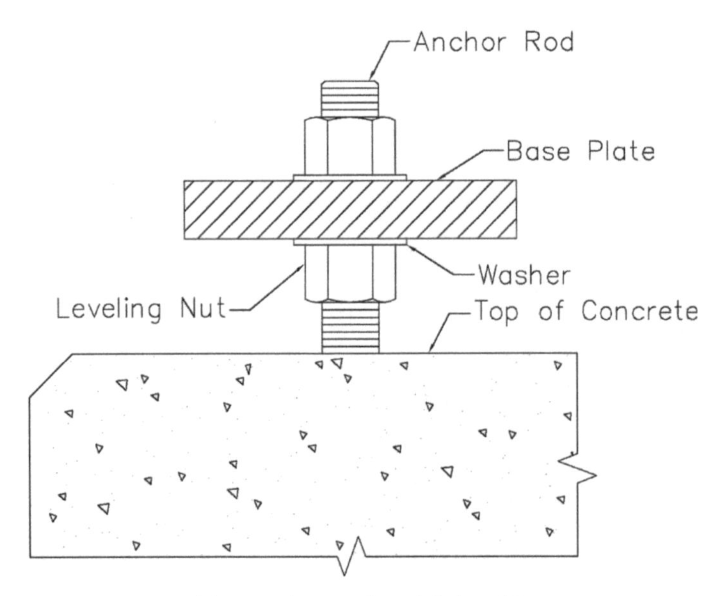

Figure D-35. A base plate supported by anchor rods with leveling nuts.

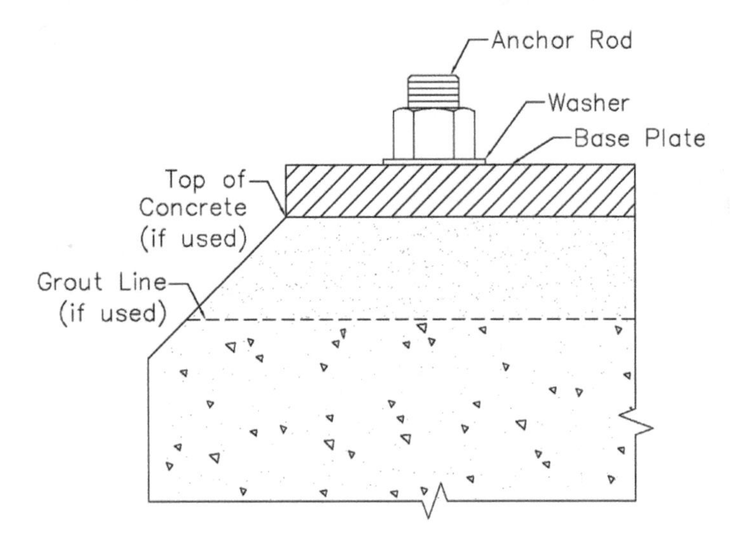

Figure D-36. Anchor rods with a base plate on concrete or grout.

Ductile design of anchors resisting seismic forces shall be in accordance with the provisions of ACI CODE-318 (ACI 2019), 17.10.5.3(a). Nonductile design of anchor rods in seismic regions with $S_{DS} > 0.167$ and $S_{D1} > 0.067$ shall follow the provisions of ACI CODE-318 (ACI 2019), 17.10.5.3 (b), (c), or (d). See ACI CODE-318 (ACI 2019), 17.10.5.3(d), where seismic loads are amplified by an overstrength factor Ω_0.

The overstrength factor for substation structures is provided in Section D.6.9.4, "Seismic Design Coefficients and Factors," Table D-15. The overstrength factor Ω_{PL} for all IEEE 693 (IEEE 2018b) qualified substation equipment is provided in Section D.6.9.10.1, "IEEE 693 (IEEE 2018b) Qualified Equipment and Supports Anchorage Design Forces."

Design level for ductile anchorage designs, Ω_{PL}, shall be 1.0 and nonductile anchorage designs, Ω_{PL}, shall be 1.5.

Performance level for ductile anchorage designs, Ω_{PL}, shall be 1.0 and nonductile anchorage designs, Ω_{PL}, shall be 1.0.

D.9.4 Anchors Cast in Place

D.9.4.1 Types of Anchors. The three types of anchors that are most commonly used are headed rods, deformed reinforcing bar rods, and hooked rods (Figure D-37).

D.9.4.1.1 Headed Rods. Headed anchor rods consist of a rod with a head or nut at the end of the anchor embedded in concrete, where tensile forces (and compressive forces as in the case of base plates supported on leveling nuts) are resisted by a concrete breakout cone or anchor reinforcement provided adjacent to the anchor.

D.9.4.1.2 Deformed Reinforcing Bar Rods. The design of deformed anchor rods shall be in accordance with the provisions of ACI CODE-318 (ACI 2019), Chapter 25, and ASCE 48 (ASCE 2019a). The shear failure modes are not addressed in ASCE 48 (ASCE 2019a) and shall be addressed as previously described in ACI CODE-318 (ACI 2019) and Sections D.9.1.1 and D.9.1.2 of this draft pre-standard.

D.9.4.1.3 Hooked Rods. Smooth-bar hooked rods are not permitted.

D.9.4.2 Design Considerations for Anchor Steel

D.9.4.2.1 Anchor Rods with a Base Plate on Concrete or Grout. The following equations shall be used to determine the maximum strength of the anchor rod in accordance with the provisions of ACI CODE-318 (ACI 2019), Chapter 17:

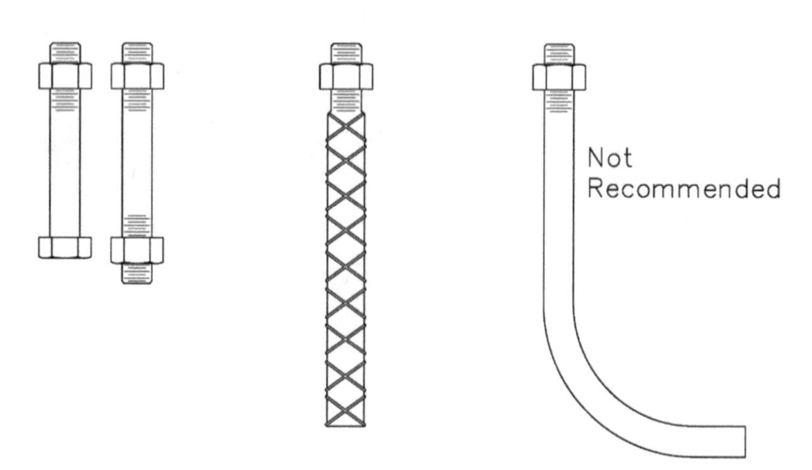

Figure D-37. Types of anchors (headed rod, threaded rebar, hooked smooth rod).

$$\frac{N_{ua}}{\phi N_n} + \frac{V_{ua}}{\phi V_n} \leq 1.2 \quad \text{(D-56)}$$

$$\frac{N_{ua}}{\phi N_n} \leq 1.0 \quad \text{(D-57)}$$

$$\frac{V_{ua}}{\phi V_n} \leq 1.0 \quad \text{(D-58)}$$

where

N_{ua} = Ultimate tensile force (factored design tensile force) per anchor rod (kips or kN),
V_{ua} = Ultimate shear force (factored design shear force) per anchor rod (kips or kN),
ϕ = Strength reduction factor,
N_n = Nominal shear strength of each anchor rod (kips or kN),
V_n = Nominal shear strength of each anchor rod (kips or kN),
ϕN_n = Design tensile strength (available capacity) of each anchor (kips or kN), and
ϕV_n = Design shear strength (available capacity) of each anchor (kips or kN).

The design tensile strength (ϕN_n) shall be determined by

$$\phi N_n = \phi f_{uta} A_{se,N} \quad \text{(D-59)}$$

where f_{uta} is the lesser of f_{ua}, the specified tensile strength of the anchor; 1.9 f_{ya}, where f_{ya} is the specified yield strength of the anchor; and 125,000 psi (860 MPa). $A_{se,N}$ is the effective cross-sectional area of the anchor rod in tension (in.^{2} or mm^2). $\phi = 0.75$ for the ductile steel element or 0.65 for the brittle steel element.

For threaded rods and headed rods, the effective cross-sectional area $A_{se,N}$ (in.^{2} or mm^2) shall be determined by

$$A_{se,N} = \frac{\pi}{4}\left(d_a - \frac{0.9743}{n_t}\right)^2 \quad \text{(D-60)}$$

where d_a is the nominal anchor rod diameter (inch or mm), and n_t is the number of threads per inch or millimeter of the anchor rod.

For anchors with reduced cross-sectional areas anywhere along the length, the effective cross-sectional area shall be provided by the anchor manufacturer.

The ultimate shear force per anchor rod V_{ua} shall be determined by

$$V_{ua} = \frac{V - \mu N_{\text{cm}}}{n_a} \quad \text{(D-61)}$$

where

V = Total shear force at the base plate (kips or kN);
N_{cm} = Compressive force on the base plate (kips or kN) concurrent with the shear force V (*Note*: $N_{\text{cm}} \geq 0$. If the base plate is subjected to net uplift, $N_{\text{cm}} = 0$);
$\mu = 0.4$ the static coefficient of friction between the base plate and the concrete (Figure D-36). For shear force V that includes seismic load combinations, $\mu = 0$;
n_a = Number of anchor rods at the base plate
$\mu^* N_{\text{cm}} < 0.2 f'_c A_c \times 10^{-3}$ or $800A_c \times 10^{-3}$ (where A_c is in in.^{2}) or $5.5A_c \times 10^{-3}$ (when A_c is in mm^2)

where f_c' is the specified compressive strength of concrete, psi or MPa, and A_c is the area of concrete section–resisting shear transfer (in.² or mm²).

The design shear strength (ϕV_n) shall be determined by

For cast-in headed stud anchors with a base plate on concrete without grout,

$$\phi V_n = \phi f_{uta} A_{se,V} \quad \text{(D-62)}$$

These types of anchors use an embedded plate in the concrete where equipment (such as transformers or circuit breakers) is welded to the embedded plate and the studs are anchored to the concrete foundation by the heads on the studs. Headed studs welded to an embedded plate shall not be used on grout.

For cast-in headed rod anchors with a base plate on concrete without grout,

$$\phi V_n = \phi 0.6\, f_{uta} A_{se,V} \quad \text{(D-63)}$$

For cast-in headed rod anchors with a base plate on grout,

$$\phi V_n = \phi 0.48\, f_{uta} A_{se,V} \quad \text{(D-64)}$$

where ϕ is 0.65 for ductile steel elements and 0.60 for brittle steel elements, and $A_{se,V}$ is the effective cross-sectional area of an anchor rod in shear (in.² or mm²)

For threaded rods and headed rods, the effective cross-sectional area $A_{se,V}$ (in.² or mm²) is given by

$$A_{se,V} = \frac{\pi}{4}\left(d_a - \frac{0.9743}{n_t}\right)^2 \quad \text{(D-65)}$$

D.9.4.2.2 Base Plate Supported by Anchor Rods with Leveling Nuts. The methodology for design of anchor rods with leveling nuts is described in this section. The moment arm for anchors in bending shall be determined on the basis of the fixity of the rod to the concrete, as shown in Figure D-38 or D-39. The following equations shall be used to design the anchorage:

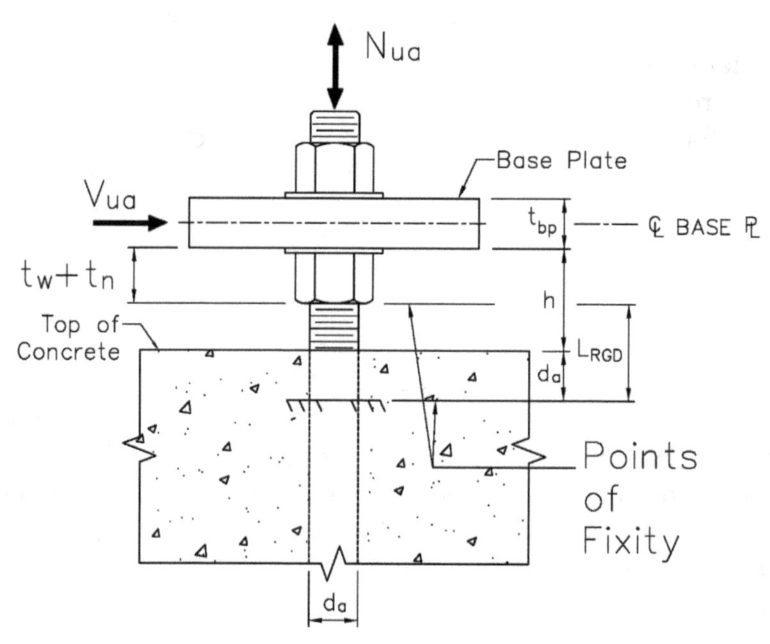

Figure D-38. Moment arm without a clamping nut.

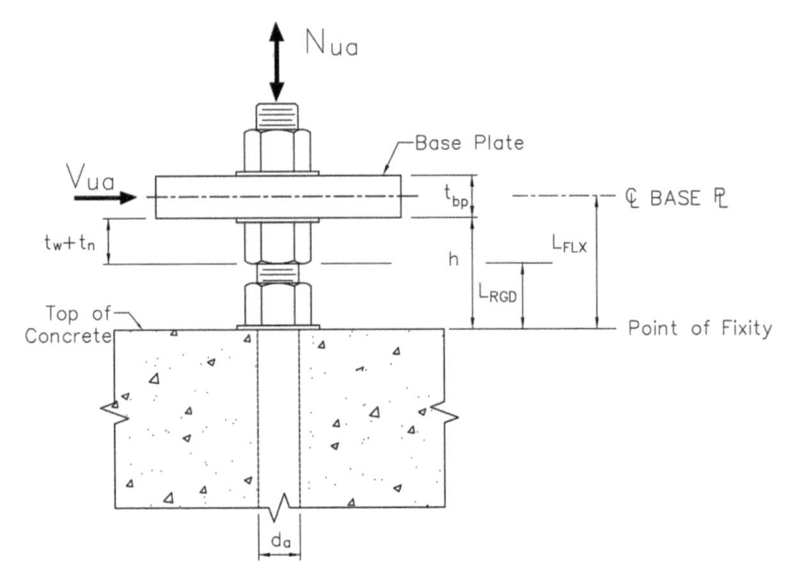

Figure D-39. Moment arm with a clamping nut.

$$L_{\mathrm{RGD}} = h - t_w - t_n + d_a \quad \text{(Figure D-22)} \tag{D-66}$$

$$L_{\mathrm{RGD}} = h - t_w - t_n \quad \text{(Figure D-23)} \tag{D-67}$$

$$L_{\mathrm{FLX}} = h + \frac{t_{bp}}{2} + d_a \quad \text{(Figure D-22)} \tag{D-68}$$

$$L_{\mathrm{FLX}} = h + \frac{t_{bp}}{2} \quad \text{(Figure D-23)} \tag{D-69}$$

where

L_{FLX} = Moment arm for flexible base plate behavior (Figure D-40),
L_{RGD} = Moment arm for rigid base plate behavior (Figure D-41),
h = Distance from the bottom of the base plate to the top of the concrete,
t_w = Thickness of the washer,
t_n = Thickness of the nut,
t_{bp} = Thickness of the base plate, and
d_a = Diameter or the anchor rod.

The bending capacity of an individual anchor is given by the following equation:

$$M_0 = 1.2 S f_u \quad \text{(only when the anchor cross section is constant)} \tag{D-70}$$

where M_0 is the bending capacity of an individual anchor.

$$S = \frac{\pi}{32}\left(d_a - \frac{0.9743}{n_t}\right)^3 \quad \text{(section modulus of the anchor rod)} \tag{D-71}$$

d_a = Nominal anchor diameter,
n_t = Number of threads per unit of measure, and
f_u = Specified tensile strength of a steel anchor.

The bending capacity is reduced because of an axial force in the anchor, which is given by the following equation:

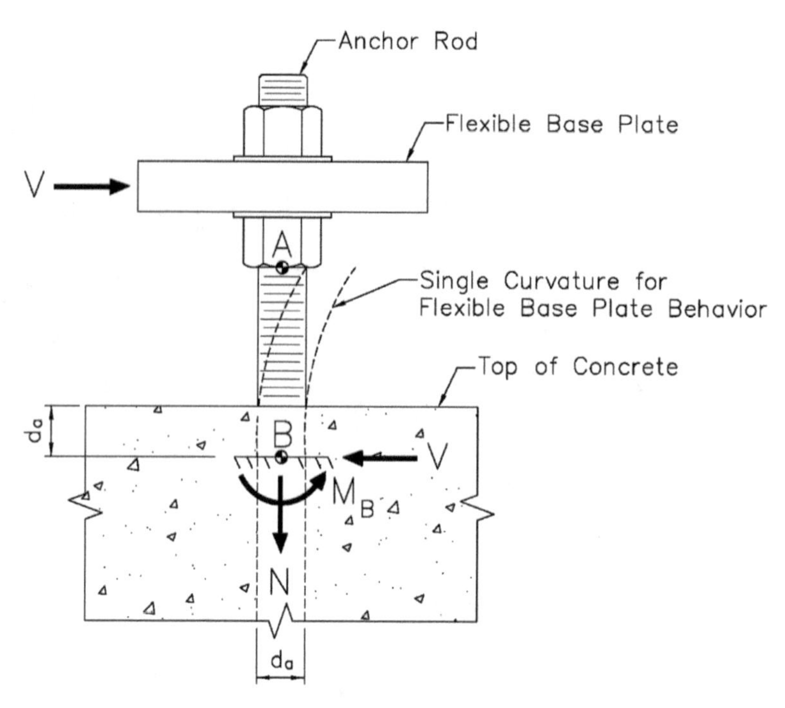

Figure D-40. Flexible base plate—rotates freely.

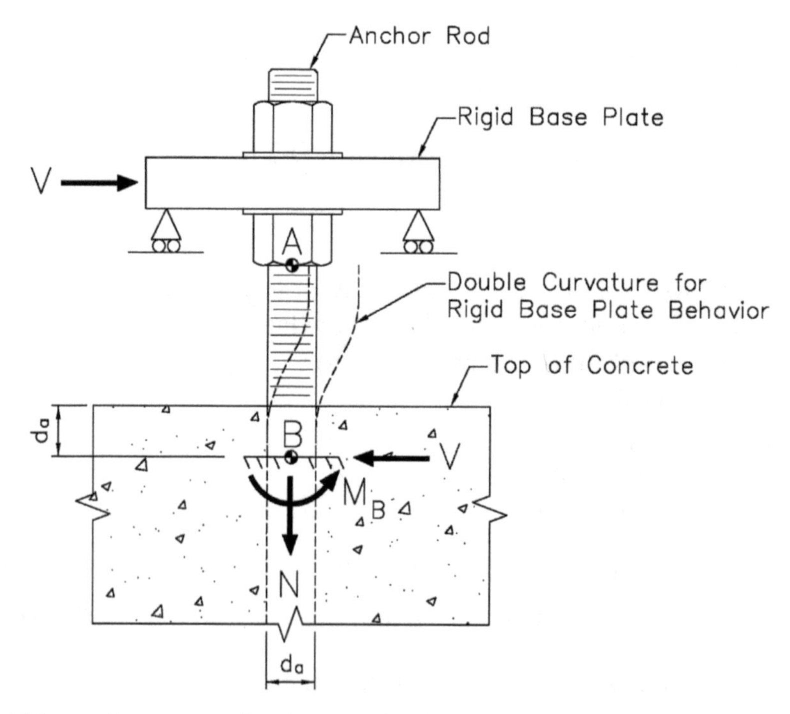

Figure D-41. Rigid base plate—rotation is restrained.

$$M_n = M_0\left[1 - \frac{N_{ua}}{\phi_a N_n}\right] \qquad \text{(D-72)}$$

where

M_n = Reduced bending capacity with an axial force,
N_{ua} = Ultimate axial force in an individual anchor (compression or tension), and

ϕ_a = Strength reduction factor, 0.75, for the ductile steel element and 0.65 for the brittle steel element

$$N_n = A_{se,N} f_{uta} \tag{D-73}$$

(nominal axial capacity of an individual anchor)

$$A_{se,N} = \frac{\pi}{4}\left(d_a - \frac{0.9743}{n_t}\right)^2 \tag{D-74}$$

f_{uta} is the lesser of f_{ua}, the specified tensile strength of the anchor; 1.9 f_{ya}, where f_{ya} is the specified yield strength of the anchor; and 125,000 psi (860 MPa).

The nominal shear force per anchor V_{nm} as limited by the bending moment shall be determined by using the following equation:

$$V_{nm} = \frac{M_n}{L} \tag{D-75}$$

where L is 1/2 L_{RGD} for rigid base plate behavior, and L is L_{FLX} for flexible base plate behavior.

The ultimate shear force V_{ua} as limited by the bending moment shall be determined as

$$V_{ua} \leq \phi_v V_{nm} = \frac{\phi_v M_n}{L} \tag{D-76}$$

where ϕ_v is the strength reduction factor, 0.65, for the ductile steel element, and 0.60 for the brittle steel element

$$V_{ua} \leq \frac{\phi_v M_n}{L} = \frac{\phi_v M_0 \left[1 - \frac{N_{ua}}{\phi_a N_n}\right]}{L} \tag{D-77}$$

Substituting the reduced bending capacity from (D-72) and solving for unity results in Equation (D-78).

$$\frac{N_{ua}}{\phi_a N_n} + \frac{V_{ua} L}{\phi_v M_0} \leq 1 \tag{D-78}$$

$$\frac{N_{ua}}{\phi_a N_n} + \frac{V_{ua}}{\phi_v V_n} \leq 1.2 \tag{D-79}$$

Equation (D-79) is the same as Equation (D-56), and it is provided here as an additional check for adequacy of the anchor rod, where

$$V_n = 0.6 f_{uta} A_{se,V} \tag{D-80}$$

$$A_{se,V} = \frac{\pi}{4}\left(d_a - \frac{0.9743}{n_t}\right)^2 \tag{D-81}$$

A rigid base plate shall be designed such that the material remains elastic, and the yield stress is not exceeded for the ultimate loads (factored or strength level loads). Design of rigid base plates shall be in accordance with Section D.7.7.

Consideration shall be given for flexural buckling of anchors on leveling nuts under compressive loads. The procedures outlined in AISC 360 (AISC 2016b) for flexural buckling of members without slender elements shall be used to verify the buckling capacity of anchor rods.

D.9.4.3 Design Considerations for Concrete. The capacity of a cast-in-place anchor rod is limited by either the capacity of a steel anchor rod or concrete. In Section D.9.4.2, the limitations of the strength of anchor rods are discussed. In this section, the limitations of concrete are also considered.

D.9.4.3.1 Capacity of Concrete. The capacity of concrete shall be in accordance with the provisions of Chapter 17 of ACI CODE-318 (ACI 2019).

D.9.4.3.2 Design of Minimum Edge Distance and Anchor Spacing. Design of minimum edge distance and anchor spacing shall be in accordance with Chapter 17 of ACI CODE-318 (ACI 2019).

D.9.4.3.3 Anchor Rod Embedment Length. The embedment for anchor rods shall be designed to resist concrete tension breakout or have adequate length to transfer the applied tension or compression to the foundation reinforcement.

Design of vertical reinforcement as anchor reinforcement shall follow the provisions of ACI CODE-318 (ACI 2019). Sufficient adjacent reinforcement shall be provided to transfer the full anchor rod tension force using a strength reduction factor of 0.75. In addition to providing adequate anchor rod embedment, the foundation vertical reinforcement shall have sufficient length above and below the failure plane extending from the head of the anchor rod to develop the reinforcement in the case of headed rods or to develop the non-contact lap splice with the adjacent rebar in the case of deformed bar anchor rods (Figure D-42). The required development length (l_d) of the vertical reinforcement in drilled shafts and spread footings is provided by ACI CODE-318 (ACI 2019), Chapter 25.

D.9.4.3.4 Concrete Punch Out from Anchor Rods. Anchor rods on leveling nuts without grout shall be designed so that they can transfer the downward compressive force by the headed anchor. If there is insufficient concrete shear capacity beneath the headed anchor that is in compression, a punching shear failure may occur (Figure D-43). If the concrete shear capacity is insufficient, anchor reinforcement that intercepts the potential failure surface created by the downward force on the anchor rod shall be provided beneath the headed anchor and extending upward into the concrete. Such anchor reinforcement shall be designed in a manner similar to an anchor rod in tension.

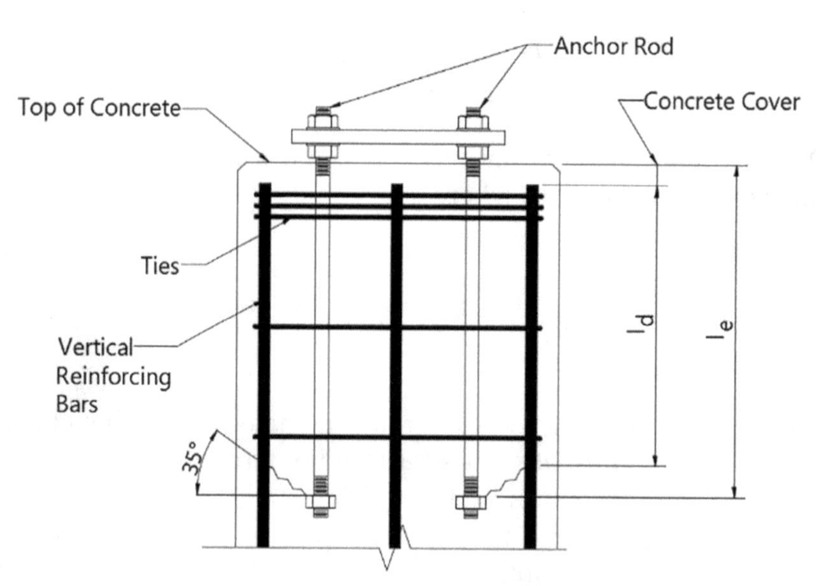

Figure D-42. Development length of a vertical reinforcing steel in drilled shafts and spread footings.

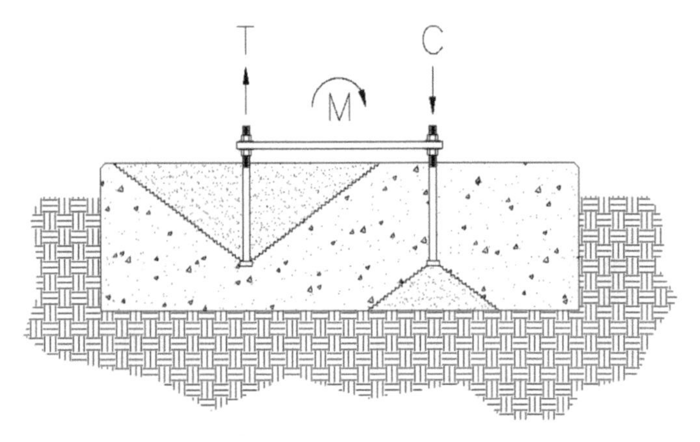

Figure D-43. Concrete breakout from tension and compression.

D.9.4.3.5 Localized Bearing Failure. Design of anchors in tension or for anchors in compression as in the case of base plates supported on leveling nuts shall consider localized bearing capacity. An anchor/bearing plate, where required, shall be designed in accordance with ACI CODE-318 (ACI 2019), Chapter 17, with a thickness not less than the unsupported edge of the plate. The nominal concrete shear strength, V_c, shall consider the impact of anchor/bearing plates.

D.9.5 Post-installed Anchors in Concrete

D.9.5.1 Types and Application. Commonly used post-installed anchor types include expansion, screw, undercut, and adhesive anchors.

Post-installed anchors shall be installed by strictly adhering to the manufacturer's instructions and in applications for which they are intended.

D.9.5.2 Design. Post-installed anchors shall be designed in accordance with the requirements of ACI CODE-318 (ACI 2019), Chapter 17, and the applicable portions of Section D.9.4.2 of this standard. For anchors on leveling nuts, the procedure of Section D.9.4.2.2 shall be applied to post-installed anchors.

D.10 FOUNDATIONS

Substation foundations shall be designed to safely transfer the vertical and lateral loads from the electrical equipment and structures to the supporting soil.

This section identifies various types of foundations that are commonly used to support substation equipment and structures, and geotechnical considerations for the design of substation foundations, and provides design considerations for various foundations.

D.10.1 Definitions

Shallow Foundations: Foundation able to distribute loading effectively through bearing pressure transferred to the soil.

Spread Footings: Single pedestal supported on a single enlarged reinforced concrete footing

Combined Footings: Spread footing foundation having two or more pedestals supported on a single reinforced concrete footing.

Mat Foundations: Also known as a slab foundation; a shallow reinforced concrete footing with no pedestals placed at or near grade level.

Grade Beams: Shallow foundation that consists of a reinforced concrete beam used to transmit the load from a bearing wall, or other similar continuous loads, to the foundation.

Figure D-44 shows different types of shallow foundations used in a substation.

Deep Foundations: Foundation with a depth-to-width ratio (*D/B*) exceeding 5.

Drilled Shafts: A common type of deep foundation that is concrete cast-in-place, constructed in drilled holes. Each foundation is usually considered rigid and thus capable of transferring a substantial amount of overturning moments into the soil, in addition to axial and lateral loads.

Figure D-45 shows a typical drilled shaft foundation used in a substation.

Piles: Type of deep foundation that is considered slender and can only transfer a limited amount of overturning moment and shear into the soil.

Direct Embedment: Poles that are installed directly into an open excavation with backfill placed in the void space around the pole.

Helical Screw Anchor Piles: A manufactured deep foundation system consisting of varying sizes of tubular hollow steel shafts with one or more helical plates welded around it.

D.10.2 Geotechnical Subsurface Explorations

A site-specific subsurface exploration and laboratory testing program shall be used for new substations, major expansions, in areas of poor subsurface conditions, or as otherwise determined by the engineer.

The subsurface explorations shall include an evaluation of the site-specific seismic design parameters and seismic hazards, where applicable. The geotechnical report shall identify whether the soils at the site are prone to liquefaction.

The results of the subsurface exploration and testing program shall be presented in a report format and certified by the geotechnical engineer of record. The contents of the report shall provide the engineer with the information and recommendations needed to select and design foundations of the appropriate type and size on the basis of the subsurface conditions.

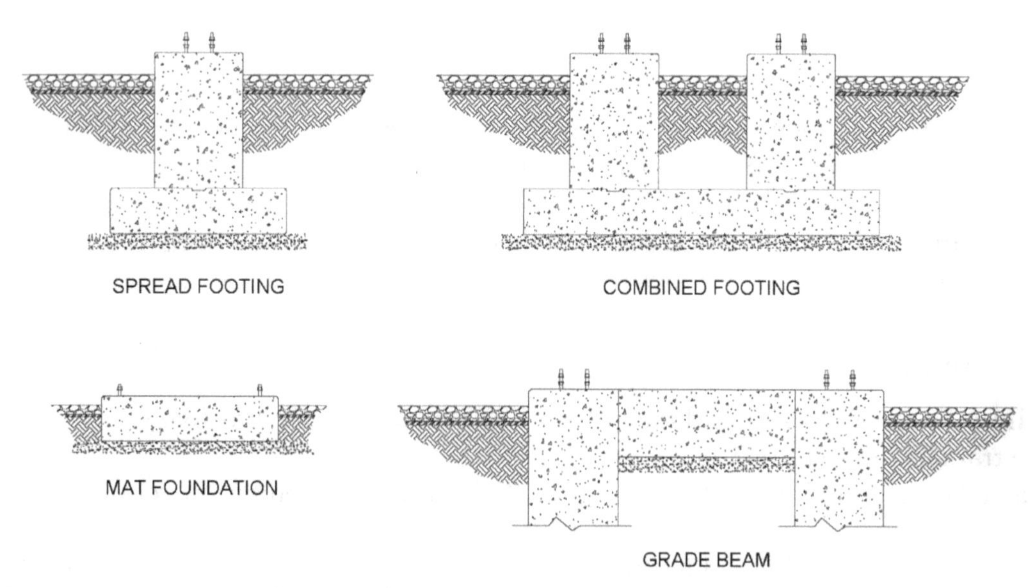

Figure D-44. Various types of shallow foundations.

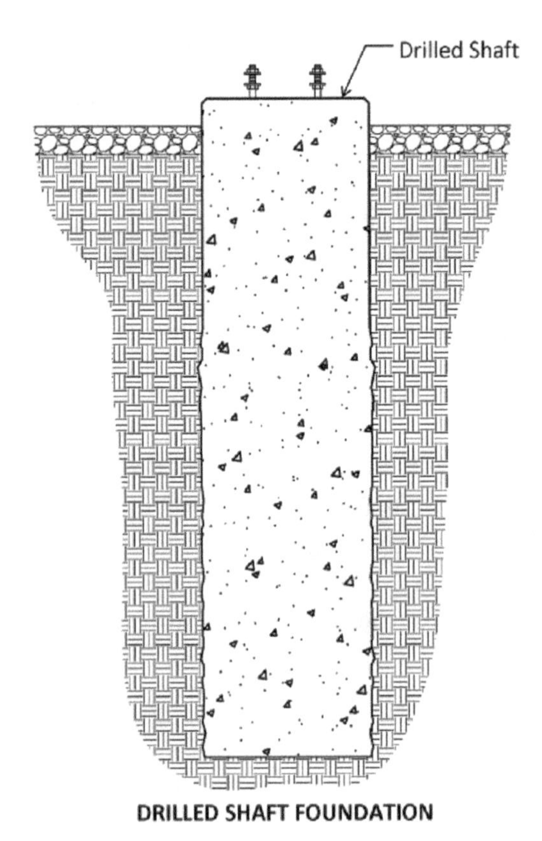

Figure D-45. Drilled shaft foundation.

D.10.3 Environmental Considerations

D.10.3.1 Frost Depth. At a minimum, foundations shall extend to a depth below the anticipated frost line. A frost penetration map is available in IEEE 691 (IEEE 2001), Figure 38, for the continental United States. In many localities, the frost depth is determined by the building official.

D.10.3.2 Soil Movement. Foundations that are placed within an active zone of expansive soils shall be designed to resist differential volume changes and resulting forces on the foundation and structure. Refer to IEEE 691, Figure 40, for a graph of Plasticity Index to Soil Swell Potential.

D.10.3.3 Corrosion. The geotechnical report shall provide any applicable recommendations for mitigation against soil corrosion.

D.10.4 Durability

Concrete used for foundations shall be able to resist weathering exposure, chemical attacks, and erosion, while still maintaining the desired properties in accordance with the recommendations of ACI CODE-318 (ACI 2019) and ACI PRC-201.2 (ACI 2023).

The quality of concrete slabs shall follow the recommendations of ACI PRC-302.1 (ACI 2015a) for contraction joints.

D.10.5 Foundation Load Combinations

Stability, bearing, and soil–structure interaction shall use either unfactored loads (ASD) with load combinations from Table D-2 or USD load combinations from Table D-1. Design of concrete, reinforcing steel, and other structural components shall use factored (USD) load combinations from Table D-1.

D.11 QUALITY ASSURANCE AND CONTROL

D.11.1 Introduction

To ensure quality of the product, good quality control (QC) and quality assurance (QA) programs shall be instituted by the fabricator and verified by the owner.

D.11.2 Material

The purchaser shall review and agree on the fabricator's material specifications, supply sources, material identification procedures, storage, and traceability procedures. Mill test reports for all material shall be reviewed for compliance with the appropriate ASTM specification by the fabricator. The fabricator shall maintain copies of all mill test reports.

Prestressed and concrete pole structure inspection shall be done according to the guidelines provided by ASCE MOP 123 (ASCE 2012).

D.11.3 Welding

The fabricators' welding procedures and welder certifications shall be in accordance with the latest revision of the American Welding Society's *Structural Welding Code for Steel*, AWS D1.1/D1.1M, D1.2/D1.2M. All welders performing work shall be AWS-certified for the process and position that they are welding.

D.11.4 Structure Coating

Where coating is specified by the owner, the system, procedures, and methods of application shall be acceptable to both the owner and the fabricator.

Where bare weathering steel is specified, the need for blast-cleaning the steel shall be decided and agreed on by the owner and the fabricator.

D.11.5 Fabrication Inspection

The fabricator shall have a detailed procedure for contract and project specification review that includes all necessary information for the owner to assure contract compliance, including a system for requests for information necessary to resolve discrepancies or variations from contract requirements.

D.11.6 Visual Inspection

Visual inspections shall address the following typical areas:

- Dimensional correctness,
- Fabrication straightness,
- Cleanliness of cuts and welds,
- Surface integrity at bends,
- Condition of punched and drilled holes,

- Hardware fit and length,
- Weld size and appearance,
- Overall product workmanship, and
- Coating thickness and quality.

D.11.7 Inspection of Welds

Weld inspection for steel structures shall be performed according to the requirement of AWS D1.1, Section 6, Inspection, Part C. Personnel qualification for non-destructive testing shall be in accordance with Section 6.14.6.1 of AWS D1.1.

D.11.8 AISC Compliance

The steel fabricator shall establish and maintain quality control procedures and perform inspections to ensure that the work is performed and inspected in accordance with the requirements of AISC 303 (AISC 2022) and AISC 360 (AISC 2016b).

D.11.9 Inspection Reports

Inspection reports shall be generated and maintained for all inspection activities, including dimensional correctness, weld quality and size, and cleanliness of cuts. Reports shall also be generated for all non-destructive inspection activities. The reports shall include the assembly part number, date of inspection, and the inspector's name.

D.11.10 Shipping

At a minimum, the fabricator shall comply with the shipping procedures listed as follows:

- Check packaging to minimize shipping damage.
- Check items to ensure that they have completed specified inspections.
- Check that specified items are included with the shipment.

Before starting fabrication, the owner shall review the fabricator's methods and procedures for packaging and shipping and agree to the mode of transportation. When receiving materials, all products shall be inspected for any shipping damage before accepting delivery.

D.11.11 Handling and Storage

The fabricator shall provide written procedures for handling and storing materials to prevent damage, loss, or deterioration of the structure. The owner shall review and approve these procedures before shipping any materials.

D.12 CONSTRUCTION AND MAINTENANCE

D.12.1 Introduction

Loading from construction and maintenance activities shall be considered in the design of substation facilities.

Note: For additional information on installation procedures, refer to the following documents:

- IEEE Standard 951-96, *IEEE Guide to the Assembly and Erection of Metal Transmission Structures*
- IEEE Standard 1025-12, *IEEE Guide to the Assembly and Erection of Concrete Pole Structures*
- IEEE Standard 524-03, *IEEE Guidelines to the Installation of Overhead Transmission Line Conductors*

Substation owner's construction specifications.

D.12.2 Construction and Maintenance Access

The design engineer shall anticipate construction loads imposed on substation structures and ensure that proper construction methods and quality materials are used to prevent undue stresses.

Inspections shall be performed during various stages of construction such as grading, placement of reinforcement and embedment, placement of concrete, structure erection, and structure loading.

The design engineer shall consider accessibility of equipment for initial installation and future maintenance and operations. Equipment shall be designed with provisions for access including crossover platforms or working platforms, especially around large transformers with coolant and fire protection piping and at locations where the equipment is raised above the design flood elevation. All structures or equipment that will be inaccessible with bucket trucks or smaller ladders shall be designed to meet all access requirements in accordance with OSHA, state, and local codes.

All structures, including temporary and permanent buildings, shall be inspected on a regular basis, but at a minimum of each time the equipment being supported is inspected or maintained. Where inspection detects signs of loss of strength, damage, corrosion, loose members, missing members, and deficient connections, corrective measures shall be taken, as necessary.

Foundations shall be inspected for signs of loss of strength, settlement, cracking, unusual corrosion of anchor rods, and deterioration, particularly caused by alkali-aggregate reactivity (AAR) and alkali-silica reactivity (ASR), and corrective measures shall be taken, as necessary.

In addition, when equipment is upgraded, structures shall be carefully checked for structural adequacy, electrical clearances, and deterioration caused by corrosion.

REFERENCES

AA (Aluminum Association). 1981. *Aluminum construction manual.* Arlington, VA: AA.

AA. 2020. *Aluminum design manual, including specification for aluminum structures.* Arlington, VA: AA.

ACI (American Concrete Institute). 2015a. *Guide to concrete floor and slab construction.* ACI PRC-302.1. Detroit: ACI.

ACI. 2015b. *Guide for the design and construction of concrete reinforced with FRP bars.* ACI PRC-440.1. Detroit: ACI.

ACI. 2019. *Building code requirements for structural concrete (with commentary).* ACI CODE- 318. Detroit: ACI.

ACI. 2023. *Guide to durable concrete.* ACI PRC-201.2. Detroit: ACI.

AISC (American Institute of Steel Construction). 2016a. *Seismic provisions for structural steel buildings*. AISC 341. Chicago: AISC.

AISC. 2016b. *Specification for structural steel buildings*. AISC 360. Chicago: AISC.

AISC. 2022. *Code of standard practice for steel buildings and bridges*. AISC 303. Chicago: AISC.

ANSI (American National Standards Institute). 2017. *American National Standard for wet-process porcelain insulators-apparatus, post type*. ANSI/NEMA C29.9. New York: ANSI.

ANSI. 2020. *American National Standard for composite insulators—Station post type*. ANSI/NEMA C29.19. New York: ANSI.

ANSI. 2022. *Wood poles-specifications and dimensions*. ANSI-O5.1. New York: ANSI.

ASCE. 2012. *Prestressed concrete transmission pole structures: Recommended practice for design and installation*. MOP 123. Reston, VA: ASCE.

ASCE. 2015. *Design of latticed steel transmission structures*. ASCE 10-15. Reston, VA: ASCE.

ASCE. 2019a. *Design of steel transmission pole structures*. ASCE 48-19. Reston, VA: ASCE.

ASCE. 2019b. *Wood pole structures for electrical transmission lines: Recommended practice for design and use*. MOP 141. Reston, VA: ASCE.

ASCE. 2020. *Guidelines for electrical transmission line structural loading*. MOP 74. 4th ed. Reston, VA: ASCE.

ASCE. 2022. *Minimum design loads and associated criteria for buildings and other structures*. ASCE 7-22. Reston, VA: ASCE.

ASTM International. 2022. *Standard specification for aluminum and aluminum-alloy seamless pipe and seamless extruded tube*. ASTM B241/B241M. West Conshohocken, PA: ASTM.

ASTM. 2015a. *Standard specification for aluminum-alloy extruded bar, rod, tube, pipe, structural profiles, and profiles for electrical purposes (bus conductor)*. ASTM B317/B317M. West Conshohocken, PA: ASTM.

ASTM. 2015b. *Standard specification for zinc coating, hot-dip, requirements for application to carbon and alloy steel bolts, screws, washers, nuts, and special threaded fasteners*. ASTM F2329/F2329M. West Conshohocken, PA: ASTM.

ASTM. 2016. *Standard specification for zinc (hot-dip) on iron and steel hardware*. ASTM A153/A153M. West Conshohocken, PA: ASTM.

ASTM. 2017. *Standard specification for sampling procedure for impact testing of structural steel*. ASTM A673/A673M. West Conshohocken, PA: ASTM.

ASTM. 2019a. *Standard specification for carbon structural steel*. ASTM A36/A36M. West Conshohocken, PA: ASTM.

ASTM. 2019b. *Standard specification for zinc-coated (galvanized) steel bars for concrete reinforcement*. ASTM A767/A767M. West Conshohocken, PA: ASTM.

ASTM. 2020a. *Standard specification for anchor bolts, steel, 36, 55, and 105-ksi yield strength*. ASTM F1554. West Conshohocken, PA: ASTM.

ASTM. 2020b. *Standard practice for safeguarding against embrittlement of hot-dip galvanized structural steel products and procedure for detecting embrittlement*. ASTM A143/A143M-07. West Conshohocken, PA: ASTM.

ASTM. 2022a. *Standard test methods and definitions for mechanical testing of steel products*. ASTM A370. West Conshohocken, PA: ASTM.

ASTM. 2022b. *Standard specification for deformed and plain carbon-steel bars for concrete reinforcement*. ASTM A615/A615M. West Conshohocken, PA: ASTM.

ASTM. 2022c. *Standard specification for low-alloy steel deformed and plain bars for concrete reinforcement*. ASTM A706/A706M. West Conshohocken, PA: ASTM.

AWS (American Welding Society). 2014. *Structural welding code—Aluminum*. AWS D1.2/D1.2M. Miami: AWS.

AWS. 2020. *Structural welding code—Steel*. AWS D1.1/D1.1M. Miami: AWS.

Bleich, F. 1952. *Buckling strength of metal structures*. 1st ed. New York: McGraw-Hill.

IEC (International Electrotechnical Commission). 2011. *Short-circuit currents—Calculation of effects—Part 1: Definitions and calculation methods*. IEC 60865, ed.3.0. Geneva: IEC.

IEEE (Institute of Electrical and Electronics Engineers). 2000. *Standard performance characteristics and dimensions for outdoor apparatus bushings*. IEEE C57.19.01. Piscataway, NJ: IEEE.

IEEE. 2001. *Guide for transmission structure foundation design and testing*. IEEE 691. Piscataway, NJ: IEEE.

IEEE. 2008. *Guide for design of substation rigid-bus structures*. IEEE 605. Piscataway, NJ: IEEE.

IEEE. 2011. *IEEE standard requirements for AC high-voltage air switches rated above 1000 V.* IEEE C37.30.1. Piscataway, NJ: IEEE.

IEEE. 2012. *Guide for application of power apparatus bushings*. IEEE C57.19.100. Piscataway, NJ: IEEE.

IEEE. 2018a. *IEEE standard for ratings and requirements for AC high-voltage circuit breakers with rated maximum voltage above 1000 V.* IEEE C37.04. Piscataway, NJ: IEEE.

IEEE. 2018b. *Recommended practice for seismic design of substations*. IEEE 693. Piscataway, NJ: IEEE.

IEEE. 2018c. *Recommended practice for the design of flexible buswork located in seismically active areas*. IEEE 1527. Piscataway, NJ: IEEE.

IEEE. 2023. *National electrical safety code* (ANSI C2). Piscataway, NJ: IEEE.

Norris, C. H., and J. B. Wilbur. 1960. *Elementary structural analysis*. 2nd ed. New York: McGraw-Hill.

PCI (Precast/Prestressed Concrete Institute). 2017. *Design handbook, precast and prestressed concrete*. 8th ed. PCI MNL-120. Chicago: PCI.

PCI. 2021. *Manual for quality control for plants and production of structural precast concrete products*. 5th ed. PCI MNL-116. Chicago: PCI.

USGS (US Geological Survey). 2018. "Relative seismic hazard map." Reston, VA: USGS.

INDEX